BIO-ENGINEERING SYMPOSIUM FOR FISH CULTURE

Proceedings

Bio-Engineering Symposium for Fish Culture

Steering Committee

Harry Westers, Chairman
Robert G. Piper, Program Chairman
Cecil L. Fox, Program
Charles E. Hicks, Publicity
Douglas Morgan, Arrangements

Arden H. Trandahl, Vice-Chairman
Edward R. Miller, Finance
Jerald Williamson, Program
Jack R. Snow, Publicity
Keith M. Pratt, Arrangements

Sponsors

Fish Culture Section of the
American Fisheries Society
Northeast Society of Conservation
Engineers
U.S. Fish and Wildlife Service
National Marine Fisheries Service
Great Lakes Fisheries Commission

North Central Division of the
American Fisheries Society
Southern Division of the
American Fisheries Society
Western Division of the
American Fisheries Society

Contributors

Coffelt Electronics Co., Inc.
Garon Company
Heath Tecna
Johnson & Anderson, Inc.
Kramer, Chin & Mayo, Inc.
National Aquaculture Information
System

North Pacific Aquaculture Development
Company
Peterson & Matz, Inc.
Peterson Fiberglas Laminates, Inc.
S.F. Sonk Associates, Inc.

Exhibitors

Aquafine Corporation
Aquaculture Research/Environmental
Associates
Baker Filtration Company
Clevepak Corporation
Frigid Units, Inc.
Heath Tecna

Hinde Engineering Company
Honeywell, Inc.
Hydrolab Corporation
North Pacific Aquaculture Development Co.
Peterson Fiberglass Laminates, Inc.
Shelby Reinforced Plastics Corporation

Symposium

held in Traverse City, Michigan

October 16-18, 1979

Proceedings of the Bio-Engineering Symposium for Fish Culture

Edited by

Lochie Jo Allen

and

Edward C. Kinney

FCS Publ. 1

FISH CULTURE SECTION
of the American Fisheries Society
5410 Grosvenor Lane
Bethesda, Maryland 20014
and the
NORTHEAST SOCIETY OF CONSERVATION ENGINEERS

Library of Congress No. 80-68383

Published by the Fish Culture Section of the American Fisheries Society

© 1981 by the Fish Culture Section of the American Fisheries Society

Printed in the United States of America by Kirby Lithographic Co., Inc.,
  Washington, D.C.

CONTENTS

SESSION III *Water Conditioning for Fish Rearing*

SESSION IV *Instrumentation and Automation in Fish Culture*

SESSION V *Hatchery Effluent Treatment*

SESSION VI *Fish Production Facilities*

OFFICE OF THE GOVERNOR
LANSING

WILLIAM G. MILLIKEN
GOVERNOR

October 12, 1979

Bio-Engineering Symposium
Park Place Motor Inn
Traverse City, Michigan

Greetings:

I regret that it is not possible for me to be with you at the
Bio-Engineering Symposium on October 16, because of prior commit-
ments.

However, I do want to take this opportunity to welcome each of
the participants in this important conference to Michigan and
to the Traverse City area.

It is appropriate that this conference is being held in Michigan
on the shores of one of the Great Lakes. The revival of the
Great Lakes fishery is—in large measure—directly attributable
to the work of people like yourselves, and that revival has
meant a great deal to Michigan, both economically and recrea-
tionally. We are proud to be host to this important gathering
of bio-engineers from around the world. I look forward with
anticipation to the results of this symposium as you discuss
ways in which to improve our abilities to deal with the problems
associated with intensive aquaculture.

Michigan has benefited greatly from your expertise and the improved
technologies you have developed, and I believe that the world can
benefit from your continuing efforts at improving the provision
of protein at reasonable costs and with minimal impact on our
environment.

Best wishes to all participants at this important event.

Sincerely,

William G. Milliken
Governor

Preface

The symposium meeting, attended by over three hundred people representing eleven different countries, is now history. These Proceedings are the tangible and lasting fruits of three days of interaction by engineers, biologists, fish culturists, and other interested individuals who met in Traverse City, Michigan.

The aim of the organizers was to bring together an international asemblage of expertise to communicate and develop new concepts in fish culture facility design, and identify critical research needs.

Although this volume represents the meat of the formal program, it is incapable to convey the enthusiasm displayed and even less so the excitement and intense discussions generated during long after-hours. The Steering Committee looks back with much satisfaction to the many hours spent on planning, execution, and completion of a most successful undertaking.

We recognize and acknowledge that this Bio-Engineering Symposium for Fish Culture and this publication would not have been possible without the attendees, program contributors, supporters, and sponsors. Although the supporters and sponsors are listed elsewhere in this volume, we want to thank them here again for their generous contributions. We also want to express our great appreciation to the Michigan Department of Natural Resources, who hosted the sym-posium and made generous contributions of manpower and equipment. In particular, we thank Jack Hammond, Jim Copeland, John Hnath, Vernon Bennett, Walter Houghton, Charles Pecor, Roger Martin, Richard Poynter, John Driver, and Warren Yoder for their role in local arrangements and especially for a highly successful and memorable field trip, culminating with a delicious salmon boil dinner at Ranch Rudolph. The registration and numerous other details were very ably handled by my secretary, Bonnie Menovske, and her assistants: Nita Simon, Marcia Matthews, Linda Johnson, and Sue Holtom.

Finally, as Chairman of the Steering Committee, I would like to express a most sincere and well-deserved appreciation to the Committee members, who worked hard and diligently, off and on, for over two years. Their organizational talents, professionalism, cooperative spirit, and keen interest in the project were instantly at my disposal throughout the entire planning period. No chairman could be more fortunate than this one with a Steering Committee such as this. Thank you.

Harry Westers

Michigan Department of Natural Resources
Fisheries Division
Lansing, Michigan 48909

Bio-Engineering Symposium for Fish Culture (FCS Publ. 1):1

SESSION I   SYMPOSIUM PERSPECTIVE

Bio-Engineering—An Overview

ROGER E. BURROWS

736 Marine View Drive
Longview, Washington 98632

It is obvious that the sport and commercial demand for salmonid fishes already exceeds the available supply. To meet the demand, fish hatcheries will have to become more efficient, more productive, larger, and more numerous. While many of the most promising sites for hatcheries are already being utilized, technology and research have progressed to the point where many other sites have become desirable and practical.

Fish culture research has developed adequate diets, methods of disease control, several pond designs, definition of some factors limiting the carrying capacity of water, methods of water reconditioning and recycling, and environmental control including disinfection, temperature control, and mineral regulation. The problem is to incorporate all or most of these developments into new hatchery designs while still maintaining simple and practical operations.

Most developments in fish culture are derived from either fish culturists or research biologists. The practical application is dependent on the ability of the engineers. This field of engineering requires capability in the mechanical, structural, hydraulical, electrical, and chemical disciplines. Unfortunately most engineering firms do not have all these capabilities confined to one person or a few persons who are willing to work together and listen to the biologists and fish culturists. As a result, only a few engineering firms are qualified for hatchery design. In defense of the engineers, there are several fields of thought among fish culturists and biologists as to just what the requirements of a hatchery are and how they can best be achieved. Add the fact that each discipline speaks a different language, and the result can be chaos.

Communication, then, becomes the principal problem in bio-engineering. An intermediary becomes necessary if communication is to be established. Usually such intermediary has a biological background with some engineering experience or mechanical aptitude. He then needs an engineer or engineers who will listen. Once communication is established the design of the hatchery can proceed.

Another problem particularly frustrating to engineers is that no two hatchery designs should be exactly alike. Fish cultural facilities have developed enough versatility that variations are possible to meet specific requirements. Water flow, water source, climate, production required, species reared, and rearing period are some of the factors affecting hatchery design.

The degree of automation can present problems to both engineers and fish culturists. Murphy's Law, that if anything can fail it will, is particularly applicable to highly automated hatchery systems. The tracing of failures requires time, which usually is not available in fish cultural systems. Mechanical overrides then become a necessity if catastrophic losses are to be avoided. In any mechanized system automatic failsafes are required, but I do not feel that the science of fish culture has advanced to the point where highly computerized systems are applicable. Fish culturists are still an essential ingredient in fish culture.

In conclusion, I would like to say that fish culturists, research biologists, and engineers need each other. By their working together, more efficient, more productive fish hatcheries can be produced.

Bio-Engineering Symposium for Fish Culture (FCS Publ. 1):2-4
© 1981 by the Fish Culture Section of the American Fisheries Society

A Commercial Fish Farmer's Viewpoint

LEO RAY

Fish Breeders of Idaho
Rt. 3, Box 193
Buhl, Idaho 83316

The purpose of the Bio-Engineering Symposium for Fish Culture is to get biologists, engineers, fish farmers, and anyone else interested in fish culture together to exchange ideas with the hope of developing lines of communication of benefit to all people in the fish industry.

When this symposium was being conceived I was asked if there were a need for a bio-engineering symposium for fish culture. After reviewing the problems and questions I had encountered in the last ten years designing, building, and operating three fish farms, I had no doubt the symposium was needed. If the farms designed had been poultry farms, the questions could have been answered by any competent person in the poultry industry. The farms were not poultry farms. They were fish farms and answers to questions that arose were not available. Simple questions did not have ready answers. Fish culture today is an art and not a science. If fish culture is to become a science, fish culturists must work with members of the other sciences. Many of the questions the fish farmer has will be found in the engineering community. Finding them will require cooperation between biologists and engineers.

Many of the problems that plague fish culture today plagued other industries in their developmental stages. Answers to many of these questions have been found in other industries. Lack of communication has kept this information from the fish industry. Engineers in other fields hold the answers or at least the principles needed to find many solutions to the fish culturists' problems. This information needs to be applied to fish culture.

In looking at the engineering problems of fish culture I am reminded of the story of Dr. Jeager, who is noted for first discovering hibernation in birds. Dr. Jeager was shocked when he found a poorwill hibernating in the Four Corners area. When he asked an assistant, a native Indian of the area, if he had ever heard of a poorwill hibernating, the Indian boy replied, "Sure, the Indian name of the poorwill is 'one who sleeps all winter.'"

Many of the answers to the fish culturists' problems will be this simple to members of another science.

If answers to fish culturists' problems are known by engineers of other fields, why haven't these answers been used by fish farmers? Why do gaps exist between the thinking of the biologist and that of the engineers? Why did fish breeders design their own fish farms without the assistance of an engineering firm? Why is it that all of the fish farms in the Magic Valley of Idaho were designed by fish farmers? Why have so many fish farms designed by engineers been failures? Why do we see so many State and Federal hatcheries designed by engineering firms, supposedly competent in fish farm designing, producing a small percentage of their capacity and doing so at such a high cost that a private hatchery would have gone broke years ago?

Some of the answers to these questions lie in the fact that an engineer's dream has too often been a biologist's nightmare. It is easy for an engineer to design an elaborate facility with valves and switches. It is impossible to build a fish facility that is fool proof. The more one tries, the more complicated and elaborate the facility becomes. The more complicated and elaborate the facility becomes, the more danger there is of a segment's failing. The

2

whole chain is only as strong as the weakest link. The more links, the greater the chances of failure.

The first engineering rule to follow in designing a fish farm is the KISS rule, "Keep it simple, Stupid." It is very easy to name many State and Federal hatcheries that violated this rule. Elaborate, complicated facilities work on paper but do not work in practice. Private industry has also built its share of colossal nightmares. In private industry these facilities fail, go broke, and disappear. The State and Federal nightmares are not destroyed but go on forever consuming tax dollars. For this reason a government facility is a poor place to visit when looking for a model fish farm.

I did not come to this symposium with the idea of telling you how to raise fish; instead, I came with a lot of questions. So if I can refrain from taking cuts at the bureaucrats, I will ask a few. Let's start with the first factor affecting production, oxygen.

Oxygen is easily absorbed into water and if elevation is present, waterfalls are the economical way to aerate water. But what type of waterfall? Are two two-foot drops better than one four-foot drop? Is a heavy spill four inches deep better than a shallow spill one-half-inch deep? Do splash boards increase or decrease oxygen intake into the water? When, where, how does oxygen enter the water? In short, how should a waterfall be designed to give maximum efficiency in aeration? The answer to this would increase the production of most fish farms by at least 25 percent.

If elevation is not available and waterfalls are not possible, what is the best means of aeration? Is it cheaper to blow air into water or pump the water into the air? Are the paddle wheels used in oriental eel culture for aeration more economical than pumps and blowers? If they are, what is the best paddle design — a blade, a perforated blade, or a wire brush?

How can oxygen be injected into water to obtain 100% absorption? If 100% of liquid oxygen could be obtained, this would probably be the most economical production increase that could be made.

The second factor affecting production is usually ammonia. How can we get rid of it? When is it economical to aerate ammonia out of water? At high pH, at high temperatures, and in soft water, ammonia goes to the gaseous stage where it can be aerated off. How can this be done economically?

If two fish farms are located on a stream, one below the other, what can be done to improve the water quality to the second farm? If moss is allowed to grow in this ditch, the moss consumes oxygen at night and can cause severe oxygen depletion in the hours before sunrise. If the moss is removed, there is nothing to remove the ammonia and the ammonia travels to the second farm. Would watercress or another type of vegetation that takes its nutrients from the water but its oxygen from the air solve the problem?

Most of the good springs and water sources in the United States have been developed for fish culture. Additional production is now coming from reusing this water. What can be done to improve the quality of this reuse water? Concrete ditches that move water from one point to the next with minimal moss and silt problems may not be the answer. The engineering of the stream must be coordinated with the biology of the stream to obtain those characteristics of water quality needed by the second farm.

What is the ideal velocity in a raceway, and what is the ideal pattern of movement for that velocity? Velocity is used to carry away the waste particles and move fresh water to the fish. How can a raceway be designed to give the maximum benefit from the velocity and provide for the removal of the unwanted with minimum effect on the fish?

Density — if all the answers were known concerning density one would find that twice as much concrete was poured in building most fish farms than was needed. Density has two dimensions, weight of fish per cubic measure of space and weight of fish per flow of water. Social as

3

well as physical factors control stocking densities. Why can fish be raised in jars where there are more fish than water and good conversion obtained; but when they are grown in raceways only 2 or 3 pounds can be grown per cubic feet of space?

Another very important aspect of fish farm design is cost. How can an expenditure for a fish farm be justified in private industry or in a government facility? Is the interest on the money to build the farm sufficient to buy more fish than the farm can raise? There have been too many government hatcheries built that fit this description. The interest on the capital expenditure of some of the colossal nightmares would buy more fish than the facility can produce.

Hatcheries can be built with a capital cost that can be justified. It is time for the government to stop spending more money for hatcheries than they are worth. Let us hope that from this symposium we can contribute effort and knowledge to the biologist and the engineer and reduce the cost of fish farm construction. It doesn't take much of an engineer to design a million-dollar fish farm that costs a million dollars to build. An engineer who can design the million-dollar fish farm that can be built for one hundred thousand dollars is a genius.

SESSION II   REQUIREMENTS OF THE FISH

Introduction

ROGER LEE HERMAN

National Fisheries Research and Development Laboratory
U.S. Fish and Wildlife Service
R.D. #4, Box 47
Wellsboro, Pennsylvania 16901

Hatcheries have been built and are being built for the convenience of hatchery personnel, not necessarily for the comfort of the fish. It is being recognized with increasing frequency that fish have needs that must be accommodated. It is the purpose of this panel and the papers to bring your attention to some of the more important requirements of the fishes.

Bio-Engineering Symposium for Fish Culture (FCS Publ. 1):6-20

The Hatchery Environment Required to Optimize Smoltification in the Artificial Propagation of Anadromous Salmonids

GARY A. WEDEMEYER

U.S. Fish and Wildlife Service
National Fisheries Research Center
Building 204, Naval Support Activity
Seattle, Washington 98115

RICHARD L. SAUNDERS

Biological Station
Fisheries and Environmental Sciences
Fisheries and Oceans, Canada
St. Andrews, New Brunswick, Canada E0G 2X0

W. CRAIG CLARKE

Pacific Biological Station
Resource Services Branch
Department of Fisheries and Oceans
Nanaimo, B.C., Canada V9R 5K6

ABSTRACT

A high priority requirement in designing hatcheries for the production of fish for salmon enhancement programs is that the resulting smolts be fully prepared behaviorally and physiologically to migrate to sea and continue to grow and develop normally. However, the environmental requirements of smolts are exacting and must be used as the bio-engineering foundation of hatchery design and operation. First, an understanding of the physiology of normal smolt development is needed so that biological monitoring can be carried out to give an early warning that adverse environmental conditions or hatchery practices are reducing smoltification success. Second, a knowledge of the effects of environmental changes on smolt physiology is needed so that bio-engineering equipment can be designed to alter smoltification timing as changing resource management needs develop.

On the basis of our present understanding of the physiology of the parr-smolt transformation and the influence of hatchery practices and environmental alterations during rearing, guidelines are given for providing the aquatic environment needed to increase efficiency in the hatchery production of anadromous salmonids.

Introduction

A perennial problem in the efficient use of artificially propagated anadromous salmonids for enhancement programs or for aquaculture is that although hatcheries can consistently out-perform nature in survival through the hatching and early development stages, river and ocean survival of hatchery smolts is too often below the estimated survival of naturally produced smolts. Hatchery experience has generally shown that, within limits, sur-vival through to adult returns tends to be directly proportional to size at release. Accordingly, most hatcheries release large smolts in an attempt to increase the contribution to the fishery and returns to the hatchery. However, the results of even this attempt frequently fall short of those desired. As an example, releases of very large coho salmon smolts (*Oncorhynchus kisutch*) may merely result in an increased incidence of precocious males (jacks) and thus an actual reduction in adult returns (Bilton 1978).

There is reason to suspect that many of the apparently healthy hatchery "smolts" that are released, though large and silvery, are physiologically unprepared to go to the sea and therefore produce a limited contribution to the fishery, even when released into the same rivers from which their parents were taken. This failure to produce good quality smolts arises both from an incomplete understanding of smolt physiology by biologists, and from a lack of understanding of the hatchery environment required for a successful parr-smolt transformation by hatchery design engineers.

This review paper will describe the physiological and behavioral criteria for smoltification of Atlantic salmon *(Salmo salar)*, Pacific salmon *(Oncorhynchus* sp.*)*, and steelhead trout *(Salmo gairdneri)*; discuss the influences of hatchery environmental conditions on the parr-smolt transformation; and conclude with guidelines for fish cultural practices to help maximize early marine survival and thus the contribution to the fishery.

Physiology of Smoltification

Body Silvering, Fin Darkening

A summary of the currently understood physiological changes occurring during the parr-smolt transformation of anadromous salmonids is given in Table 1. Among the visible signs that occur, the external body's silvering and fin margin's blackening (in Atlantic salmon, coho salmon, and steelhead trout) are among the most dramatic characteristics used by biologists to distinguish smolts from parr. The silvery color results from deposition of guanine and hypoxanthine crystals in the skin and scales. It has long been known that thyroid hormones are involved, since feeding beef thyroid gland or adding thyroxine to the water results in body silvering (Landgrebe 1941; Piggins 1962; Dodd and Matty 1964). In naturally produced Atlantic salmon smolts particularly, the caudal and pectoral fin margins also become black because of melanization. This characteristic helps to differentiate silvery parr — often seen in

TABLE 1. — Physiological changes that must occur in a coordinated fashion during the parr-smolt transformation of anadromous salmonids.

Physiological characteristic	Level in smolts compared with parr
1. Body silvering, fin margin blackening	Increases
2. Hypoosmotic regulatory capability	Increases
3. Salinity tolerance and preference	Increases
4. Weight per unit length (condition factor)	Decreases
5. Growth rate	Increases
6. Body total lipid content	Decreases
7. Oxygen consumption	Increases
8. Ammonia production	Increases
9. Liver glycogen	Decreases
10. Blood glucose	Increases
11. Endocrine activity (thyroid, T_4, interrenal, pituitary, growth hormone)	Increases
12. Gill microsome, Na^+, K^+ – ATPase enzyme activity	Increases
13. Ability to grow well in full-strength seawater (35 °/oo)	Increases
14. Buoyancy (Atlantic salmon)	Increases
15. Migratory behavior	Increases

the early part of the outmigrant run — from true smolts. Though silvering and fin blackening, however, may be a good indication of smolt status in wild fish, these colors often develop in hatchery parr when the fish are not otherwise smolt-like. Thus additional physiological criteria are necessary.

Seawater Tolerance

Another major physiological aspect of the smolting process is the increase in hypoosmoregulatory ability, which preadapts the fish to life in seawater. This is frequently associated with a dramatic increase in gill chloride cell numbers and in gill Na^+, K^+-ATPase enzyme activity — the so-called salt pump. Both of these can be monitored in hatcheries to follow the development of smoltification. Short-term ability to survive in seawater, however, cannot by itself be used to distinguish smolts from parr. Juvenile steelhead trout and coho and Atlantic salmon develop seawater tolerance before the smolting process is complete (Conte and Wagner 1965; Conte et al. 1966; Wagner 1974b; Farmer et al. 1978). Nevertheless, the hypoosmoregulatory ability of parr can be distinguished physiologically from that of true smolts during the process of adaptation to seawater. For example, high salinity (40 º/oo) challenge reveals salinity tolerance in Atlantic salmon smolts that is undeveloped before smoltification and therefore is an excellent indicator of smolt status and preparedness for hypoosmotic regulation (Saunders and Henderson 1978).

Salinity tolerance is also influenced by fish size. In species such as Atlantic salmon, steelhead trout, and coho salmon hypoosmoregulatory capacity increases with growth until smolting occurs. However, if the smolts are prevented from entering seawater, their ability to adapt to it then decreases despite continued growth (Evropeitseva 1957; Koch 1968; Zaugg and McLain 1970). This phenomenon (desmoltification or parr-reversion) has been suggested as one reason for the frequently lower growth rates of landlocked compared with sea-run salmon. However, Saunders and Henderson (1969) concluded that stunting of landlocked Atlantic salmon was more likely due to inadequate food supply in fresh water than to salinity *per se*. In "non-smolting" species such as non-anadromous rainbow trout *(Salmo gairdneri),* continued growth appears to confer increased seawater adaptive capability (Houston 1961; Landless and Jackson 1976).

Because of the seasonal increase in salinity tolerance during the freshwater rearing phase, periodic seawater challenge tests can be used to determine if hypoosmoregulatory ability is developing normally. Salinity tolerance generally peaks at the time of smolting, and physiological monitoring via the seawater challenge test can help prevent the dramatic suppression of feeding behavior, growth and resultant stunting, and en-docrine abnormalities that can occur when salmon are transferred to seawater before smoltification is complete (Clarke and Nagahama 1977).

Proximate Composition, Lipid-Moisture Dynamics, Carbohydrate Metabolism

Another physiological change that could be used in hatchery monitoring of smoltification development is the reduction in total body lipid that occurs. For example, Komourdjian et al. (1976b) have demonstrated that the total lipid content of Atlantic salmon is significantly lower than in nonsmolted controls. Coincident with falling lipid levels, total moisture content tends to increase. Both of these effects are probably the result of metabolic changes such as the increased oxygen consumption rates that accompany smolting.

Carbohydrate metabolism is also altered during the parr-smolt transformation. Blood glucose is elevated at the expense of liver glycogen reserves, and protein and lipid catabolism is accelerated (Wendt and Saunders 1973). At the same time of year, salmon parr have high liver glycogen levels and low blood glucose, and unlike smolts survive transport without sharp reverses in these carbohydrate levels. Wedemeyer (1972) showed that handling stress typical of hatchery operations caused severe hyperglycemia, hypochloremia, and the activation of subclinical bacterial kidney disease infections in coho salmon smolts. Carbohydrate metabolism of parr was much less affected.

Endocrine System

A number of endocrine changes have been observed histologically during the parr-smolt transformation, such as an activation of thyroid, interrenal, and pituitary growth hormone cells. Recently, Dickhoff et al. (1978) demonstrated as well that circulating thyroid hormone levels are very high in coho salmon smolts. As a hatchery application, oral administration of thyroid hormone, particularly triiodothyronine, shows considerable promise as a physiological technique for stimulating growth and smolting of cultured salmon (Higgs et al. 1977).

That the pituitary growth hormone is also involved in the smolting process is indicated in several ways. First of all, there is a period of rapid growth during the smolting process, and secondly, there is a strong correlation between growth rate itself and smolting tendency. Histologically, hypertrophy as well as hyperplasia of the pituitary growth hormone cells is seen (Clarke and Nagahama 1977). Also, injections of mammalian growth hormone enable brown trout *(Salmo trutta)* or Atlantic salmon parr to survive in seawater (Smith 1956; Komourdjian et al. 1976a). Similarly, Clarke et al. (1977) demonstrated that growth hormone injections improve the hypoosmoregulatory performance of underyearling sockeye salmon *(Oncorhynchus nerka)*. However, much additional work will be needed to clarify the functions of growth hormone fully, and other hormones in coordinating the smolting process. Particularly intriguing is the possible role of prolactin, which is known to influence the osmoregulatory capability of salmon in fresh water. Little is known about its function in smolts (Zambrano et al. 1972; Clarke and Nagahama 1977; Nagahama et al. 1977).

Gill Na⁺, K⁺-ATPase Activity

From a physiological monitoring point of view, one of the important developmental changes taking place during salmonid smolting is an increase in the Na⁺, K⁺-activated ATPase enzyme activity of the gill microsomal system (Zaugg and McLain 1970; Zaugg and Wagner 1973; McCartney 1976; Giles and Vanstone 1976; Saunders and Henderson 1978). The gill ATPase system is involved, directly or indirectly, in the "physiological distillation" of seawater that salmonids drink to replace the water lost osmotically. Other euryhaline fishes also show increased ATPase and ion pump activity before and after adaptation to seawater. In salmonids, the increase in ATPase activity begins in fresh water during smoltification, and peaks near the

time the fish show typical active migratory behavior (Fig. 1). If the smolts enter seawater, gill ATPase activity continues to rise and stabilizes at an elevated level. If the smolts are not allowed to migrate, the ATPase activity will gradually decline to initial levels as parr-reversion occurs.

Behavioral Changes

In addition to the physiological aspects of smolting, several behavioral changes distinguish smolts from parr. Atlantic salmon parr are territorial and live on or near the stream bottom, whereas smolts tend to develop schooling behavior and swim at mid-depth. Salmonids are physostomous and smolts maintain larger volumes of air in their swimbladders than do parr. The resulting increased buoyancy (despite the reduced lipid content of smolts) probably influences the initiation of downstream migration.

Another behavioral pattern is the development of preference for increased salinities. Baggerman (1960) demonstrated that salinity preference was also affected by photoperiod, and suggested that endocrine activity, controlled by environmental priming factors, was involved.

Perhaps the most noticeable behavioral change shown by smolts is downstream migration. Migratory activity has long

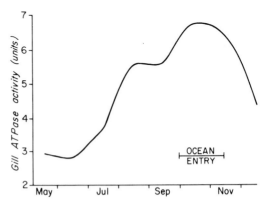

FIGURE 1.—Gill Na⁺, K⁺-ATPase activity in juvenile spring chinook salmon over a 2-year period. This chinook race normally enters the ocean from the Rogue River (Oregon) in September-October (redrawn from Ewing et al. 1979).

been used in hatcheries as an index of smolt development. However, it has limitations when used by itself. Baggerman (1960) also postulated the model of diadromous fish migrations in which environmental priming factors, such as daylength and temperature, condition smolts physiologically, mainly through the pituitary, to prepare for migration. External releasing factors, such as light intensity and river flow rates, finally trigger migrations. From this and many case history observations, Koch (1968) concluded that smoltification and migration may have separate causes.

A practical application of the relation between smolting and migratory behavior has been the use of delayed smolt releases from hatcheries to obtain "nonmigrating" populations of Pacific salmon in Puget Sound (Mahnken and Joyner 1973). As an example, chinook salmon (*Oncorhynchus tshawytscha*) have been induced to form residual populations in Puget Sound (Washington) by delaying their release until parr reversion begins.

Size and Time at Release

A final criterion to consider is the influence of body size on smolting. In wild Atlantic salmon, the minimum size threshold for smolting appears to be 12-13 cm, although most smolts are in the 14-17-cm range. However, smolts may range in age from two, to as much as five years. Although smaller salmon may be silvery, have a certain degree of salinity tolerance, and show increased buoyancy at smolting time, other aspects, especially migratory behavior, do not develop until after the threshold size has been reached. It thus appears that, although several of the components of smolting exist in very small salmon, the needed physiological coordination is lacking. For Pacific salmon and steelhead, the general rule governing size at release has been that, within limits, "bigger is better." However, recent work has shown that larger smolts do not necessarily have the highest gill ATPase activities. For spring chinook salmon, at least, there is a minimum and a maximum fish size beyond which parr cannot enter

the gill ATPase development cycle (Ewing et al. 1979). For spring chinook salmon, these lower and upper limits are approximately 80-90 mm and 140-150 mm (Fig. 2).

It is difficult to separate the effects of smolt size from the effects of time of release on the number of returning adult salmon. Peterson (1973) found that releasing 2-year-old Atlantic salmon smolts smaller than about 14 cm gave low adult returns. Greater numbers of returning adults were obtained as smolt length increased from 114 to 18 cm, while, above this, returns again tapered off. In contrast, yearling smolts showed good adult returns at release lengths of 12.5 cm to 14 cm. This difference between the two age classes indicated that size is not an absolute factor in determining marine survival. Perhaps growth rate is more important. For Pacific (coho) salmon, time of release for a given smolt size also has a profound effect on marine survival. Bilton (1978) quantified this phenomenon by releasing smolts of several size categories over a period of four months. He found that the optimal size varied with time of release and that maximum return of adult salmon resulted from the release of smolts of approximately 20 g, just prior to the summer solstice. Releasing very large smolts resulted in more precocious males (jacks) in the population and thus a reduced total adult biomass.

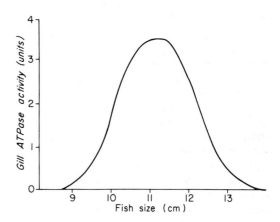

FIGURE 2.—The influence of fish size on development of gill Na⁺, K⁺-ATPase activity; spring chinook salmon. Minimum size is 8-9 cm (redrawn from Ewing et al. 1979).

Presently, most salmon culture facilities are managed such that release sizes are about 90/lb or 200/kg (5 g) for fall chinook salmon, 15-17/lb or 33-38/kg (26-30 g) for coho salmon, and 7-8/lb or 16-18/kg (56-63 g) for steelhead trout and spring chinook salmon.

In summary, considerable progress has been made in understanding the physiology of smoltification, but much additional information is still needed before biologists will be able to control it to increase efficiency in the use of hatchery-propagated juveniles for anadromous fisheries enhancement programs. The former belief that several separate physiological processes must take place before smolting is complete has been replaced by the more functional view that smolting is the coordination of several separate physiological and behavioral processes, including downstream migration, which results in the ability of the fish to continue to grow and develop normally in the ocean. Some of these changes are in progress well before smolting time, some occur in response to recognizable environmental stimuli, and perhaps all of them have an endogenous rhythm. In Atlantic and Pacific salmon, photoperiod has been identified as the major environmental priming factor and coordinator that brings these endogenous physiological processes together on a temporal basis. Water temperature controls the range within which these processes can go on and determines their rates of reaction. Hatchery biologists can thus activate, postpone, or coordinate such aspects of smoltification as body silvering, salinity tolerance and preference, growth rate, and gill ATPase activity, through the use of bio-engineering equipment for photoperiod and temperature control.

Effect of Hatchery Environment and Fish Cultural Methods on Smoltification Success

Water Quality

It has recently become apparent that the water chemistry requirements for smoltification are exacting (Table 2).

TABLE 2. — Aquatic environmental quality recommended during the freshwater (hatchery) phase of the anadromous salmonid life cycle.

Water chemistry parameter	Environmental limits
Acidity	pH 6-9
Alkalinity	at least 20 ppm (as $CaCO_3$)
Ammonia (NH_3)[1]	0.02 ppm (un-ionized form)
Cadmium[2]	0.4 ppb in soft water (<100 ppm as $CaCO_3$
Cadmium[3]	3 ppb in hard water (>100 ppm)
Chloride[4]	2-3 ppm in soft water (<100 ppm)
Copper[4]	less than 5 ppb in soft water, 30 ppb in hard water
Herbicides	(See text for recommendation.)
Lead	0.03 ppm
Mercury (organic or inorganic)	0.2 ppb maximum, 0.05 ppb average
Nitrogen[5]	maximum total gas pressure 110% of saturation
Nitrite	100 ppb in soft water, 200 ppb in hard (30 and 60 ppb nitrite, nitrogen)
Oxygen	at least 75% of saturation at 5 °C, 90% at 15 °C
Ozone residual[6]	3 ppb
Polychlorinated biphenyls (as Arochlor 1254)	0.002 ppm
Total suspended and settleable solids	80 ppm or less

[1]0.005 ppm is probably healthier for salmonids.

[2]To protect salmonid eggs, fry, and normal smoltification. For nonsalmonids, 4 ppb.

[3]To protect salmonid eggs and fry. For nonsalmonids, 0.03 ppm is acceptable.

[4]Minimum to prevent NO_2 toxicity in soft water re-use hatcheries and assure normal smolt development.

[5]Cu above ppb may suppress gill ATPase activity and compromise smoltification in anadromous salmonids.

[6]Maximum safe residual when ozone is used for fish disease control in hatcheries or is discharged to receiving water when used as a replacement biocide for chlorine in power plant cooling systems or sewage treatment plant effluents.

Quite small alterations can have a major deleterious effect on subsequent seawater survival. One example of this that has particularly serious hatchery management consequences is the effect of trace heavy metal exposure during rearing. This would normally be due to hatchery design faults or to exposure from mineral deposit drainage or nonpoint source industrial pollution that stems from siting problems. Unfortunately, the gill ATPase enzyme system of smolts and pre-smolts is sensitive to levels of dissolved heavy metals well within the current maximum safe exposure limits recommended for freshwater fish populations. For example, chronic copper exposure during the parr-smolt transformation of coho salmon, at only 20-30 $\mu g/l$, partly or completely inactivates the gill ATPase system (Lorz and McPherson 1976). The biological damage is not apparent at the hatchery but when the fish enter salt water, severe mortalities begin. An equally devastating consequence is the suppression of normal migratory behavior. Similar results were reported by Davis and Shand (1978) for sockeye salmon *(Oncorhynchus nerka)*. Exposing smolts for 144 h to copper at 30 $\mu g/l$ in fresh water impaired subsequent hypoosmoregulatory performance, as revealed by elevated blood electrolytes during the seawater challenge test and a later seawater mortality.

Cadmium levels of more than 4 $\mu g/l$ in fresh water also result in a dose-dependent mortality when previously exposed coho salmon smolts are transferred directly into 30 o/oo seawater (Lorz et al. 1978a). However, if a 5-day recovery period in fresh water is allowed, seawater survival returns to normal. In contrast to copper, sublethal exposure to cadmium or zinc during rearing apparently does not adversely affect migratory behavior. However, if even very low levels of copper (10 $\mu g/l$) are simultaneously present, migratory behavior is inhibited and normal gill ATPase activity is suppressed. Unfortunately, few apparent alterations in freshwater growth or in feeding behavior occur to act as a warning to the fish culturist and the smolts appear to be "normal" when released. One biological consequence of smolts' being unable to migrate directly into the ocean would be an increased residence time in the river and estuary, with consequent increased exposure to predation and to diseases such as

vibrio and viral erythrocytic necrosis (VEN).

Nickel or chromium exposure at up to 5 mg/l for 96 hours in fresh water apparently does not inhibit migratory behavior or capability for ocean survival, but sublethal exposure to mercury results in a dose-dependent seawater mortality (Lorz et al. 1978a).

Less is known about the effects of low-level heavy metal exposure on smoltification and migration of salmon such as sockeye, pink *(O. gorbuscha)*, chum *(O. keta)*, or fall chinook, which have much different freshwater behavior patterns. Servizi and Martens (1978), who studied the effects of copper, cadmium, and mercury, reported that mortality, hatching, and growth of sockeye salmon during the egg-to-fry stage were not affected by continuous exposure to 5.7 $\mu g/l$ cadmium. For copper, the incipient lethal level during the egg-to-fry stage was within the range of 37 to 78 $\mu g/l$ for sockeye salmon and 25 to 55 $\mu g/l$ for pink salmon. Copper inhibited egg capsule softening as well, but mortality at hatching occurred only at concentrations also lethal to eggs and alevins. Dissolved copper was concentrated by eggs, alevins, and fry in proportion to exposure concentrations. For pink salmon, mortalities occurred when tissue copper concentrations reached levels of 105 mg/kg in eyed eggs and 7 mg/kg in fry.

For (inorganic) mercury exposure, concentrations of only 2.5 $\mu g/l$ during egg incubation caused development of malformed embryos. Hatching success, fry mortality, and growth were less sensitive to mercury exposure than was malformation. Mercury was also concentrated by sockeye and pink salmon eggs, and embryo abnormalities occurred when concentrations reached about 1.9 mg/kg.

In addition to the heavy metal exposure problem, increasingly intensive forest, range, and agricultural practices are now occurring that may eventually result in chronic low-level herbicide concentrations in juvenile salmon rearing waters. Coho salmon smolts exposed for only 96 h to the herbicide Tordon 101 at 0.6-1.8 mg/l just prior to release did not migrate as successfully as did those in the control group (Lorz et al. 1978b). Tordon 101 is a formulation of picloram and the dimethylamine salt of 2,4-D used for brush, weed, and vine control on non-crop lands, including rights of way. Other 2,4-D and 2,4,5-T formulations, such as the esters used for the control of Eurasian watermilfoil, may also inhibit smolt functions and migratory behavior. Low level 2,4-D exposure is known to cause an avoidance response in nonanadromous trout (Folmar 1976).

Water Temperature

To reduce costs, some anadromous fish are now being reared at elevated water temperatures in re-use hatcheries to accelerate growth and shorten the normal time needed to produce smolts. However, such artificial temperature regimes must be used with care because rearing temperature has a strong influence on the pattern of gill ATPase and hypoosmoregulatory development as well as on growth. Guidelines for hatcheries are given in Table 3. In some anadromous species, elevated rearing temperatures will not only accelerate the onset of smolting but will also hasten parr-reversion so that the duration of the smolting period itself is shortened (Fig. 3). For example, coho salmon show a slow, sustained rise in gill ATPase

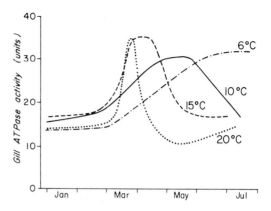

FIGURE 3.—Summary of the effect of water temperature on the pattern of gill Na⁺, K⁺-ATPase activity development of juvenile coho salmon (redrawn from Zaugg and McLain 1976).

activity at 6°C, a more normal pattern at 10°C, and a precocious development pattern at temperatures up to 20°C (Zaugg and McLain 1976). However, temperatures as high as 15°C have been used to rear coho salmon so that 200-mm "smolts" can be released at the end of their first year. Unfortunately, as shown in Figure 3, parr-reversion is also accelerated so that there is very little latitude in release time (Novotny 1975; Donaldson and Brannon 1976; Clarke and Shelbourn 1977). Juvenile fall chinook salmon also undergo a more rapid reversion to the parr condition at elevated holding temperatures (Clarke and Blackburn 1977). If elevated rearing temperatures must be used, it may be possible to overcome the problem of accelerated parr-reversion by holding sockeye, coho, or chinook salmon smolts in dilute seawater of 10-20 ⁰/oo salinity (W.C. Clarke, 1978, unpublished data).

In some salmonids, temperature acceleration of growth may cause inhibition of smoltification. Steelhead trout are particularly sensitive (Fig. 4), and ATPase development is strongly inhibited at 13°C and above (Adams et al. 1973, 1975; Zaugg and Wagner 1973). In the case of Atlantic salmon, laboratory experiments have shown apparent smoltification at temperatures as high as 15°C (Saunders and Henderson 1970; Komourdjian et al. 1976b). However, until their temperature

TABLE 3.—Maximum water temperatures recommended for the hatchery production of anadromous salmonids to optimize egg development, hatching, growth, smoltification, and migratory behavior.

Temperature		Physiological aspect considered
°C	°F	
15	59	Growth, smoltification, and migration of anadromous salmonids (except steelhead trout and Atlantic salmon).
13	55	Spawning, hatching, and egg development of Pacific salmon and trout; normal smoltification growth and migration of steelhead trout and Atlantic salmon.
9	48	Spawning, hatching, and egg development of Atlantic salmon.

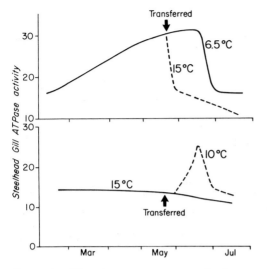

FIGURE 4.—Effect of water temperature on development of gill Na$^+$, K$^+$-ATPase activity of steelhead trout. A. Fish held at 6.5°C, and transferred to 15°C (arrow). B. Held at 15°C and transferred to 10°C (arrow); the ATPase development phase of smoltification then occurred (redrawn from Zaugg et al. 1972).

dependence has been fully characterized, it is prudent to assume that Atlantic salmon are not unlike steelhead trout. In this regard, it may be significant that native runs of Atlantic salmon show their greatest downstream migrant activity as the temperature rises to 10°C and smolt migration is normally completed before the water warms to 15°C.

Photoperiod

The seasonal cycle of growth and smolting in juvenile salmon has a strong endogenous component that is apparently synchronized by the yearly photoperiod cycle (Hoar 1965, 1976; Poston 1978). Experiments with juvenile Atlantic salmon have demonstrated that the spring growth acceleration, as well as smolting, can be advanced by several months if an increasing or long photoperiod is applied during winter (Saunders and Henderson 1970; Knutsson and Grav 1976; Komourdjian et al. 1976b). Photoperiod also has a marked effect on the growth and hypoosmoregulatory ability of underyearling sockeye and coho salmon (Clarke and Shelbourn 1977;

Clarke et al. 1978). Wagner (1970, 1974a) demonstrated that the photoperiod cycle synchronized the development of smolt characteristics and migratory behavior in juvenile steelhead trout. The rate of change of photoperiod is an important cue, and the timing of smolting can be modified more successfully by shifting the phase of the photoperiod cycle than by altering its frequency. Prolonged exposure to a long photoperiod inhibits growth and smolting. Again, since water temperature controls the rate of physiological response to photoperiod, effects are apparent sooner at elevated temperatures.

Hatchery Practices, Fish Disease Treatments, Handling, Scale Loss

The smolt performance problems identified thus far suggest that improvements in present hatchery design and operation would contribute to the production of better quality smolts, increased ocean survival, adult returns, and total contribution to the fishery. Two major areas need to be considered in some detail. First, the fish cultural methods themselves. For example, normal ATPase development in juvenile chinook salmon can apparently be suppressed by crowding stress at high population densities (Strange et al. 1978). Few data are yet available, but these preliminary results have extremely important implications for bio-engineering design. Second, certain fish disease therapeutants, including several drugs and chemicals commonly used pre-release, are now known to reduce the ability of smolts to tolerate seawater as a side effect of the treatment (Bouck and Johnson 1979).

In the case of crowding stress at high hatchery raceway loadings, guidelines, which are used by the Washington Department of Fisheries and give good results under the indicated limitations, are summarized in Table 4. Raceways using recirculated water, however, must be operated at reduced loadings. For example, maximum loadings of spring chinook salmon in one type of deep recirculating raceway are about $0.9 \text{ kg·l}^{-1}\text{·min}^{-1}$ with a space density requirement of $0.05 \text{ kg·m}^{-3}\text{·mm}^{-1}$ fish length. Increasing the loadings to 0.9

TABLE 4.—Recommended maximum hatchery pond loadings based on experience by the Washington Department of Fisheries (WDF) expressed as flow index (kg/l·min⁻¹·mm⁻¹ length) for coho, and fall and spring chinook salmon (re-calculated from Wedemeyer and Wood 1974). Information is not available for other temperatures, sizes, or species. Recirculating raceways require reduced loadings (J. Wood, WDF, personal communication).

Water temperature	Fish size (number per kg)						
	2222	1111	222	111	73	56	33
Coho salmon	Fish length (mm						
	36	43	74	94	107	117	140
3	0.06	0.014	0.014	0.015	0.018	0.021	0.022
9	0.01	0.012	0.011	0.014	0.017	0.018	0.017
14	0.008	0.0095	0.008	0.013	0.012	0.013	0.014
17	—	0.006	0.006	0.007	0.0085	0.01	0.009
20	—	—	0.0025	0.0025	0.0035	0.0035	0.0035
Chinook salmon	Fish length (mm)						
	38	48	81	104	119	130	155
3	0.01	0.011	0.0095	0.01	0.0115	0.012	0.011
9	0.0085	0.008	0.008	0.008	0.0095	0.01	0.009
14	0.0065	0.006	0.0055	0.0055	0.007	0.0075	0.0075
17	—	0.003	0.0045	0.004	0.004	0.0045	0.0045

NOTES: The values represent final values at time of smolt release for fish sizes of 222 fish per kg or larger. Loadings should not exceed ¾ of the table value before the final pond divisions so that the above loadings will not be exceeded at time of release. In addition, chinook salmon should not be fed at levels in excess of 0.012 kg of food per liter per minute inflow before they are 444/kg; 0.018 kg/l·min⁻¹ at 222 fish/kg, and 0.024 kg/l·min⁻¹ at larger fish sizes.

The maximum space density for chinook appears to be about 0.18 kg/m³·mm⁻¹ body length.

The raceway measurements should approximate 1 unit depth : 3 units width : 10 units length with 1.5 to 2.0 exchanges per hour. Recirculation raceways require that reduced loadings be used.

kg·l⁻¹·min⁻¹ and 0.07 kg·m⁻³·mm⁻¹ fish length will not actually increase adult returns even though more "smolts" are produced (J. Wood, 1977, personal communication).

In the case of fish disease treatments, Bouck and Johnson (1979) found that high seawater mortality rates followed treatments of clinically healthy coho salmon smolts with copper sulfate, hyamine 1622, potassium permanganate, malachite green or MS-222. About 10% mortality occurred in groups treated with formalin and nifurpirinol. Little or no seawater mortality resulted from treatment with trichloron, simazine, or nifurpirinol. If a 4-day freshwater recovery period was allowed, seawater mortality was reduced after treatments with copper sulfate, potassium permanganate, MS-222, and malachite green; greatly reduced after treatments with hyamine 1622; and reduced to nil for the other agents. Thus the present emphasis on releasing hatchery fish free of external parasites and subclinical bacterial infections can easily result in the release of disease-free "smolts," which have difficulty converting to seawater. These findings, which are summarized in Table 5, suggest that it would be good practice to allow at least a 1-week freshwater recovery period before smolts are released to migrate to the ocean. For hatcheries sufficiently distant from the sea, the period of river migration might suffice.

Another hatchery practice that can adversely affect survival is the handling or hauling of smolts. Two physiological factors must be considered. First, smolts are much more sensitive to handling stress itself than are parr or post-smolts (Wed-

TABLE 5.—Summary of seawater survival of coho salmon smolts following standard treatments with drugs and chemicals commonly used in fish culture (Bouck and Johnson 1979).

Chemical	Concentration of active ingredient (mg/l)	Length of daily exposure (min)	Consecutive days treatment was given	Total mortality (%) during 10 d in seawater	
				Direct transfer into seawater following treatment	Treatment, then 4 d in freshwater, then 4-h acclimation to seawater
Controls	none	none	none	0	0
				4	
Copper sulfate	37.0	20	1	100	20
Endothal	5.0	60	1	100	4
Formalin	167.0	60	1	12	0
Hyamine 1622	2.0	60	4	68	4
Malachite green	1.0	60	1	44	12
MS-222	100.0	6	1	100	12
Nifurpirinol	1.5	60	4	8	0
Oxytetracycline	1.0	60	1	20	12
Potassium permanganate	2.0	60	3	80	12
Quinaldine	2.5	10	1	0	0
Simazine	2.5	60	1	4	0
Trichlorofon	0.5	60	1	0	0

emeyer 1972). Problems with reduced survival , including that due to subsequent disease epizootics, have led to the development of stress mitigation procedures. Mineral salt additions of several kinds have been developed to help minimize handling stress in both smolts and nonanadromous fishes. These include NaCl at 3-5 °/oo (Wedemeyer 1972; Hattingh et al. 1974) or $CaCl_2$ at 50 mg/l (Wedemeyer and Wood 1974). For smolt hauling, salt (NaCl) at up to 15 °/oo has found considerable use both for stress mitigation and for *Saprolegnia* control, a major disease-related cause of delayed smolt mortality (Long et al. 1977). Normally, some potassium must also be added if NaCl at 10 °/oo or more is to be used. If easily available, diluted seawater itself (5-15 °/oo) is very effective in mitigating handling stress.

Unfortunately, in addition to the physiological stress itself, various degrees of scale loss commonly occur when smolts are handled or hauled which, in turn, causes additional stress. Bouck and Smith (1979)

showed that considerable seawater mortality occurred after smolts were experimentally descaled as little as 10% of the body surface. The estimated 10-day TL_m was 50% mortality for a 10% scale loss on the ventral surface. These results have serious implications for smolt-hauling programs involving stocking directly into seawater or for interpreting high seas tagging results because of the delayed aspect of the mortality pattern. However, for smolt hauling involving release into rivers, the consequences would potentially be lessened by the fact that a 5-day recovery period in fresh water almost completely restored full tolerance to seawater, as judged by lack of mortality. However, partially descaled smolts released near a river mouth would probably show an avoidance reaction to salinity and would tend to remain in the estuarine area longer than normal, thus increasing exposure to diseases such as VEN and vibrio, as well as to predation.

Another physiological consequence of scale loss to consider is that partially descaled smolts can absorb harmful amounts of Mg^{++} and K^+ from seawater or from the stress-mitigating salt mixtures themselves (Wedemeyer, unpublished data). As little as 10% (dorsal) scale loss can result in a life-threatening hyperkalemia and hypermagnesemia in coho salmon smolts. Even if immediate mortalities do not occur, blood electrolyte imbalances of this magnitude are debilitating and would be expected to markedly reduce the ability of smolts to survive further stress or to escape predation.

Recommendations for Hatchery Bio-Engineering Design and Operations

On the basis of the present understanding of the need for coordination of the physiological processes of smoltification, the following environmental requirements and hatchery design and operations criteria are suggested for the freshwater rearing phase of anadromous salmonids.

1. Water temperature should ideally follow a natural seasonal pattern if production constraints allow it. If elevated temperatures must be used to accelerate growth, this should be done during October through December. However, temperature should not be elevated too quickly or to more than about 10 °C in late winter unless accelerated smolting is desired. Temperature should be held below 13 °C at least 60 days prior to release of Atlantic salmon or steelhead trout. Similarly, for coho and chinook salmon, water temperatures should be held below about 12 °C, if possible, in order to prevent premature smolting and parr-reversion. For other fishes, the temperature regimens needed to produce the most successful smolting are unknown.

2. In the absence of complicating factors such as altered river and estuarine ecology, the time, age, and size-at-release strategies should coincide as nearly as possible with the historical seaward migration of naturally produced fish in the recipient stream if genetic strains are similar. At headwater production sites, much earlier release may be called for. The

desired result is that hatchery-reared smolts that are genetically similar to wild smolts enter the sea at or near the same time. Planting large parr in the fall would not be unreasonable if observation shows naturally produced parr move to down-river sites at that time.

3. The photoperiod is probably one of the easier environmental factors for bio-engineers to control in production facilities. However, this environmental priming factor should be altered cautiously, since it is easily misused. Unless accelerated or delayed smolting is required, the parr should be held in outdoor ponds with no artificial light. This constraint may preclude the use of the night floodlighting often needed for hatchery security unless intensities are sufficiently low. Indoor rearing should also be done under natural light, but if artificial light must be used, it should be timed to simulate natural photoperiod cycle, phase, and intensity.

4. Pre-release fish disease treatment should ideally be limited to medication known to have no effect on smolt performance. When this is not possible, the smolts should be held in fresh water long enough to allow recovery of their hypoosmoregulatory ability.

5. When environmental control is used to alter time, age, or size at release because of changed resource management requirements, physiological testing should be used to monitor smolt development and detect any adverse effects of the hatchery practices or the environmental factors that were used. Gill Na^+, K^+-ATPase, plasma thyroxine (T_4) levels, or the seawater challenge blood sodium test is recommended.

6. Finally, in evaluating present or proposed bio-engineering design concepts, the major criterion of success should be a benefit-cost ratio based on the summation of hatchery returns and contribution to the fishery.

References

ADAMS, B.L., W.S. ZAUGG, AND L.R. McLAIN. 1973. Temperature effect on parr-smolt transformation in steelhead trout (Salmo gairdneri) as measured by gill sodium-potassium stimulated adenosinetriphosphatase. Comp. Biochem. Physiol. 4A:1333-1339.

———. 1975. Inhibition of salt water survival and Na-K-ATPase elevation in steelhead trout *(Salmo gairdneri)* by moderate water temperatures. Trans. Am. Fish. Soc. 104(4):766-769.

BAGGERMAN, B. 1960. Salinity preference, thyroid activity, and the seaward migration of four species of Pacific salmon *(Oncorhynchus)*. J. Fish. Res. Board Can. 17:295-322.

BILTON, H.T. 1978. Returns of adult coho salmon in relation to mean size and time at release of juveniles. Fish. Mar. Serv. Tech. Rep. No. 832. 27 pp.

BOUCK, G., AND D. JOHNSON. 1979. Medication inhibits tolerance to seawater in coho salmon smolts. Trans. Am. Fish. Soc. 108:63-66.

BOUCK, G.R., AND S.D. SMITH. 1979. Mortality of experimentally descaled smolts of coho salmon *(Oncorhynchus kisutch)* in fresh and salt water. Trans. Am. Fish. Soc. 108:67-69.

CLARKE, W.C., AND J. BLACKBURN. 1977. A seawater challenge test to measure smolting of juvenile salmon. Fish. Mar. Serv. Tech. Rep. 705. 11 pp.

CLARKE, W.C., S.W. FARMER, AND K.M. HARTWELL. 1977. Effect of teleost pituitary growth hormone on growth of *Tilapia mossambica* and on growth and seawater adaptation of sockeye salmon *(Oncorhynchus nerka)*. Gen. Comp. Endocrinol. 33:174-178.

CLARKE, W.C., AND Y. NAGAHAMA. 1977. Effect of premature transfer to sea water on growth and morphology of the pituitary, thyroid, pancreas, and interrenal in juvenile coho salmon *(Oncorhynchus kisutch)*. Can. J. Zool. 55:1620-1630.

CLARKE, W.C., AND J.E. SHELBOURN. 1977. Effect of temperature, photoperiod and salinity on growth and smolting of underyearling coho salmon. Am. Zool. 17(4):957 (Abstr.).

CLARKE, W.C., J.E. SHELBOURN, AND J.R. BRETT. 1978. Growth and adaptation to seawater in underyearling sockeye *(Oncorhynchus nerka)* and coho *(O. kisutch)* salmon subjected to regimes of constant or changing temperature and daylength. Can. J. Zool. 56:2413-2421.

CONTE, F.P., AND H.H. WAGNER. 1965. Development of osmotic and ionic regulation in juvenile steelhead trout *Salmo gairdneri*. Comp. Biochem. Physiol. 14:603-620.

CONTE, F.P., H.H. WAGNER, J. FESSLER, AND C. GNOSE. 1966. Development of osmotic and ionic regulation in juvenile coho salmon *(Oncorhynchus kisutch)*. Comp. Biochem. Physiol. 18:1-15.

DAVIS, J.C., AND I.G. SHAND. 1978. Acute and sublethal copper sensitivity, growth, and saltwater survival in young Babine Lake sockeye salmon. Fish. Mar. Serv. Tech. Rep. 847. 55 pp.

DICKHOFF, W., L. FOLMAR, AND A. GORBMAN. 1978. Changes in plasma thyroxine during smoltification of coho salmon *Oncorhynchus kisutch*. Gen. Comp. Endocrinol. 36:229-232.

DODD, J.M., AND A.J. MATTY. 1964. Comparative aspects of thyroid function. Pages 303-356 *in* R.P. Rivers and W.R. Trotter, eds. The thyroid gland. Butterworth, Land. Washington, D.C.

DONALDSON, L.R., AND E.L. BRANNON. 1976. The use of warm water to accelerate the production of coho salmon. Fisheries 1(4):93-100.

EVROPEITSEVA, N.V. 1957. Transformation of smolt stage and downstream migration of young salmon. Uch. Zap. Leningr. Gos. Univ. 228. Ser. Biol. Nauk. 44:117-154. (Fish. Res. Board Can. transl. Ser. No. 234).

EWING, R., S. JOHNSON, H. PRIBBLE, AND J. LLICHATOWICH. 1979 Temperature and photoperiod effects on gill (Na $^+$K) – ATPase activity in chinook salmon *(Oncorhynchus tshawytscha)*. J. Fish. Res. Board Can. 36:1347-1353.

FARMER, G.J., J.A. RITTER, AND D. ASHFIELD. 1978. Seawater adaptation and parr-smolt transformation of juvenile Atlantic salmon, *Salmo salar*. J. Fish. Res. Board Can. 35(1):93-100.

FOLMAR, L. 1976. Overt avoidance response of rainbow trout fry to nine herbicides. Bull. Environ. Contam. Toxicol. 15:509-513.

GILES, M.A., AND W.E. VANSTONE. 1976. Changes in oubain sensitive adenosine triphosphatase activity in gills of coho salmon *(Oncorhynchus kisutch)* during parr-smolt transformation. J. Fish. Res. Board Can. 33:54-62.

HATTINGH, J., F. LEROUX FOURIE, AND J.H.J. VAN VUREN. 1974. The transport of freshwater fish. J. Fish Biol. 7(4):447-449.

HIGGS, D.A., U.H.M. FAGERLUND, J.R. MCBRIDE, H.M. DYE, AND E.M. DONALDSON. 1977. Influence of combinations of bovine growth hormone, 17-methyltestosterone, and L-thyroxine on growth of yearling coho salmon *(Oncorhynchus kitsutch)*. Can. J. Zool. 55(6):1048-1056.

HOAR, W.S. 1965. The endocrine system as a chemical link between the organism and its environment. Trans. R. Soc. Can. Vol. III, Ser. IV:175-200.

———. 1976. Smolt transformation: Evolution, behavior, and physiology. J. Fish. Res. Board Can. 33:1233-1352.

HOUSTON, A.H. 1961. Influence of size upon the adaptation of steelhead trout *(Salmo gairdneri)* and chum salmon *(Oncorhynchus keta)* to sea water. J. Fish. Res. Board Can. 18:401-415.

KNUTSSON, S., AND T. GRAV. 1976. Seawater adaptation in Atlantic salmon *(Salmo salar L.)* at different experimental temperatures and photoperiods. Aquaculture 8:169-187.

KOCH, H.J.A. 1968. Migration. Pages 4-22 *in* E.J.W. Barrington and C. Barker Jorgenson, ed. Perspectives in Endocrinology. Academic Press Inc., London and New York.

KOMOURDJIAN, M.P., R.L. SAUNDERS, AND J.C. FENWICK. 1976a. The effect of porcine somatotropin on growth and survival in sea water of Atlantic salmon *(Salmo salar)* parr. Can. J. Zool. 54(4):531-535.

_____. 1976b. Evidence for the role of growth hormone as a part of a 'light-pituitary axis' in growth and smoltification of Atlantic salmon (Salmo salar). Can. J. Zool. 54(4):544-551.

LANDGREBE, F.W. 1941. The role of the pituitary and thyroid in the development of teleosts. J. Exp. Biol. 18:162-169.

LANDLESS, P.J., AND A.J. JACKSON. 1976 Acclimatising young salmon to seawater. Fish Farming Int. 3(2):15-17.

LONG, C.W., J.R. McCOMAS, AND B.H. MONK. 1977. Use of salt (NaCl) water to reduce mortality of chinook salmon (Oncorhynchus tshawytscha) during handling and hauling. Mar. Fish. Rev. 39(7):6-9.

LORZ, H.W., AND B.P. McPHERSON. 1976. Effects of copper or zinc in freshwater on the adaptation to seawater and ATPase activity and the effects of copper on migratory disposition of coho salmon (Oncorhynchus kisutch). J. Fish. Res. Board Can. 33(9):2023-2030.

LORZ, H.W., R.H. WILLIAMS, AND C.A. FUSTISH. 1978a. Effects of several metals on smolting in coho salmon. U.S. Environmental Protection Agency, Grant Rep. R-804283. Oregon Dep. Fish Wildl., Corvallis. 187 pp.

LORZ, H.S., R.H. WILLIAMS, D. KUNKEL, L. NORRIS, AND B. LOPER. 1978b. Effect of selected herbicides on smolting of coho salmon. U.S. Environmental Protection Agency, EPA-600/3-79-071. Environ. Res. Lab., Corvallis, Oregon. 138 pp.

MAHNKEN, C., AND T. JOYNER. 1973. Salmon for New England fisheries. III. Developing a coastal fishery for Pacific salmon. Mar. Fish. Rev. 35:9-13.

McCARTNEY, T.H. 1976. Sodium-potassium dependent adenosine triphosphatase activity in gills and kidneys of Atlantic salmon (Salmo salar). Comp. Biochem. Physiol. 53(4A):351-353.

NAGAHAMA, Y., W.C. CLARKE, AND W.S. HOAR. 1977. Influence of salinity on ultrastructure of the secretory cells of the adenohypophyseal pars distalis in yearling coho salmon (Oncorhynchus kisutch). Can. J. Zool. 55(1):183-198.

NOVOTNY, A.J. 1975. Net-pen culture of Pacific salmon in marine waters. Mar. Fish. Res. 37:36-47.

PETERSON, H.H. 1973. Adult returns to date from hatchery reared one-year old smolts. Pages 219-226 in M.V. Smith and W.M. Carter, eds. Int. Atlantic Salmon Foundation Spec. Publ. Ser. 4.

PIGGINS, D.J. 1962. Thyroid feeding of salmon parr. Nature (Lond.) 195:1017-1018.

POSTON, H.A. 1978. Neuroendocrine mediation of photoperiod and other environmental influences on physiological responses in salmonids: A review. U.S. Fish Wildl. Serv. Tech. Pap. 96. 25 pp.

SAUNDERS, R.L., AND E.B. HENDERSON. 1969. Growth of Atlantic salmon smolts and post-smolts in relation to salinity, temperature and diet. Fish. Res. Board Can. Tech. Rep. 149. 20 pp.

_____. 1970. Influences of photoperiod on smolt development and growth of Atlantic salmon (Salmo salar).

J. Fish. Res. Board Can. 27:1295-1311.

SAUNDERS, R.L., AND E.B. HENDERSON. 1978. Changes in gill ATPase activity and smolt status of Atlantic salmon (Salmo salar). J. Fish. Res. Board Can. 35(12):1542-1546.

SERVIZI, J.A., AND D.W. MARTENS. 1978. Effects of selected heavy metals on early life of sockeye and pink salmon. Int. Pac. Salmon Comm., Prog. Rep. 39, New Westminster, B.C., Canada. 39 pp.

SMITH, D.C.W. 1956. The role of the endocrine organs in the salinity tolerance of trout. Mem. Soc. Endocrinol. 5:83-101.

STRANGE, R.J., C.B. SCHRECK, AND R.D. EWING. 1978. Cortisol concentrations in confined juvenile chinook salmon (Oncorhynchus tshawytscha). Trans. Am. Fish. Soc. 107:812-819.

WAGNER, H.H. 1970. The parr-smolt metamorphosis in steelhead trout as affected by photoperiod and temperature. Ph.D. thesis, Oregon State Univ., Corvallis. 175 pp.

_____. 1974a. Photoperiod and temperature regulation of smolting in steelhead trout Salmo gairdneri. Can J. Zool. 52:219-234.

_____. 1974b. Seawater adaptation independent of photoperiod in steelhead trout (Salmo gairdneri). Can. J. Zool. 52:805-812.

WEDEMEYER, G. 1972. Some physiological consequences of handling stress in the juvenile coho salmon (Oncorhynchus kisutch) and steelhead trout (Salmo gairdneri). J. Fish. Res. Board Can. 29:1780-1783.

WEDEMEYER, G., AND J. WOOD. 1974. Stress as a predisposing factor in fish diseases. U.S. Fish Wildl. Serv. Fish Dis. Leafl. No. 38. 8 pp.

WENDT, C.A.G., AND R.L. SAUNDERS. 1973. Changes in carbohydrate metabolism in young Atlantic salmon in response to various forms of stress. Int. Atl. Salmon Sympos. Spec. Pub. Ser. 4(1):55-82.

ZAMBRANO, D., R.S. NISHIOKA, AND H.A. BERN. 1972. The innervation of the pituitary gland of teleost fishes. Pages 50-66 in K.M. Knigge, D.E. Scott, and A. Weindl, eds. Brain endocrine interaction. Median eminence: Structure and function. Karger, Basel.

ZAUGG, W.S., B.L. ADAMS, AND L.R. McLAIN. 1972. Steelhead migration, potential temperature effects as indicated by gill adenosine triphosphatase activities. Science 176:415-416.

ZAUGG, W.S., AND L.R. McLAIN. 1970. Adenosine triphosphatase activity in gills of salmonids: Seasonal variations and salt water influence in coho salmon, Oncorhynchus kisutch. Comp. Biochem. Physiol. 35:587-596.

_____. 1976. Influence of water temperature on gill sodium, potassium-stimulated ATPase activity in juvenile coho salmon (Oncorhynchus kisutch). Comp. Biochem. Physiol. 54A:419-421.

ZAUGG, W.S., AND H.H. WAGNER. 1973. Gill ATPase activity related to parr-smolt transformation and migration in steelhead trout (Salmo gairdneri): Influence of photoperiod and temperature. Comp. Biochem. Physiol. 45B:955-965.

19

QUESTIONER: What physiological parameter do you consider to be the best indicator of smolt quality?

GARY WEDEMEYER: I don't think that there is any one best indicator. Smoltification is a pattern of the development of several things. You need to consider a variety of things such as migratory behavior, the growth spurt that occurs during the last stage of smoltification, the development of hypoosmoregulatory ability, the ATPase test, and perhaps the thyroid hormone levels. However, since we all live in the real world, you can't do all of these things. If I were to choose one, I would use the seawater challenge blood sodium test that was developed at the Canadian laboratory at Nanaimo. The test is simple. You put what you think are smolts into full-strength seawater and leave them for 24 hours. Then you take a blood sample. If they can keep salt out of their blood, they are smolts. If they can't, they are not. Survival has nothing to do with it. As you saw on my second slide, in order to find out for sure if the fish are going to survive, you have to observe them for several months. The salt level in the blood after 24 hours in salt water is diagnostic. If the blood sodium level is below about 170, they are smolts and will probably grow and develop normally. If it rises much above that, they are not smolts. They are just parr. That has a predictive value too. You can predict that they will not grow and develop normally.

QUESTIONER: If you were to concentrate on growing efficiency at either a warmwater or coldwater hatchery, which of your parameters would you like to look at? Which would you concentrate on?

GARY WEDEMEYER: If you did not want to use growth, which is probably the least sensitive, I don't know. I do not have a good answer. I don't think that there is any one thing that you can follow. You are going to have to develop a pattern of things that are practical. Perhaps things as simple as blood hemoglobin or something like that; a pattern of things which are practical for you to do and which have some meaning at your particular station. I will tell you some of the things that I looked at the Dworshak National Fish Hatchery, where we were concerned with steelhead trout. We found that the steelhead were going out to the ocean dehydrated. Blood water is an easy thing to measure. They were going out dehydrated with a depressed white blood cell count and with an electrolyte imbalance in their blood. If they start out that way at the hatchery, you can imagine what would happen after they battle all of those dams in the Columbia River. Dehydration, white blood cell count, and electrolytes would provide good indicators.

Bio-Engineering Symposium for Fish Culture (FCS Publ. 1):21-28

Requirements of Warmwater Fish

NICK C. PARKER AND KENNETH B. DAVIS

Southeastern Fish Cultural Laboratory
U.S. Fish and Wildlife Service
Marion, Alabama 36756

ABSTRACT

Most warmwater fish survive under a wider range of environmental conditions than do either coldwater or tropical species. Factors controlling life processes are physical, chemical, and biological and operate at three distinct levels for survival, growth, and reproduction. The requirements for each of these components are broadest for survival, more exacting for growth, and even more so for reproduction.

There are wide differences among fish generally considered to be warmwater species and these differences must be recognized to optimize fish culture systems. Suboptimum conditions may induce sublethal responses affecting feeding, growth, and reproduction. Warmwater fish may survive and even grow under suboptimum conditions but the most successful culture requires an aquatic environment of high quality.

The broad data base on which coldwater bio-engineers rely is sorely lacking for most warmwater species. We are only now investigating basic physiology and defining biological parameters necessary to design efficient and economical intensive culture systems. This paper draws attention to those physical, chemical, and biological components that bio-engineers must consider to meet the requirements of warmwater fish.

Introduction

The basic requirements of fish may be classified as physical, chemical, and biological. Physical factors include temperature, photoperiod, water velocity, and substrate. Typically, physical factors change more slowly than do chemical factors and are not often responsible for rapid reductions in an established population. These factors usually limit an organism to a particular aquatic habitat. Chemical factors include dissolved gases, pH, nitrogenous metabolites, inorganic ions, hardness, alkalinity, and environmental contaminants. The chemical environment can rapidly shift beyond the tolerance range of some aquatic organisms and produce sudden mortalities. Biological factors may be either intrinsic or extrinsic. Intrinsic factors include physiology of the individual and the response of individuals to their environment. The major extrinsic biological factors are pathogens, predators, and competing species. Either physical, chemical, or biological factors may be limiting in the aquatic environment.

There are three different conditions under which limiting factors can be defined. First, immediate needs of fish are those necessary for survival of the individual. Second, requirements for growth are more narrowly defined than are requirements for survival. Third, requirements for successful reproduction are more narrowly defined than are requirements for growth.

Climatic conditions are more variable in the temperate zones than in either colder or more tropical regions of the world. Consequently, most temperate zone or warmwater fish have a relatively wide tolerance range for both physical and chemical factors (Fig. 1). For example, the thermal range for warmwater fish (4-25 °C) is usually much wider than that for coldwater (4-15 °C) and tropical fish (25-35 °C). For the purpose of this paper we define warmwater fish as those species commonly found in soth temperate zones where water temperature may vary annually from less than 4 °C to over 35 °C. Tolerance to high temperatures separates warmwater fish from coldwater fish and tolerance to low temperatures separates warmwater fish from tropical fish.

Historically warmwater fish have been reared at much lower densities than have coldwater fish (U.S. Dept. Interior 1972).

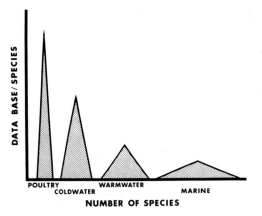

FIGURE 1. — Warmwater fish can be defined by their ability to tolerate a wide range of environmental conditions; whereas, coldwater and tropical species are more environmentally restricted.

These lower densities and the wide tolerance of most warmwater fish to environmental change are probably responsible for their successful culture. Many excellent fish culturists acquired their "know how" by trial and error and not by formal training. Our knowledge of warmwater fish culture has basically developed by this trial-and-error procedure.

With the birth of the catfish industry in the early 1960's there was a public demand to increase our basic knowledge of warmwater fish, specifically catfish. Today we probably know more about catfish, carp, and goldfish than we do about any other warmwater species.

Requirements for Survival

Physical

Temperature

The effects of water temperature are probably the most far reaching of any single component of the environment. It affects the amount of oxygen that can be dissolved in water, the diffusion rate of gases, the metabolic rate of fish, and the life history of potential pathogenic organisms (Brett 1960).

Warmwater fish may be exposed to winter temperatures close to 0 °C and to summer temperatures of 35 °C. Some fish tolerate higher temperatures than others. For instance, catfish are frequently ac-

climated to water about 30 °C (U.S. Dept. Interior 1972) while striped bass are stressed when exposed to temperatures much above 25 °C (Bonn et al. 1976).

Warmwater fish tolerate wide ranges of temperature; however, the greatest negative effect on fish health results from rapid changes in temperature. Exceeding the temperature transfer tolerance results in thermal shock, which may kill fish by affecting the central nervous system. Catfish have been reported to tolerate transfer to water as much as 12 °C less than the acclimation temperature without mortality (McCraren and Millard 1978). Complete acclimation to a given temperature may require up to two weeks and likely involves changes in enzymes, lipids, hemoglobin, and hormone levels (Hoar and Randall 1971).

Photoperiod

Photoperiod has not been shown to affect survival of warmwater fish dramatically. However, lower light intensity seems to encourage feeding by some species and reduces the fright response to extraneous activity around culture tanks.

Water velocity

Water velocity considerations have not been thoroughly studied. Fish tend to orient into a flow of water and a too-rapid flow can result in interference with feeding activity and fatigue if continued for long periods. Low velocity current causes a slow constant swimming pattern, whereas high currents result in short, very intense bursts of swimming activity (Hoar and Randall 1971). Water velocity in culture conditions should possibly duplicate velocities found in the species' natural habitat. Forcing fish to swim in a current might be an additional advantage, particularly if the fish are to be stocked in natural conditions. The exercise should cause an increase in body musculature and stamina, better preparing the fish to survive in the wild.

Water velocity determines the type of substrate that can be maintained and different species require different substrates (Alexander 1974). Burrowing species require a soft substrate, which can only be maintained at low water velocities. Eggs and larval stages of some species may not survive when the proper substrate is not available. Even adult fish may not survive in culture units with improper substrates. Skin abrasion resulting from body contact with rough substrates will increase disease susceptibility.

Chemical

Dissolved oxygen

The maintenance of adequate dissolved oxygen (D.O.) is also essential for survival. At very low D.O., below about 2 mg/l, most fish cease feeding, reduce locomotor activity and use the available oxygen to maintain the basic life support systems, primarily the circulatory and nervous systems. The severity of low oxygen is intensified at higher temperatures since the metabolic rate and oxygen consumption are greater than at colder temperatures. Maintaining adequate D.O. levels is especially important during feeding, since feeding activity in a closed system can rapidly lower oxygen levels (Simco 1976).

Nitrogen

Supersaturation of water with N_2 gas is often produced by entrainment of atmospheric gas in water under pressure or be elevating temperature of water saturated with N_2 gas (Wolke et al. 1975). Nitrogen supersaturation greater than 110% is considered detrimental to most fish and levels of 104% may be lethal to larval stages. Dissolved N_2 gas enters the body fluids of fish and equilibrates with the environment. If the fish rapidly moves to an area of lower N_2 gas saturation, gas can move out of solution and gas emboli form. These bubbles can interrupt blood flow and induce blood clot formation. The effect on the fish depends on the location in the body where emboli form. If no critical tissue is affected, the fish will survive but may have bubbles in the fins or protruding eyes characteristic of "popeye" disease. Gas bubble disease may not be lethal but is a sure sign of a gas problem and will render affected fish more susceptible to predation and disease.

Carbon dioxide, and pH

Metabolic production of energy produces carbon dioxide (CO_2) as a waste gas, which diffuses into the water. Excretion of CO_2 into the water lowers the pH because CO_2 reacts with H_2O to form H_2CO_3, which dissociates to form H^+ and HCO_3^-.

Slightly alkaline water is considered the most productive water probably because it provides a buffer against the constant CO_2 excretion of the fish. Alkalinity levels of around 200 mg/l are considered desirable for many cultured species (Bardach et al. 1972).

Nitrogenous metabolites

Maintenance of the proper pH is also involved in the toxicity of nitrogenous waste products of fish. Ammonia is produced by deamination of amino acids and is excreted into the water primarily by the gills. The toxicity of ammonia is due primarily to the un-ionized form (NH_3) and as pH and temperature increase, the percent of NH_3 also increases (pK = 9.4 at 20°C). The 24-h median lethal concentrations (24-h LC_{50}) of total ammonia nitrogen to channel catfish fingerlings are 264 mg/l at pH 7, 39 mg/l at pH 8, and 4.5 mg/l at pH 9; these levels of total ammonia nitrogen produce un-ionized ammonia nitrogen levels of about 1.6 mg/l. Sublethal concentrations of ammonia reduce growth and are detrimental to fish health (Colt and Tchobanoglous 1978). A linear relationship has recently been found between ammonia levels and plasma corticosteroid hormones (Tomasso, personal communication). Corticosteroids may provide an estimate of the general nonspecific stress response in fish as in other vertebrates.

Ammonia and other metabolic wastes are either diluted with fresh water or detoxified by bacterial nitrification. Ammonia in pond water or in the biological filter of a recirculating system is first converted by *Nitrosomonas* to the even more toxic compound, nitrite. Nitrite is converted by *Nitrobacter* to the relatively nontoxic compound nitrate. Nitrite toxicity increases with decreasing pH and causes the oxygen-carrying pigment hemoglobin to be converted to methemoglobin, which cannot bind oxygen (Russo and Thurston 1977). Exposure to channel catfish fingerlings to as little as 1 mg/l nitrite at pH 7 for 24 h caused 20% of the hemoglobin to be converted to methemoglobin (Tomasso et al. 1979).

Inorganic ions

A protective effect of certain inorganic ions has been found against ammonia and nitrite toxicity. Increasing the total hardness of water by addition of $CaSO_4$ decreases the toxicity of ammonia. A dramatic protective effect of chloride ions on nitrite toxicity has also been described in channel catfish. An ionic ratio of 16 Cl^- to 1 NO_2^- is capable of complete suppression of nitrite-induced methemoglobin formation (Tomasso et al. 1979). Sodium chloride and calcium chloride were equally effective when equivalent weights of Cl^- were tested. Other small monovalent anions also afforded protection against nitrite toxicity (Crawford and Allen 1977).

A further desirable effect of salts is likely derived from their buffering capacity and in decreasing osmoregulatory problems. Disturbances of fish in fresh water frequently cause a decrease in plasma electrolytes, and the advantage of transporting salmon (Smith 1978) and striped bass (Bonn et al. 1976) in water to which sodium chloride has been added is well known. Increased buffering capacity of the water suppresses wide fluctuations in pH.

Contaminants

A water supply free of environmental contaminants is essential to successful aquaculture. This frequently means the use of well water rather than surface water, especially in areas with heavy agricultural use (Liao and Mayo 1974). Although pesticide levels in agricultural run-off are generally low, the high partition coefficient of pesticides in lipids allows pesticide concentrations in fish to build up to several hundred times that of the environment.

Heavy metal contamination whether natural or added by industrial run-off is also potentially damaging to culture operations. Heavy metals are a particular problem in water with low D.O., alkalinity, and pH because the metals do not precipitate and therefore enter the body of the fish.

The heavy metals most frequently of concern to aquaculturists are copper and iron. Copper sulfate is used to control plankton, parasites, and algal growth; copper inhibits many enzyme systems and is very toxic to fish. Iron, often found in well water, can interfere with O_2 exchange by precipitating on gill and egg surfaces.

Biological

Pathogens

Fish are reared in hatcheries and commercial production facilities at population densities greater than those found in nature, and therefore cultured fish are very likely to contact pathogenic organisms transmitted within the system (Plumb 1979). The spread of an infectious organism is very rapid in intensively stocked culture units. Further, many diseases are referred to as "stress mediated" to indicate they occur when the resistance of the fish is decreased (Wedemeyer et al. 1976). Lowered resistance is thought to be caused by secretion of corticosteroid hormones in response to a nonspecific stressor such as detrimental environmental conditions.

Predation

Cannibalism is frequently encountered in culture conditions and may be explained as either agonistic behavior or a feeding response. Predator fish, such as striped bass, are particularly notorious for eating other fish. When there is a wide variation of fish size, cannibalism increases. It is unclear whether cannibalism is due to a behavioral phenomenon or dietary deficiencies. Agonistic behavior, as evidenced by biting in catfish, produces large open wounds on individuals and is more frequently encountered in catfish held at low densities.

Requirements for Growth

Environmental

Temperature

Each species has an optimum temperature for growth likely determined by the optimum temperature of critical enzyme activity. Since enzymes control the rate of every biological event including feeding, digestion, food conversion, and energy production, the optimum temperature for anabolic enzyme activity may be the optimum temperature for growth. Warmwater species generally have optimum temperatures for growth between 20 °C and 30 °C. The higher temperatures associated with increased metabolic rate in warmwater fish, as compared to coldwater species, also increase the rate of microbial waste product degradation (Plumb 1979). Elevated temperatures also result in the rapid multiplication of pathogenic organisms.

Dissolved oxygen

The chemical composition of the water can also affect growth rate. Growth of channel catfish has been shown to be proportional to D.O. concentration in tanks with controlled D.O. levels (Raible 1975). A possible explanation of growth responses might be related to feeding activity. The feeding activity of catfish at a density of 213 kg/m^3 in a closed recirculating system induced a sharp decrease in D.O.; feeding activity decreased when D.O. levels declined to 3 mg/l and ceased at 2 mg/l regardless of the amount eaten (Simco 1976). Additional aeration at feeding might increase the amount of food taken and, as a result, growth of the fish.

Metabolites

Low, nontoxic levels of nitrogenous waste products are also detrimental to growth. Ammonia may be an uncoupler of oxidative phosphorylation, and reduced growth may result from the inability of the animal to convert food energy to ATP (Russo and Thurston 1977). Corticosteroid hormones are also released proportionally to sublethal ammonia exposure. These hormones cause a negative nitrogen balance by deaminating amino acids. Such deaminated amino acids are not available for protein synthesis essential for growth. Thus, chronic stress of any type may inhibit growth.

Biological

Feeding schedule

Several studies have shown that feeding fish twice a day rather than once increased the amount of food taken and the efficiency of its utilization (Stockney and Lovell 1977). When fish are at warm temperatures in systems with continuous aeration, the time of day of the two feedings is probably best eight to ten hours apart. In a system without temperature control, feeding rate should be reduced and be done during the warm part of the day at a time when D.O. is highest.

Multiple feedings are advantageous for young fish. Striped bass fingerlings fed seven times a day at 2-h intervals feed aggressively at each feeding (Parker, unpublished data). This frequent feed application may reduce incidence of cannibalism.

Warmwater fish show sharply reduced feeding and digestion rates at temperatures below 15 °C. Catfish grow best when held at 28 °C and fed at 3% of the body weight per day. When the temperature is 15 °C or lower, some feeding and growth can be achieved by feeding at 1% (Stickney and Lovell 1977). However, such feeding must be adjusted to the severity of the winter. A rapid decrease in temperature following heavy feeding may result in decreased digestion rate and lead to bloating due to bacterial action on food in the gut.

Nutrition

Pellet size and acceptability (taste) must be matched to the size and age of the fish. Nutrition of predaceous fish is a particular problem because little is known of their nutritional requirements and acceptance of a pelleted food is sometimes difficult. Until satisfactory pelleted food is developed for all predaceous fish at all life stages, live forage fish must be cultured. Culturing large amounts of fish for forage is time consuming and expensive and may require as much effort as the species of interest. Culture begets culture.

Pelleted food can be formulated as either a floating or sinking chow. Floating food is not lost through underwater drains as rapidly as sinking chow but may be lost more rapidly in systems where water is removed from the surface. Floating pellets allow the aquaculturist to observe feeding behavior easily; however, some species prefer not to eat from the surface.

Therefore, sinking feed may be the best choice for some fish regardless of system drainage.

Ammonia production by fish has been shown to be quantitatively related to amount of food eaten and temporally related to feeding time (Simco 1976). In a recirculation system the rate of ammonia production increased almost immediately after feeding and reached a peak approximately 6 h after feeding. In a closed system with a biofilter for nitrification this means that pulses of ammonia are expected to be highest during the 6 h following feeding.

Requirements for Reproduction

Spawning

At the present time many fish can be induced to spawn by injections of either dried fish pituitaries or preparations of human chorionic gonadotropins (Smitherman et al. 1978). Induced spawning is successful only when the fish are reproductively mature. It is not common practice to bring fish to reproductive readiness out of season. This limitation decreases fingerling availability and off-season production of marketable fish. Gonadal maturation is determined by age, size, nutritional state, and environmental history of fish. With proper environmental manipulation it should be possible to induce out-of-season gonadal development routinely. A number of species have been experimentally spawned out of season by manipulating photoperiod and temperature (Schreck 1974). Very little systematic research has been done on off-season spawning of warmwater species. Further development of off-season spawning techniques requires fish culture systems with complete environmental control including all of the physical, chemical, and biological factors. To induce off-season spawning systematically, a fish culture system must provide the important environmental cues found in diel and circannual rhythms.

Nutrition

Nutrition of warmwater broodstock is an area that has similarly received little at-

tention. It must be determined if prepared feeds are formulated properly for gonadal development. Recently 5-year-old striped bass, raised from 4-day-old fry, were successfully spawned at the Southeastern Fish Cultural Laboratory. The fish had been raised on a combination of forage and pelleted trout food and were spawned at the same time of the year as were wild-caught striped bass.

Genetics

The success of selective genetic development to produce salmonid strains with proved growth, survival, and food conversion rates suggests that similar programs would be beneficial for warmwater species (Burrows and Combs 1968; Donaldson 1970). Domestication and selective breeding of some salmonids have produced strains more adaptable to intensive culture conditions than to conditions in the wild. Only recently have warmwater culturists attempted the systematic evaluation and selection of strains suitable for intensive culture (Kirpichnikov 1970; Cherfas 1972).

Warmwater vs Coldwater Culture Requirements

Coldwater fish are typically reared in raceways or other culture units with high water-exchange rates. Water exchange flushes waste metabolites from the system and simultaneously dilutes concentrations of pathogenic organisms. Similar water-exchange rates are not common in warmwater fish culture, which is still conducted primarily in shallow earthern ponds.

Warmwater fish require water of high quality for most successful culture; however, many warmwater species successfully tolerate wide fluctuations of environmental conditions. Warmwater fish adapt to a much broader thermal range and to lower D.O. than do coldwater fish. Since the best temperature for warmwater culture is 10 °C to 20 °C higher than the optimum temperature for coldwater fish and since warm water holds less oxygen than cold water, D.O. can quickly become limiting in warmwater systems. Furthermore, the reproductive rate of pathogenic organisms and the course of disease are more rapid in warm water than in cold water. At elevated temperatures increased disease problems, higher metabolic rates, and higher excretion rates of metabolic wastes may help explain why warmwater culture is not performed at densities as high as those in cold water. Furthermore, warmwater fish are often cultured in ponds with limited freshwater inflow as opposed to coldwater species, which are typically cultured in raceways. Since warmwater fish survive under a wide range of environmental variables, perhaps not enough attention has been directed to sublethal responses that may affect feeding, growth, and reproduction.

The data base on which coldwater bioengineers rely is sorely lacking for most warmwater species. We are only now investigating basic physiology and defining biological parameters necessary to design efficient and economical intensive culture systems. In contrast to the poultry industry, which is based on a limited number of species with a large data base for each one, the aquaculture industry has a large number of species with only a limited data base for most of them (Fig. 2). Intensive culture systems for warmwater fish have not been standardized; systems are custom tailored to fit the requirement of the species

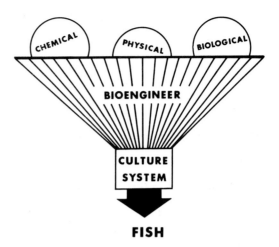

FIGURE 3. — The bio-engineer serves as a funnel to blend chemical, physical, and biological components into an efficient aquaculture system.

in each particular location. The large number of species thus impedes development of a standard system appropriate for all locations.

As the name of this symposium, Bio-Engineering, implies, the biological requirements of fish must be the first consideration when designing aquaculture systems whether for warmwater or coldwater species. A successful bio-engineer must blend the chemical, physical, and biological components (Fig. 3) to produce a system that achieves the desired goal of survival, growth, or reproduction. The interdisciplinary approach combining the expertise of biologists and engineers will be mandatory for the development of efficient, practical aquaculture systems.

References

ALEXANDER, R. McN. 1974. Functional design in fishes. Hutchinson Univ. Library, Hutchinson Co., London. 160 pp.

BARDACH, J.E., J.H. RYTHER, AND W.O. McLARNEY. 1972. Aquaculture: The farming and husbandry of freshwater and marine organisms. Wiley—Interscience, New York, N.Y. 868 pp.

BONN, E.W., W.M. BAILEY, J.D. BAYLESS, K.E. ERICKSON, AND R.E. STEVENS. 1976. Guidelines for striped bass culture. Striped Bass Comm., Southern Div. Am. Fish. Soc. 103 pp.+ Appendix 14 pp.

BRETT, J.R. 1960. Thermal requirements of fish — Three decades of study, 1940-1970. Pages 110-117 in

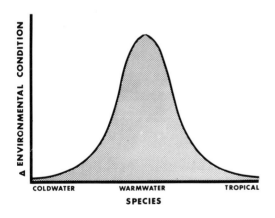

FIGURE 2. — The more limited the number of species of interest to culturists the less diverse the research interest and, thus, the greater the data base per species. System design for successful culture of any species is enhanced by the magnitude of the available data base.

Biological problems in water pollution. Robt. A. Taft Sanit. Eng. Cent. Tech. Rep. W60-3.

BURROWS, R.E., AND B.D. COMBS. 1968. Controlled environments for salmon propagation. Prog. Fish-Cult. 30:123-136.

CHERFAS, B.I., ed. 1972. Genetics, selection, and hybridization of fish. Translated from Russian. U.S. Dept. Comm. National Tech. Info. Serv., Springfield, Va. #TT 71-50112. 269 pp.

COLT, J., AND G. TCHOBANOGLOUS. 1978. Chronic exposure of channel catfish, *Ictalurus punctatus,* to ammonia: Effects on growth and survival. Aquaculture 15:353-372.

CRAWFORD, R.E., AND G.H. ALLEN. 1977. Seawater inhibition of nitrite toxicity to chinook salmon. Trans. Am. Fish. Soc. 106:106-109.

DONALDSON, L.R. 1970. Selective breeding of salmonid fishes. Pages 65-74 *in* W.J. McNeil, ed. Marine Aquaculture. Oregon State Univ. Press, Corvallis.

HOAR, W.S., AND D.J. RANDALL, eds. 1971. Fish physiology. Vol. 6. Environmental relations and behavior. Academic Press, New York. 559 pp.

KIRPICHNIKOV, V.S., ed. 1970. Selective breeding of carp and intensification of fish breeding in ponds. Translated from Russian. U.S. Dept. Comm. Clearinghouse Fed. Sci. Tech. Info., Springfield, Va. #TT 70-50064. 249 pp.

LIAO, P.B., AND R.D. MAYO. 1974. Intensified fish culture combining water reconditioning with pollution abatement. Aquaculture 3:61-85.

McCRAREN, J.P., AND J.L. MILLARD. 1978. Transportation of warmwater fishes. Pages 43-88 *in* Manual of fish culture, Sec. G.: Fish transportation. U.S. Fish Wildl. Serv., Washington, D.C.

PLUMB, J.A., ed. 1979. Principal diseases of farm-raised catfish. Southern Cooperative Series No. 225. Auburn Univ., Auburn, Ala. 92 pp.

RAIBLE, R.W. 1975. Survival and growth rate of channel catfish as a function of dissolved-oxygen concentration.

Water Resources Research Center Publ. No. 33. Univ. Ark., Fayetteville. 35 pp.

RUSSO, R.C., AND R.U. THURSTON. 1977. The acute toxicity of nitrite to fishes. Pages 118-131 *in* R.A. Tubb, ed. Recent advances in toxicology. E.P.A. 600/3-77-085.

SCHRECK, C.B., ed. 1974. Control of sex in fishes. Va. Polytech. Inst. St. Univ., Blacksburg, Va. UPI-SG-74-01. 106 pp.

SIMCO, B.A. 1976. The reuse of water in commercial raising of catfish: Phase two. Univ. Tenn. Water Res. Ctr. Res. Rep. No. 52. 66 pp.

SMITH C.E. 1978. Transportation of salmonid fishes. Pages 9-41 *in* Manual of fish culture, Sec. G.: Fish transportation. U.S. Fish Wildl. Serv., Washington, D.C.

SMITHERMAN, R.O., H.M. EL-IBIARY, AND R.E. REAGAN, eds. 1978. Genetics and breeding of channel catfish. South. Coop. Ser. Bull. No. 223, Auburn Univ., Auburn, Ala. 34 pp.

STICKNEY, R.R., AND R.T. LOVELL, eds. 1977. Nutrition and feeding of channel catfish. South. Coop. Ser. Bull. No. 218. Auburn Univ., Auburn, Ala. 67 pp.

TOMASSO, J.R., B.A. SIMCO, AND K.B. DAVIS. 1979. Chloride inhibition of nitrite-induced methemoglobinemia in channel catfish *(Ictalurus punctatus).* J. Fish. Res. Board Can. 36:1141-1144.

U.S. DEPT. INTERIOR, FISH AND WILDLIFE SERVICE. 1972. Factors affecting the growth and production of channel catfish in raceways. U.S. Dep. Comm. Tech. Assist. Proj. No. 14-16-0008-571. 98 pp.

WEDEMEYER, G.A., F.P. MEYER, AND L. SMITH. 1976. Environmental stress and fish disease. Diseases of fishes, Book 5. S.F. Snieszko and H.R. Axelrod, eds. T.F.H. Publ., Neptune City, N.J. 192 pp.

WOLKE, R.E., G.R. BOUCK, AND R.K. STROUD. 1975. Gas-bubble disease: A review in relation to modern energy production. Pages 239-265 *in* S.B. Saila, ed. Fisheries and energy production: A symposium. D.C. Heath Co., Lexington, Mass.

Bio-Engineering Symposium for Fish Culture (FCS Publ. 1):29-33

Relationships of Trout Behavior and Management: Hatchery Production and Construction

ROBERT L. BUTLER

Pennsylvania Cooperative Fishery Research Unit
The Pennsylvania State University
University Park, Pennsylvania 16801

Introduction

Hatchery trout are different from wild trout. I did not say "inferior," because the inferior features of hatchery trout most often have more to do with how the hatchery products are managed. The general goal for the production of trout used in short-term recreational management and the production of food are both worthwhile objectives and can be achieved (Butler and Borgeson 1965).

It is well and good that the Bio-Engineering Symposium is most specifically directed to more effective and economical production of fish. That is where the future lies. I could accept this as the only objective if we were selling all our fish at the hatchery rather than making large contributions to lakes and streams for recreational benefit. A missing realistic objective is how to increase survival or, in alternative terms, how to decrease the well-recognized high natural mortality rates in trout from hatcheries. In light of the recent work at the Pennsylvania Cooperative Fishery Research Unit, we may be able to suggest engineering features for hatcheries of brown trout as well as other trout hatcheries that could improve salmonid management.

Hatchery Trout

The present hatchery product has been developed both from purposeful as well as from inadvertent selection of genetic stock. Most of the stocks have been developed for rapid growth in the hatchery, low mortality in the hatchery, ease of hatchery operation, which includes ease of sorting, prophylactic treatment, and transportation. These trout are exposed to all stimuli of the hatchery operation including people and equipment without apparent stress or decline in production. Great effort has been made at the hatcheries to reduce disease and improve food quality and feeding techniques to achieve maximum growth and healthy fish accompanied with low cost.

Behavior

Behavioral features of such stock, referred to above, are limited to observations of nonerratic types of swimming and vigorous response to feeding, the only obvious characteristics of "normal" behavior related to the hatchery operation. The product appears to be one of great uniformity with predictable features of production at each hatchery installation. Unfortunately, under such systems the product may not necessarily meet some of the demands made by fishermen or fisheries managers.

Discussion

It appears to me that much of a hatchery production and genetic selection for wild characteristics in the hatchery environment are mutually exclusive. There may be opportunities, however, to modify the hatchery environment for trout with more wild-like behavior. We must first accept some of the basic facts of culture in humans as well as fish. Fish are like humans in many ways. First, we both are victims of genetic and cultural background (Hall 1977). The template for selection of wild trout in the natural stream or environment is quite different from that template for selection in a hatchery. Butler's (1975) review of the AAAS 1970 symposium on

29

the Use of Space by Animals and Man (Esser 1971) found that fish behavior was not represented among the 50 or more participants nor was any citation given to fish behavior. Jenkins (1969) made the first significant contribution to an understanding of the behavior of brown and rainbow trout in a seminatural stream at Convict Creek, California. However, some of his conclusions should be held in abeyance since the fish were not resident. They may have suffered what Welker and Welker (1958) refer to as "novelty stimulus effect." For several days the fish tried to escape the enclosed stream sections. A strange place is not home. What is important is that Jenkins was the first to examine carefully the behavior of our common species of trout.

If we are to produce a more "natural" product, what should be considered? Earl Leitritz in the 1950's and 1960's as head of hatcheries in California attempted hatchery automation. It seems that this offers a very important point of development, not only from the standpoint of a more "natural" fish, but also in reducing the stress commonly referred to as GAS, the general adaptation syndrome (Selye 1973). More recently Mazeaud et al. (1977) and Strange et al. (1977) have shown that handling fish causes an elevation in catecholamines and cortiocosteroids. Secondary effects of such handling causes also an increase in blood glucose. Some of these changes are not obliterated after the passage of two to three weeks.

There is a host of studies carried on since the 1940's to demonstrate that stocked hatchery trout have an excessively high natural mortality rate compared to wild trout of the same size. Many biologists of the past as well as present consider natural mortality primarily caused by bird and mammal predation. There are no substantive data supporting this idea.

The thought has occurred to me that under hatchery conditions the Mauthner cell may be serving a maladaptive function in the hatchery. In the wild it appears to serve a specific response for survival that is characterized by a tail-flip from a source of impacting sound such as might be made by a diving fish predator (Diamond 1971). Is it possible that the response to pellets discharged on the water surface provides reinforcement for attraction rather than repulsion? This may be particularly deleterious to fingerling trout that may be more subject to bird predation than are adult fish.

Psychologists at the University of Michigan some 20-odd years ago, working on the above premise, approached the subject of high natural mortality in planted fish through an effort of conditioning hatchery trout with electrical shock when they swam near a simulated heron. The approach to this method of conditioning was unsuccessful. More recently Fraser (1974) attempted to train hatchery brook trout to avoid predation by the common loon in a similar fashion. Again this approach did not achieve the principal objective.

From our work it appears that the high natural mortality may be more closely related to excessive activity caused by a lack of response for proper site selection and establishment of home range in the natural stream. The problems of anchor ice encountered in much of the mid-latitudes may also be a factor in over winter mortality of hatchery fish not selected for or exposed to those conditions (Hawthorne and Butler 1979).

In some recent work of the Unit, James McLaren (1979) has related behavior of wild and hatchery-reared brown trout to the population dynamics involved in stocking a trout stream. His first study compared the behavior of wild and hatchery-reared trout during daylight in a seminatural stream environment at Lamar National Fish Hatchery, Lamar, Pennsylvania. A special statistical approach was used to evaluate the effects of origin (hatchery versus wild), environment (test sections and season), density, and prior residency on six behavioral categories (overall activity, cover use, offensive acts, defensive acts, total agonistic activity, and foraging). In his second study he observed the migration and mortality of hatchery-reared trout superimposed on wild brown

30

trout populations residing in Spruce Creek, Huntingdon County, Pennsylvania. In the seminatural stream, hatchery-reared trout were more active than wild trout, used cover less and were frequently involved in foraging and agonistic behavior. Hatchery-reared trout appeared less responsive to the effects of physical habitat and seasons than wild trout. When density was doubled by superimposing one group upon the other, there was an increase in activity, foraging, and agonistic behavior, and a decrease in cover use by both groups. Despite the greater frequency of agonistic behavior in hatchery-reared trout, they held no advantage in dominance relationships with wild trout. Dominance was more closely related to size.

Behavioral differences during these studies between wild and hatchery-reared trout were evaluated in terms of survival as observed at Spruce Creek. Nine weeks after stocking hatchery-reared brown trout into Spruce Creek, emigration and mortality had reduced the total standing crop to a level nearly equal to that present before stocking. Total losses of hatchery-reared trout were 1.5 to 2 times greater than for wild trout. In the McLaren studies all the fish were displaced, marked or tagged, and placed in a constrained testing situation.

Agonistic behavior of hatchery Atlantic salmon when planted in densities greater than that occurring under natural conditions has been found to be higher than that of landlocked wild Atlantic salmon under the same densities (Fenderson and Carpenter 1971). However, when planted in densities comparable to wild condition, wild salmon had higher agonistic levels.

Demand feeding controlled by the fish could avoid maladaptive behavior previously mentioned with respect to the Mauthner cell "tail-flip." Adron et al. (1973) found that rainbow trout could learn to actuate a trigger for self-regulation of feeding. He found growth faster with less food under demand feeding. Undoubtedly the principles discovered by Bitterman (1975) relate to the results of Adron. Fish can learn as well as other

vertebrates in what is known as 100 percent probability learning; that is, learning can be induced very rapidly with any reinforced stimulus that is consistently favorable. Those that do learn a proper response in conditioning training have followers that learn through social facilitation (Zajonc 1964). Greenburg (1947) demonstrated the ability of a population of green sunfish to be socially facilitated and conditioned with the turning on of a light and feeding. One hundred percent probability learning seems to be a characteristic of all vertebrates.

Recently at Spruce Creek and through the help of Del Graff of the Pennsylvania Fish Commission, we planted an equal number of mature hatchery brown trout on top of a wild brown trout population. Robert Bachman, doctoral candidate of the Unit, has been studying this population for three years. He knows his brown trout as individuals without ever having electrofished, seined, touched, marked, mangled, or manipulated them in any way. They are living in their original unconstrained environment. Each individual has accepted 4 to 5 specific sites for feeding and resting during the daylight period. Bachman and his assistants have well over 2,000 hours of observation on site utilization, foraging, activity, and agonistic behavior.

When the hatchery brown trout were superimposed on the resident wild brown trout population, Bachman recorded on video-tape the vigorous interaction of the two groups. All wild brown trout regardless of size, challenged in agonistic behavior the introduced members. Within a few days only a small number of hatchery brown trout had become established. The interesting aspect of their establishment was their selection of specific sites that had been used by wild brown trout, not a selection of new sites. The work will be repeated in 1980. This tells us something about the capacity of at least a few hatchery brown trout to recognize the hydraulic features of the site — the subject of another doctoral study on Spruce Creek currently underway by Gregory Pierce of the Unit. From information in the recently completed doctoral thesis of James McLaren on the interaction of tagged wild brown trout and hatchery trout at Lamar National Fish Hatchery and their survival in Spruce Creek coupled with the more recent work of Robert Bachman, it appears worthwhile to consider in your bio-engineering development of salmonid hatcheries the concept of conditioning raceways.

Conditioning raceways would be built on the basis of hydraulic features utilized by wild brown trout. The practice would be to expose the hatchery fish to hydraulic conditions they would meet in the environment. At present we can only assume that a large portion of the planted fish would recognize and become accustomed to these hydraulic features, establish a home range, and thus reduce their high bioenergetic costs of swimming. We know now that some hatchery trout can recognize these features and will utilize appropriate sites, and it's obvious from analysis of tailbeat that they thereby conserve considerable energy.

We have an imbalance in research on trout and especially trout as a hatchery product. We are learning more about the behavior of wild trout in their natural environment but virtually nothing is known about the behavior of hatchery fish in their "natural" environment or in the hatchery itself, or about what conditioning raceways can do to reduce natural mortality. As stated earlier, I am convinced that the high natural mortalities of adult fish are not related to bird predation, but are related to behavioral features missing in hatchery fish.

Furthermore, I would suggest installation at some hatcheries of underwater windows for observation of behavior as developed in the hatchery environment. At this stage I cannot predict what could be learned from this type of facility.

Conclusion

Managers have insisted on using the hatchery product in situations not compatible with the genetic background nor with the system in which trout are reared. If we know more about the behavioral

repertoire of trout in the hatchery and their capacity to respond to a natural environment, we could make better use of them and could design changes in hatchery operation for improvement of quality for specific management goals.

References

ADRON, J.W., P.T. GRANT, AND C.B. COWEY. 1973. A system for the quantitative study of the learning capacity of rainbow trout and its application to the study of food preferences and behavior. J. Fish. Biol. 5:625-636.

BITTERMAN, M.E. 1975. The comparative analysis of learning. Science 188:699-709.

BUTLER, R.L. 1975. Some thoughts on the effects of stocking hatchery trout on wild trout populations. Pages 83-87 *in* Wild Trout Management Symposium, Yellowstone National Park, 1974. Trout Unlimited and U.S. Department of Interior, Fish Wildl. Serv.

BUTLER, R.L., AND D.P. BORGESON. 1965. California "catchable" trout fisheries. Calif. Dep. Fish Game, Fish Bull. 127. 47 pp.

DIAMOND, J. The Mauthner cell. Pages 265-346 *in* W.S. Hoar and D.J. Randall, eds. Fish physiology, Vol. V. Academic Press, New York.

ESSER, A.H., ed. 1971. Behavior and environment: The use of space by animals and man. Plenum Press, New York. 411 pp.

FENDERSON, O.C., AND M.R. CARPENTER. 1971. Effects of crowding on the behavior of juvenile hatchery and wild landlocked Atlantic salmon *(Salmo salar* L.). Anim. Behav. 19:439-447.

FRASER, J.M. 1974. An attempt to train hatchery-reared brook trout to avoid predation by the common loon. Trans. Am. Fish. Soc. 103(4):815-818.

GREENBURG, B. 1947. Some relations between territory, social hierarchy, and leadership in the green sunfish *(Lepomis cyanellus)*. Physiol. Zool. 20(3):267-299.

HALL, E.J. 1977. Beyond culture. Anchor Press/Double-day, Garden City, N.Y. 298 pp.

HAWTHORNE, V., AND R.L. BUTLER. 1979. A trout stream in winter. A 16-mm color and sound documentary. Pennsylvania St. Univ., Audio-Visual Serv., University Park, Pa. 18 min.

JENKINS, T.M., JR. 1969. Social structure, position choice and microdistribution of two trout species *(Salmo trutta* and *Salmo gairdneri)* resident in mountain streams. Anim. Behav. Monogr. 2(2):57-123.

MAZEAUD, M., F. MAZEAUD, AND E. DONALDSON. 1977. Primary and secondary effects of stress in fish: Some new data with a general review. Trans. Am. Fish. Soc. 106(3):201-212.

McLAREN, J. 1979. Behavioral competition between hatchery-reared and wild brown trout. Doctoral thesis,

Dep. Biology, The Pennsylvania State University. 145 pp.

SELYE, H. 1973. The evolution of the stress concept. Am. Sci. 61(6):692-699.

STRANGE, R.J., C.B. SCHRECK, AND J.T. GOLDEN. 1977. Corticoid stress responses to handling and temperature in salmonids. Trans. Am. Fish. Soc. 106(3):213-218.

WELKER, W.I., AND J. WELKER. 1958. Reactions of fish *(Eucinostomus gula)* to environmental changes. Ecology 39(2):283-288.

ZAJONC, R.B. 1964. Social facilitation. Science 149:269-274.

QUESTIONER: Those fish you planted in the stream — did it seem to make any difference what speed or what type of environment they were raised in? Do you see what I am getting at? Were they reared in raceways with fast-moving water or in ponds with very little current?

DEL GRAFF: The fish that Bob used in Spruce Creek were reared at the Big Spring Fish Hatchery in Cumberland County, Pennsylvania. The velocity and exchange in the raceways are conventional for our system, a little over three exchanges in a 10-foot wide, 100-foot section of raceway. The water quality is very similar to that of Spruce Creek, where the fish were stocked. That variable could be ruled out. The calcium, hardness, and everything else were the same. I don't know whether those fish are tough enough to make it in the water velocities they were exposed to. They were a normal representation of the catchable brown trout that we use in Pennsylvania.

ROBERT BUTLER: I am going to ask Del for another plant of similar fish next year. We need confirmation or a complete denial. I don't care. It will give us a reasonable hypothesis to proceed on.

QUESTIONER: Do you have any idea yet about the hydrological requirements of these trout? You said that you were going to have students working on this. Do you have any ideas yet on what the fish are looking for?

ROBERT BUTLER: We are confident that the trout are looking for sites with very low bienergetic costs. Each fish is doing everything to minimize his cost of living. He can't expend the energy (metabolism) as measured by physiologists in the laboratory. All of his movements are minimized. He selects a site where he gets food coming over the top. I also have a student working on whether or not the sites have difference in drift. We do not think that this is going to occur. We think that the features in a hydraulic site are the most important. It would be interesting if tests could be conducted in a raceway before the fish are stocked to find out which fish can locate these sites.

E.A. HUISMAN: I was very pleased with the lecture of Dr. Butler. In addition to that, we carried out some research in the Netherlands on the evaluation of pike stocks. Just to make a brief comment — we found that site variability was the determining factor for survival. Every year, in a certain spot, the amount of survival between stocked and wild fish was equal. We could be stocking pike earlier in the season when we had the advantage of growth or when we stocked them on a smaller migratory recruitment.

ROBERT BUTLER: In a sense, what you were doing was essentially imprinting the fish regardless of the situation for the site. They had the opportunity of familiarizing themselves with the hydraulic features. Am I correct?

E.A. HUISMAN: Yes. That is correct. What was peculiar to this situation was that we could not increase the total density of the individuals.

ROBERT BUTLER: But you did not change the substrate. That is the next step. That is what we have to do. We know that with the proper substrate the fish does not exhaust or even slightly depress the total amount of food in the stream. We know this and have published papers on it. If it is a hydraulic feature and is so specific that it is used time after time by different fish, then all we have to do is to pack that stream with hydraulic features that meet the requirements of the fish. We will increase the carrying capacity. Roger Burrows said that most streams are just barely perking along at impoverished levels. I think that it is not food related. So, we are going to build sites and we are going to pack numbers of trout beyond what were there originally.

QUESTIONER: Dr. Butler, do you see any practical application for your work in hatchery systems? Should we be devising new types of hatchery structures?

ROBERT BUTLER: I have been stunned by noting what a psychologist did in Michigan about 40 years ago. I want to be very careful that I don't open that bag of tricks and become a fish psychologist. First, I would like to find out if those individuals that came from the hatchery recognize forage sites and become part of the brown trout population. Do they have something unique? Can this occur in a hatchery situation where raceways could be provided with hydraulic features? That will be a future investigation.

Bio-Engineering Symposium for Fish Culture (FCS Publ. 1):34-47

Nitrogen Toxicity to Crustaceans, Fish, and Molluscs

JOHN E. COLT

Department of Civil Engineering
University of California
Davis, California 95616

DAVID A. ARMSTRONG

College of Fisheries
University of Washington
Seattle, Washington 98195

ABSTRACT

The toxicity of nitrogenous compounds is a serious problem in the culture of aquatic organisms. Sublethal ammonia and nitrite levels may reduce growth, damage gills and other organs, and may be a predisposing factor in several diseases. A study of the nitrogen excretion of fish, crustaceans, and molluscs in terms of culture practices and environmental conditions; of the chemistry of the excreted compounds and their decomposition products; and of the physiological mechanisms of both lethal and sublethal toxicity resulted in this paper. Since ammonia and nitrite are weak acids (or bases), particular emphasis is placed on the relative toxicity of the un-ionized and ionized form of each compound, and methods for control or reduction of their toxicity are suggested.

Introduction

Water quality management in the intensive culture of aquatic animals has been the subject of much research over the last ten years. This has been due in part to interest in the use of partially or totally recycled systems for the culture of fish, crustaceans, or mulluscs. These systems produce very high yields but require close control of environment variables such as temperature, dissolved oxygen, or ammonia levels. In conventional ponds, water quality has also become critical as production levels have been increased by increases in feeding levels, higher stocking levels, or aeration. Reduced water supplies, water pollution requirements, and higher feed costs have forced raceway and hatchery operators to optimize their production capabilities.

In all of these systems, the toxicity of excreted nitrogen compounds is the singly most limiting parameter once adequate dissolved oxygen levels are maintained. It is therefore the purpose of this paper to discuss the toxicity of the various nitrogen compounds that may occur in culture systems, the physiological basis of their sublethal and lethal toxicity, and suggest methods for control or reduction of their toxicity. Special emphasis will be placed on the effect of these compounds on the growth of the culture organism, as this is probably the most important parameter in culture.

Source of Nitrogen Compounds in Culture Systems

The major source of nitrogen compounds in culture systems is from the metabolism of protein contained in the feed. Ammonia is the major end-product of protein catabolism excreted by fish, crustaceans, and molluscs (Campbell 1973). Urea is the only other nitrogen compound excreted in significant qualities. Urea is non-toxic to aquatic animals at the concentrations present in culture systems, but is rapidly hydrolyzed to ammonia and carbon dioxide (Colt and Tchobanoglous 1976).

Ammonia is oxidized to nitrate in a two-step process by two groups of aerobic bacteria (Sharma and Ahlert 1977). Nitrification is a slow process; consequently, ammonia and urea will be the major nitrogen species in flow-through systems. In a well-aged recycled system, the rate of ammonia oxidation is equal to the rate of nitrite oxidation under steady-state conditions and nitrite levels are typically low. Because of the slow growth of these bacteria, improper start-up procedures or shock loads of ammonia can produce lethal nitrite levels (Collins et al. 1975; Liao and Mayo 1974).

The ammonia production of aquatic animals is proportional to the feeding rate. In channel catfish the ammonia production rate is 20 g ammonia-N/kg of feed·d for fish that ranged from 60 g to 940 g (Page and Andrews 1974). For other animals, the waste production rate can be estimated from the protein conversion factor (average values vary from 0.65 to 0.80), and the protein level of the feed by the following equation:

Ammonia-N Production (g/kg of food·d) =
$$(1.0 - PCF) * \frac{(PL)}{6.25} * 1000$$

where
  PCF = Protein Conversion Factor (fraction)
  PL = Protein Level (percent)

Chemistry of Nitrogenous Compounds in Culture Systems

After dealing with the nitrogenous compounds that may be found in a culture system, the chemistry of each compound will be discussed.

Urea

Because of the low toxicity of urea and its rapid conversion to ammonia and carbon dioxide, urea is not important to the design of culture systems (Colt and Tchobanoglous 1976).

Ammonia

Dissolved ammonia gas is a weak base; the equilibrium expression for this reaction can be written as

$$NH_3 + H_2O = NH_4^+ + OH^-$$

The following convention will be used: NH_3 will be referred to as un-ionized ammonia and NH_4^+ as ionized ammonia. Ammonia will be used for the sum of NH_3 + NH_4^+. The concentration of all compounds will be expressed on a nitrogen basis. These compounds will be written in the following way:

NH_3-N   un-ionized ammonia as nitrogen

NH_4^+-N   ionized ammonia as nitrogen

Ammonia-N   un-ionized + ionized ammonia as nitrogen

The un-ionized ammonia levels can be computed from the following formula:

$$(NH_3\text{-}N) = \frac{(\text{Ammonia-N})}{1 + 10.0^{(pKa-pH)}} \quad (1)$$

where

(Ammonia-N)   = the measured concentration of ammonia

pKa   = the acidity constant for the reaction

pH   = the measured pH of the solution.

The pKa value can be computed as a function of temperature from the work of Emerson et al. (1975). At 25 °C, the pKa value is equal to 9.25. Under normal pH and temperature conditions the percentage of NH_3-N may range from 0.01 percent to 15 percent. High pH and temperatures favor the un-ionized form. If all compounds are expressed on a nitrogen basis, the ionized ammonia level can be obtained from the following equation:

$$NH_4^+\text{-}N = \text{Ammonia-N} - NH_3\text{-}N$$

In some cases, it is more convenient to express the un-ionized ammonia in µg/l (1000 µg/l = 1 mg/l).

The un-ionized ammonia level computed from Equation 1 is correct in freshwater where the effects of ionic strength can be neglected (Colt 1978). The computation of un-ionized ammonia in brackish or seawater is complicated by the effects of salinity on the pKa value and the measurement of pH. The theoretical pKa

35

values over the range on 20 to 40 parts per thousand determined by Whitfield (1974) have been experimentally confirmed (Whitfield 1978; Khoo et al. 1977). A computer routine (Hampson 1977a) and a tabular list (Bower and Bidwell 1978) can be used to compute the pKa value as a function of temperature and salinity based on Whitfield's work. To use Whitfield's model or other models based on his work, it is necessary to calibrate the pH electrode with the tris-seawater buffer developed by Hansson (1973). This procedure eliminates errors due to changes in the liquid junction potential, but requires a very complex and time-consuming titration. Failure to use this procedure may result in a $\pm$ 0.16 error in the pH, which could cause a $\pm$ 40 percent error in the un-ionized ammonia level. Because of these analytical problems, the ammonia tolerance of animals in seawater should be viewed with caution. In brackish water up to 10 to 12 parts per thousand salinity, the method developed by Armstrong et al. (1978) can be used without too much error.

Nitrite

Nitrite is the ionized form of nitrous acid, a weak acid. This reaction can be written as

$$HNO_2 = H^+ + NO_2^- \qquad (2)$$

The nitrous acid level in freshwater can be computed from the following formula:

$$(HNO_2\text{-}N) = \frac{(Nitrite\text{-}N)}{1 + 10.0^{(pH - pKa)}}$$

where

(Nitrite-N)	= the measured nitrite value
pKa	= the acidity constant for this reaction
pH	= the measured pH of the solution.

The pKa as a function of temperature was determined by Tummavuori and Lumme (1968). At 25 °C, the pKa for this reaction is 3.14. At normal pH and temperatures the percentage HNO_2-N ranged from 0.0005 percent to 0.05 percent. Low pH and temperatures favor the nitrous acid

form. The computation of nitrous acid in seawater is subject to the same error discussed in the ammonia section.

Nitrate

Nitrates are in general readily soluble in water, showing slight tendencies to form coordination compounds, and for most purposes can be considered to be dissociated completely (Latimer and Hildebrand 1951).

Analytical Procedures

The analytical procedures for ammonia, nitrite, and nitrate will be covered briefly.

Ammonia

The ammonia probe (Barcia 1973) is probably the most convenient method for Ammonia-N levels higher than 0.1 mg/l and is not affected by suspended material. Several colorimetric methods (Hampson 1977b; Standard Methods 1976; Solorzano 1969) may be useful under special conditions.

Nitrite

The NO_x probe (Orion 1976) is the most convenient method for nitrite-N levels above 0.1 mg/l. Several wet chemical methods (Standard Methods 1976) can be used, but may require filtration prior to analysis.

Nitrate

Wet chemical methods can be used to determine the nitrate levels in culture systems (Standard Methods 1976).

Ammonia Toxicity

Since ammonia is the principal nitrogenous compound excreted by aquatic animals, ammonia toxicity problems may occur in all types of culture systems. Documentation on the effect of ammonia on crustaceans and molluscs is poor, so in many cases, the discussion will be biased toward the fish. The toxicity of ammonia to aquatic animals will be discussed in terms of the mechanisms of toxicity, lethal effects, and effects on growth.

Mechanisms of Toxicity

As the ammonia level of the ambient water increases, the ammonia excretion of most aquatic animals decreases and the ammonia level in the blood and tissue increases. This increased ammonia level can have serious effects on the physiology of the animal on a cellular, organ, and system level.

Effects on the cellular level

On the cellular level, the release of NH_3 into the blood from ambient water or metabolic production is converted to NH_4^+ with the release of an OH^-. The subsequent elevation of blood (and perhaps the intercellular pH) can have a pronounced effect on enzyme-catalyzed reactions and membrane stability (Campbell 1973). Also, the high levels of ammonia may cause a reversal of the glutamate dehydrogenase reaction, withdrawing α-ketoglutarate from the tricarboxylic acid cycle as well as decreasing the amount of NADH available for oxidation (Campbell 1973). Then the increased glutamate concentration would then serve to lower the cellular concentration of ATP due to an increased conversion of glutamate to glutamine. In work with coho salmon *(Oncorhynchus kisutch)*, Sousa and Meade (1977) found that exposure to ammonia decreased the blood pH because of the accumulation of acid metabolites which resulted from the enzymatic stimulation of glycolysis by NH_4^+ and the simultaneous suppression of the TCA cycle.

Effects of nitrogen excretion

Metabolic ammonia can be excreted by aquatic animals by three principal routes: 1) diffusion of HN_3 from the blood to the water through the gills, 2) exchange transport of NH_4^+ with Na^+, and 3) conversion to a nontoxic compound like urea.

Addition of ammonia to the external medium will reduce the ammonia excretion of rainbow trout, *Salmo gairdneri* (Olson and Fromm 1971), goldfish, *Carassius auratus* (Olson and Fromm 1971), crab, *Callinectes sapidus* (Mangum et al. 1976), and the freshwater shrimp, *Macrobrachium rosenbergii* (Armstrong 1978). It appears that diffusion of NH_3 is the major excretion pathway in most aquatic animals. As it becomes more difficult to excrete ammonia, the first reaction of aquatic animals may be the reduction or cessation of feeding to reduce the production of metabolic ammonia. Therefore, one of the major sublethal effects of ammonia will be a reduction in growth rate.

The exchange transport of NH_4^+ in the blood for a Na^+ ion in the external medium is an excretion pathway of questionable importance in the excretion of ammonia, but may be very important in the maintenance of ion and charge balance. This process will be discussed in detail in the next section.

When the ammonia level of the ambient water is increased, the ammonia excretion of both rainbow trout and goldfish decreases, but the goldfish is able to increase urea production and maintain nitrogen balance (Olson and Fromm 1971). The source of this urea was not from the ornithine cycle, but from purine synthesis and catabolism (Cvancara 1969). While the use of this pathway may have a significant adaptive use, the synthesis of urea from ammonia requires energy and therefore reduces the amount of energy available for other uses. The use of this pathway by other aquatic species is not documented.

Effects on osmoregulation

High ambient ammonia levels can affect the osmoregulation of aquatic species by increasing the permeability of the animal to water and thereby reducing internal ion concentrations (Lloyd and Orr 1969). This mechanism is probably more important in freshwater because in this environment aquatic species are hyperosmotic.

Addition of ammonium ions to the external medium results in inhibition of sodium absorption (Armstrong et al. 1978; Maetz et al. 1976; Mangum et al. 1976; Shaw 1960). Armstrong et al. (1978) postulated that one mode of toxicity for shrimp larvae in water of low pH (6.8) was inhibition of normal Na^+ influx due to high ambient NH_4^+ concentrations.

In rainbow trout, Lloyd and Orr (1969)

found that urine flow may increase as much as 6 times when exposed to lethal levels of ammonia. Increased urine flow could overload the readsorption mechanisms in the kidneys and result in a significant loss of NaCl, glucose, proteins, and amino acids. This effect has not been confirmed experimentally.

Effects on oxygen transport

Ammonia can have a serious effect on the ability of aquatic species to transport oxygen to the tissues. These effects include damage to the gills, reduction of the capacity of the blood to carry oxygen because of a lowered pH, an increased oxygen demand, and histological damage to the red blood cells and tissues that produce the red blood cells.

In rainbow trout, 21 μg/l NH_3-N caused significant pathological changes in gill tissue during a 6-month exposure (Smith and Piper 1975). This gill damage can reduce the surface area of the gills and therefore the ability of fish to transfer oxygen (Mayer and Kramer 1973).

Exposure of coho salmon to ammonia reduced the blood pH because of the accumulation of acid metabolites (Sousa and Meade 1977). A decrease in the blood pH reduced the ability of the blood to carry oxygen at a given oxygen level and reduced the saturation level of the blood. Coho salmon showed some ability to restore normal pH levels by renal and respiratory mechanisms.

Exposure of rainbow trout to lethal ammonia levels resulted in an increase in the oxygen consumption of approximately 3 times (Smart 1978). This increased oxygen demand may be due to increased activity, increased energy expenditure for maintenance of water and salt balance, or a disturbance in cellular metabolism. The dorsal aortic blood pO_2 decreased but no significant change in erythrocytes number, hematocrit, hemoglobin concentration, blood pH, or P50 were observed. The ammonia level in this experiment was high; survival times varied from 0.9 to 3.0 hours, and therefore the results of this experiment can be expected to vary from long-term experiments.

Reichenbach-Klinke (1967) found significant decreases in the number of erythrocytes and hemoglobin content in fish exposed to ammonia. Damage to the spleen and liver has also been reported (Reichenbach-Klinke 1976; Flis 1963).

Effects on tissue

Sublethal and lethal ammonia levels can cause histological changes in the kidneys, liver, spleen, thyroid tissue, and blood parameters of many fish (Smith and Piper 1975; Reichenbach-Klinke 1967; Flis 1963). The significance of these changes is difficult to assess, but will certainly adversely affect the animals' ability to survive if exposure to ammonia is prolonged. At some level of damage, the effects may prove irreversible.

Effects on disease

Documentation on the effects of ammonia on the incidence of diseases of aquatic animals is sparse, but it is not unreasonable to expect that an animal weakened by some of the effects discussed above may be more susceptible to disease. Burrows (1964) suggested that the extensive epithetical hyperplasia of the gills of fish caused by ammonia is a predisposing factor to bacterial gill disease. In recent work with rainbow trout sac fry, Burkhalter and Kaya (1977) found that ammonia levels of 0.19 mg/l NH_3-N and higher produced blue-sac disease.

Lethal Effects

Based on short-term lethal toxicity tests, Wuhrmann and Woker (1948) found that only un-ionized ammonia is toxic to fish and that ionized ammonia has little or no toxicity. The most direct mechanism of lethal toxicity appears to be impairment of cerebral energy metabolism due to depletion of high energy compounds in the brain (Smart 1978).

Latter work has shown that the ionized form (NH_4^+) may have a significant toxicity under low pH conditions (Armstrong et al. 1978; Tabata 1962) where the NH_3 level is very low. For an excellent discussion of the relative toxicity of weak acids and bases, see Broderius and Smith (1977). Low Na^+ levels may increase the toxicity of the ionized form (Armstrong et al. 1978; Shaw 1960).

The 96-h LC_{50} value of un-ionized ammonia (NH_3-N) ranges from 0.4 to 3.1 mg/l for fish (Colt and Tchobanoglous 1976; Ball 1967), 0.40 to 2.31 mg/l for crustaceans (Armstrong et al. 1978; Delestraty et al. 1977; Wickins 1976), and 3.3 to 6.0 mg/l for marine molluscs (Epifanio and Srna 1975). The lethal effects of ammonia and other chemicals on molluscs are difficult to assess because the time of death is difficult to determine.

Effects on Growth

There is no evidence that the sublethal effects of ammonia can be attributed solely to the concentration of NH_3. With the exception of three experiments (Armstrong et al. 1978; Tabata 1962; Wuhrmann and Woker 1948), all of the research discussed in this article was conducted at a single pH; therefore, the un-ionized ammonia, ionized, and ammonia levels are all highly correlated. Isolation of the effects of ionized and un-ionized ammonia requires that the pH be varied. In this manner, it is possible to vary the ionized ammonia level while keeping the un-ionized level constant. The hypothesis that ionized ammonia has no effect on growth is based on the results of short-term lethal bioassays (Wuhrmann and Woker 1948).

Working with larval shrimp, *Macrobrachium rosenbergii*, Armstrong et al. (1978) found significant reduction in growth at 90 μg/l NH_3-N (26.2 mg/l NH_4^+-N) in water of pH 6.8. Yet in an NH_3-N concentration of 805 μg/l, growth was not reduced (pH = 8.34, NH_4^+-N = 7.4 mg/l). Therefore, under the conditions of this experiment the effects of ammonia on growth were due to the NH_4^+ level rather than the NH_3 concentration. This effect may be due to the inhibition of Na^+ uptake as discussed previously.

In most of the sublethal research, it was assumed that NH_3 was the toxic form, and therefore, the research was reported in terms of NH_3. Because of this fact, the effects of ammonia on growth will be discussed in terms of the un-ionized ammonia, but it should be kept in mind that

the ionized ammonia may also have a significant effect.

Colt and Tchobanoglous (1978) found that un-ionized ammonia reduced the growth of juvenile channel catfish *(Ictalurus punctatus)* in a linear manner over the range of 48-989 μg/l NH_3-N (0.31-5.71 mg/l NH_4^+-N) during a 31-day growth trial. The EC_{50} (concentration causing 50 percent reduction in weight gain) value was 517 μg/l NH_3-N for wet gain, and the no-growth level was 967 μg/l NH_3-N. This experiment was conducted at a single pH, but the regression equations were computed in terms of un-ionized ammonia, ionized ammonia, and ammonia. The results of this experiment suggests that a "no-effects" level of ammonia does not exist; that is, any level of ammonia will have an effect on growth.

In work with the rainbow trout sac fry, Burkhalter and Kaya (1977) found that 50 μg/l NH_3-N had a significant effect on growth. An approximate no growth level may range from 200 to 600 μg/l NH_3-N. Blue-sac disease was common in sac dry reared in ammonia levels about 190 μg/l NH_3-N.

Levels of un-ionized ammonia ranging from 100 to 600 μg/l NH_3-N reduced the growth of the shrimp *Macrobrachium rosenbergii* by 30 percent (Armstrong et al. 1978; Wickin 1976). The EC_{50} value for the growth of *Panaeus* shrimp was approximately 450 μg/l NH_3-N (Wickin 1976).

The level of un-ionized ammonia that causes a 50 percent reduction in the filtration rate of juvenile clams, *Mercenaria mercenaria,* and juvenile oysters, *Crassostrea virginica,* are 280 and 140 μg/l NH_3-N, respectively (Epifanio and Srna 1975).

At an un-ionized ammonia level of 50-200 μg/l NH_3-N, significant growth reduction will occur in most aquatic animals. Determination of the effects of the ionized ammonia and interaction of sodium levels on the toxicity of ionized ammonia on the growth of aquatic animals will require futher research.

Nitrite Toxicity

In flow-through systems, ammonia will be the principal toxic metabolic by-product, but in recycled systems both ammonia and nitrite may occur at toxic levels. Documentation on the effects of nitrite is excellent for fish, but poor for the other aquatic animals. The toxicity of nitrite to aquatic animals will be discussed in terms of the mechanisms of toxicity, lethal effects, and effects on growth.

Mechanisms of Toxicity

The toxicity of nitrite is due to its effects on the oxygen transport, oxidation of important compounds, and tissue damage.

Effects on oxygen transport

The major effect of nitrite (or nitrous acid) is the oxidation of the iron in the hemoglobin molecule from Fe^{+2} to Fe^{+3} (Smith and Russo, 1975). Since the oxidized form of hemoglobin (ferrihemoglobin) is unable to act as an oxygen carrier, hypoxia and cyanosis may result (Kiese 1974) if sufficient ferrihemoglobin is formed. It is likely that the same reaction can occur with the copper of crustacean hemocyanin but such a conversion has not been reported for this animal class. Nitrite can cause relaxation of smooth muscle of small blood vessels which, in this dilated state, result in a pooling of blood (Gleason et al. 1969).

Oxidization of other compounds

Besides the reaction with hemoglobin, nitrite may oxidize other important compounds (Miale 1967). In work with rainbow trout and chinook salmon (Crawford and Allen 1977; Smith and Williams 1974) it was found that mortality did not correlate with the ferrihemoglobin levels. It was thought that the fish were dying from other toxic reactions of nitrite.

Effects of tissue damage

A nitrite level of 0.060 mg/l NO_2-N produced a minimal degree of hypertrophy, hyperplasia, and lamellar separation in rainbow trout after 3 weeks' exposure (Wedemeyer and Yasutake 1978). At the end of 28 weeks, most of the fish had recovered and showed little or no lamellar epithelial change. Smith and Williams (1974) found that lethal levels of nitrite

produced hemorrhages and necrotic lesions in the thymus of trout and salmon. The long-term effects of nitrite on disease and histology are not well documented.

Lethal Effects

The lethal effects of nitrite to fish is extremely variable depending on water chemistry and species. The 96-h LC_{50} value of nitrite to channel catfish, rainbow trout, and the mottled sculpin *(Cottus bairdi)* range from 12.8 to 13.1; 0.20-0.40; and > 61.0 mg/l NO_2^--N, respectively (Russo and Thurston 1977). Addition of calcium (Wedemeyer and Yasutake 1978; Crawford and Allen 1977) or chloride ions (Wedemeyer and Yasutake 1978; Perrone and Meade 1977; Russo and Thurston 1977) can increase the tolerance of salmonid fish to nitrite by a factor of 20 to 60 times. These ions may compete with nitrite for transportation across the gill and therefore reduce the effect of nitrite toxicity.

Based on the fact that un-ionized molecules can cross biological membranes faster than ionized molecules, Colt and Tchobanoglous (1976) suggested that nitrous acid rather than nitrite may be the toxicity form. If this is the case, a decrease in pH would reduce the tolerance of animals to nitrite. In recent experiments, Russo and Thurston (1977) found pH had no effect on nitrite tolerance, but Wedemeyer and Yasutake (1978) reported that pH did have an effect. Additional experiments will be necessary to clarify the toxicity of nitrite or nitrous acid.

The 96-h LC_{50} value of nitrite to shrimp ranges from 8.5 to 15.4 mg/l (Wickins 1976; Armstrong et al. 1976). The LC_{50} values of nitrite to the clam *Mercenaria mercenaria* and oysters *Crassostrea virginica* are 756 and 532 mg/l NO_2^--N, respectively (Epifanio and Srna 1975). The extremely high tolerance may be an artifact to the ability of these animals to avoid toxic chemicals by closing their valves.

Effects on Growth

The chronic exposure for six months in soft water to 0.03-0.06 mg/l NO_2^--N to steelhead trout caused no significant growth reduction, changes in gill histology, blood chemistry, or impaired ability of the smolts to adapt to seawater (Wedemeyer and Yasutake 1978). In work with crustaceans, 1.8 mg/l caused a 35 percent reduction in the growth of *Macrobrachium rosenbergii* larvae (Armstrong et al. 1976) and a value of 6.2 mg/l caused a 50-percent reduction in the growth of juvenile *Penaeus* shrimp (Wickins 1976). The filtration rate of marine molluscs *Mercenaria mercenaria* and *Crassostrea virginica* was reduced by 50 percent at approximately 280 mg/l NO_2-N. (Epifanio and Srna 1975).

Because of the large variation of both the toxic and sublethal effects of nitrite, criteria for even a single species are impossible to propose. Additional research will be necessary to clarify the effects of environmental parameters on the toxic effect of nitrite.

Nitrate Toxicity

Nitrate toxicity may only be a potential problem in recycled systems where high levels of nitrate can accumulate from the nitrification of ammonia. The toxicity of nitrate to aquatic animals will be discussed in terms of mechanisms of toxicity, lethal effects, and effects on growth.

Mechanisms of Toxicity

Nitrates are the least toxic of the inorganic nitrogen compounds. The toxicity of this compound is due to its effects on osmoregulation and possibly on oxygen transport.

Effects on osmoregulation

The 96-h LC_{50} value of nitrate to aquatic animals ranges between 1000 to 3000 mg/l NO_3-N (Wickin 1976; Colt and Tchobanoglous 1976; Epifanio and Srna 1975). In most cases, the nitrate was added as $NaNO_3$. Therefore, the NO_3-N value should be multiplied by 6.07 to obtain total solute concentration. At these levels, the toxicity of $NaNO_3$ may be due in part to the Na^+ level. In freshwater fish, the lethal $NaNO_3$ levels are comparable to the lethal $NaCl$ levels (Trama 1954). The toxicity of

nitrate seems related to the failure of the animal to maintain osmoregulation in water containing a high salt content.

Effects on oxygen transport

Nitrate can also oxidize hemoglobin to ferrihemoglobin. In mammals, nitrate can be reduced to nitrite by certain gut bacteria and then absorbed across the lower gastrointestinal tract (Gleason et al. 1969). Experimental confirmation of this mechanism has not been reported in aquatic animals.

Gradba et al. (1974) found that low levels of nitrate (5-6 mg/l NO_3-N) caused significant increases in the ferrihemoglobin content of the blood of rainbow trout. There was also serious damage in the peripheral blood and hematopoietic centers as well as serious damage to liver tissue. There is a possibility that nitrite was produced by bacteria in the culture water rather than by gut bacteria. The interpretation of this experiment is difficult because the nitrite levels in the culture water were not measured. The results of this experiment, if valid, would indicate that nitrate toxicity could be more significant than previously thought.

Lethal Effects

The 96-h LC_{50} value of NO_3-N to fish ranged from 1000 to 2000 mg/l (Colt and Tchobanoglous 1976; Westin 1974; Trama 1954). The 48-h LC_{50} to juvenile shrimp (Wickins 1976) was 3,400 mg/l. The 96-h tolerance of the marine mollusc *Crassostrea virginica* ranged from 2,600 to 3,800 mg/l NO_3-N (Epifanio and Srna 1975).

Effects on Growth

Nitrate-N levels of 90 mg/l and 200 mg/l had no effect on the growth of channel catfish (Knepp and Arkin 1972) or juvenile Penaeid shrimp (Wickins 1976), respectively. The growth of the freshwater shrimp, *Macrobrachium rosenbergii* was reduced by 50 percent in 180 mg/l NO_3-N (Wickins 1976). Nitrate levels of 450 mg/l NO_3-N caused extended larvae development in the shrimp, *Palaemontes pugio* (Hinsman 1977). The research by Gradba et al. (1974) showed significant fer-rihemoglobin production and histological damage to rainbow trout at 5 to 6 mg/l NO_3^--N, but additional research should be done to confirm those results. A 50-percent reduction in the filtration rate of *Mercenaria mercenaria* and *Crassostrea virginica* lies in the range of 2500 mg/l NO_3^--N (Epifanio and Srna).

The toxicity of nitrate to aquatic animals does not seem to be a serious problem. Additional work will be required to confirm the results of Grabda et al. (1974). Bacterial reduction of nitrite by bacteria in the culture system or in the gut of the culture animals could significantly increase the toxicity of nitrate.

Design of Aquatic Culture Facilities to Reduce Nitrogen Toxicity Problems

After discussing the source and composition of nitrogen compounds, their chemistry, and their toxicity, the design of culture systems to reduce the effects of nitrogen toxicity will be covered. Design will be based on both lethal and sublethal effects.

Reduction of Lethal Effects

The lethal effects of ammonia may be a significant design parameter in flow-through systems if the water supply is interrupted. Failure of a biofilter in a recycled system would produce lethal ammonia levels in a short time. In both systems, feeding of the animals should be stopped immediately. This would reduce the ammonia production rate to about 25 percent of the normal level (Pecar 1979; Brett and Zala 1975). A stand-by generator could be used to power emergency aeration and water supply system in either types of systems. A small emergency water system would prevent total mortality, although some sublethal damage could result.

Reduction of Effects on Growth

The design of a culture system for aquatic animals should be based on a knowledge of the effects of each toxic compound or environmental variable. It is necessary to know not only the "no-effects" level but also the functional relationship between growth and each toxic compound. The design level of ammonia (or nitrite) will depend both on the effects of ammonia on the culture organism and on the cost of maintaining a given level. In a flow-through system the costs of maintaining a given ammonia level will depend on the pumping, pretreatment, or post-treatment costs. In most systems, a maximum flow rate that exists is due to hydraulic considerations or a restricted water supply. In a recycled system, the costs to maintain a given water quality will depend on the capital and operating costs of unit processes in the system. Since there are very few quantifiable data on the effects of nitrite or nitrate on the growth of aquatic animals, the remainder of this discussion will be based on ammonia. The same methodology could be applied to the other parameters.

Growth Response of the Species

The shape of the unit growth curve as a function of ammonia will have a significant effect on the design of the system. Three cases of general interest are presented in Figure 1a-c. These cases are (1) a "no-effect" level with a linear decrease, (2) a linear decrease with no plateau, and (3) a linear decrease with a nonlinear portion at the higher ammonia levels. The unit growth rate should be measured as a function of the ammonia (or dissolved oxygen) level. Under most normal conditions, operational variables such as stocking density or flow rate have no direct effect on growth. The ammonia and dissolved oxygen levels resulting from a given stocking density and flow rate are the critical parameters.

The effect of ammonia on the total production of the system will be more important than the effects on the unit growth rate. Total production (over a given time period) is equal to the unit growth rate times the number of animals. In Figure 1, the effects of metabolic ammonia on the total production is also presented. In the first case, there is no reason to operate below the x_1 ammonia level, but there may

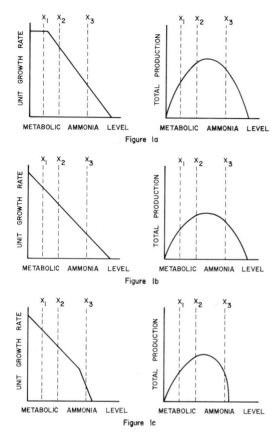

FIGURE 1. — Effects of metabolic ammonia on the unit growth and total production in an aquatic system.

be a good reason to operate at a higher level. If more fish are added to system 1a the growth rate of the fish will decrease, but the total production of the system will increase up to a point. The operating point of the system will depend on the production objectives. For an excellent discussion of the control of a culture unit to achieve rapid growth, produce the maximum weight possible, produce the cheapest fish, optimize fish quality, or combine two or more goals, see Brauhn et al. (1976).

The same type of analysis will apply to the curves presented in Figures 1b and 1c, except in Figure 1c, where an increase in stocking density from $x_2 \rightarrow x_3$ would result in a much greater decrease in total production. The only information on the effect of ammonia on growth is of the type presented in Figure 1b (Colt and Tchobanoglous 1978). If the incidence of disease is

increased at high ammonia levels, the growth response presented in Figure 1c could be produced. The proper design of culture systems should be based on the growth response of the culture animal to each toxic compound over the whole production cycle.

Hydraulic Interactions

The ammonia production will depend primarily on the mass of the animals, the feeding rate, and the protein level in the feed (Page and Andrews 1974; Harris 1971). If Q = the flow rate, M = mass of the animals, and F = ammonia production rate per unit mass of fish at a given feeding rate, then the ammonia level concentration in the effluent of the culture systems will be equal to $\dfrac{ME}{Q}$ at steady-state. This analysis assumes that the ammonia excretion rate does not depend on the ambient ammonia level — which is not entirely correct (Olson and Fromm 1971).

In fish culture, two commonly used culture systems are the circular pond and the raceway. The hydraulics of the circular pond and the raceway may be approximated by the continuous-flow stirred-tank reactor (CFSTR) and the plug flow reactor (PFR), respectively. While the ammonia level in the effluent from both of these systems will be identical, the ammonia levels within the reactors are quite different (Fig. 2).

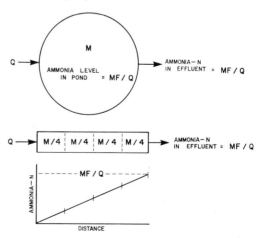

FIGURE 2. — Variation of the ammonia level in different culture systems.

In an ideal CFSTR, the contents of the reactor are fully mixed so the concentration in the reactor world also be MF/Q. In the PFR, the ammonia level would linearly increase from the level in the influent to MF/Q at the end of the raceway. Real circular ponds may not be fully mixed so that the ammonia level in some areas of the reactors may exceed $\dfrac{MF}{Q}$ (Larmoyeux et al. 1973; Burrows and Chenoweth 1970, 1955). The characteristics of raceways may be very close to that predicted from the PFR model (Westers and Pratt 1977). While the hydraulics of real culture systems may differ significantly from theory, this will not have a critical effect on the following analysis.

The total production of the two systems will be identical for an ammonia concentration less than x_1 for a type 1a growth curve. The total production of the PFR system will be larger than the CFSTR for all ammonia levels for type 1b and 1a growth curves and for ammonia levels $> x_1$ for the type 1a curve.

This type of analysis can lead to the rational design of culture systems, but depends on a knowledge of the effects of ammonia (or nitrite or dissolved oxygen) on the growth of the culture animals and hydraulic characteristics of the culture systems. The growth response curves to these important parameters are not known for most species, but the hydraulic characteristics of culture systems have been investigated for several common types (Larmoyeux et al. 1973; Burrows and Chenoweth 1970, 1955).

Addition of Chemicals

The short-term toxicity of ammonia depends strongly on the pH of the ambient water, and therefore reduction of the pH could reduce the toxic effects. If the sublethal effects of ammonia are due in part to the NH_4^+ level, addition of Na^+ may reduce the sublethal effects. Wedemeyer and Yasutake (1978) have suggested that the water supply for hatcheries using recycled systems should have a total water hardness of at least 100 mg/l (as $CaCO_3$), a pH of 7 or above, or Cl^- and Ca^{++} levels equivalent to 25 mg/l $CaCl_2$.

As an example, the cost to add 25 mg/l $CaCl_2$ will be computed for Warm Springs National Fish Hatchery, a hatchery using recycled water (Kramer, Chin, and Mayo 1971). This hatchery uses 9.4 million liters of make-up water a day. The calcium level is 6.6 mg/l, and the chloride level is 0. Therefore, it would be necessary to add 25 mg/l $CaCl_2$. This would require 235 kg a day. At $0.33/kg (80 percent $CaCl_2$) this would cost $35,000 per year without consideration of the operating and capital costs of the mixing and metering systems. The estimated annual operating costs for this hatchery is $282,000, so chemical addition would increase the operating costs by at least 12 percent. Chemical addition may be required only during the peak production period, so the cost of chemical addition would be significantly reduced. Acid or base addition would also require elaborate control systems and expensive piping and tankage for these corrosive chemicals. Delivery costs of these chemicals could also be expensive in some areas. The cost of chemical addition in flow-through systems may not be feasible because of the large flows involved.

Aeration

The effects of ammonia and nitrite depends strongly on the dissolved oxygen levels. In some cases, dissolved oxygen depletion may have a more serious effect on growth than ammonia toxicity (Larmoyeux and Piper 1973). Therefore, it is critical that adequate dissolved oxygen be maintained in culture systems. Several types of aeration devices have been used in hatcheries, but the maintenance of high levels (6-9 mg/l) of dissolved oxygen may be more economical with pure oxygen systems (Ruane et al. 1977) or U-tube aerators (Speece and Orosco 1970).

Conclusions

The control of nitrogen toxicity problems in the culture of aquatic animals will depend on a knowledge of the mechanisms of toxicity, the effect of each compound on the growth of the culture species over the whole production period, and an understanding of the hydraulics of the culture system. The effects of ammonia and nitrite on the growth of the culture animal, probably the most important sublethal effect, is unknown for most animals. The functional relationship between growth and, for example, ammonia must be known because it will probably be impossible to operate in the "no-effects" levels because of economic considerations. The design level of each toxic compound will depend on the objectives of the fish culturists, the costs of maintaining a given level, the amount of water available, or a combination of these factors. Aeration and addition of chemicals to reduce the toxicity of nitrogen compounds may also be economic under certain conditions.

References

ARMSTRONG, D.A. 1978. Toxicity and metabolism of nitrogen compounds: Effects on survival, growth and osmoregulation of the prawn *Macrobrachium rosenbergii.* Ph.D. dissertation, Ecology Group, Univ. Calif., Davis. 94 pp.

ARMSTRONG, D.A., D. CHIPPENDALE, A.W. KNIGHT, AND J.E. COLT. 1978. Interaction of ionized and unionized ammonia on short-term survival and growth of prawn larvae, *Macrobrachium rosenbergii.* Biol. Bull. 154(1):15-31.

ARMSTRONG, D.A., M.J. STEPHENSON, AND A.W. KNIGHT. 1976. Acute toxicity of nitrite to larvae of the giant Malaysian prawn, *Macrobrachium rosenbergii.* Aquaculture 9(1):39-46.

BALL, I.R. 1967. The relative susceptibility of some species of freshwater fish to poisons — I. Ammonia. Water Res. 1(11/12):767-775.

BARCIA, J. 1973. Reliability of an ammonia probe for electrometric determination of total ammonia nitrogen in fish tanks. J. Fish. Res. Board Can. 30(9):1389-1392.

BOWLER, C.E., AND J.P. BIDWELL. 1978. Ionization of ammonia in seawater: Effects of temperature, pH, and salinity. J. Fish. Res. Board Can. 35(7):1012-1016.

BRAUHN, J.L., R.C. SIMON, AND W.R. BRIDGES. 1976. Rainbow trout growth in circular tanks: Consequences of different loading densities. U.S. Fish Wildl. Serv., Tech. Pap. No. 86. 16 pp.

BRETT, J.R., AND C.A. ZALA. 1975. Daily pattern of nitrogen excretion and oxygen consumption of sockeye salmon *(Oncorhynchus nerka)* under controlled conditions. J. Fish. Res. Board Can. 32(12):2479-2486.

BRODERIUS, S.J., AND L.L. SMITH, JR. 1977. Relationship between pH and acute toxicity of free cyanide and dissolved sulfide forms to the fathead minnow. Pages 88-117 *in* R.A. Tubb, ed. Recent advances in fish toxicology. U.S. Environmental Protection Agency, EPA-600/3-77-085. U.S. Government Printing Office, Washington, D.C.

BURKHALTER, D.E., AND C.M. KAYS. 1977. Effects of prolonged exposure to ammonia on fertilized eggs and sac fry of rainbow trout *(Salmo gairdneri).* Trans. Am. Fish. Soc. 106(5):470-475.

BURROWS, R.E. 1964. Effects of accumulated excretory products on hatchery-reared salmonids. Bur. Sport Fish. Wildl., Res. Rep. No. 66. 12 pp.

BURROWS, R.E., AND H.H. CHENOWETH. 1970. The rectangular circulating pond. Prog. Fish-Cult. 32(2):67-80.

———. 1955. Evaluation of three types of fish rearing ponds. U.S. Fish Wildl. Serv., Res. Rep. No. 39. 29 pp.

CAMPBELL, J.W. 1973. Nitrogen excretion. Pages 279-316 *in* C.L. Prosser, ed. Comparative animal physiology. W.B. Saunders, Philadelphia, Pa. 966 pp.

COLLINS, M.T., J.B. GRATZEK, E.B. SHOTTS, JR., D.L. DAWE, L.M. CAMPBELL, AND D.R. SENN. 1975. Nitrification in an aquatic recirculating system. J. Fish Res. Board Can. 32(11):2025-2031.

COLT, J. 1978. The effects of ammonia on the growth of channel catfish, *Ictalurus punctatus.* Ph.D. dissertation, Dep. Civil Eng., Univ. Calif., Davis. 185 pp.

COLT, J., AND G. TCHOBANOGLOUS. 1978. Chronic exposure of channel catfish, *Ictalurus punctatus,* to ammonia: Effects on growth and survival. Aquaculture 15(4):353-372.

———. 1976. Evaluation of the short-term toxicity of nitrogenous compounds to channel catfish, *Ictalurus punctatus.* Aquaculture 8(3):209-224.

CRAWFORD, R.E., AND G.H. ALLEN. 1977. Seawater inhibition of nitrite toxicity to chinook salmon. Trans. Am. Fish. Soc. 106(1):105-109.

CVANCARA, V.A. 1969. Studies on tissue arginase and ureogenesis in freshwater teleost. Comp. Biochem. Physiol. 30(3):489-496.

DELISTRATY, D.A., J.M. CARLBERG, J.C. VAN OLST, AND R.F. FORD. 1977. Ammonia toxicity in cultured larvae of the American lobster, *Homarus americanus.* Pages 647-672 *in* J.W. Avault, ed. Proceedings of the Eighth Annual Meeting, World Mariculture Society, Louisiana St. Univ., Baton Rouge.

EMERSON, K., R.C. RUSSO, R. LUND, AND R.V. THURSTON. 1975. Aqueous ammonia equilibrium calculations: Effects of pH and temperature. J. Fish. Res. Board Can. 32(12):2379-2388.

EPIFANIO, E.C., AND R.F. SRNA. 1975. Toxicity of ammonia, nitrite ion, nitrate ion, and orthophosphate to *Mercenaria mercenaria* and *Crassostrea virginica.* Mar. Biol. 33(3):241-246.

FLIS, J. 1963. Anatomicohistopathological changes

induced in carp *(Cyprinus carpio* L.) by ammonia water. Part II. Effects of subtoxic concentrations. Acta Hydrobiol. 10(1/2):225-238.

GLEASON, M.M., R.E. GASSELIN, H. HODGE, AND R. SMITH. 1969. Clinical toxicity of commercial products: Acute poisoning. The William and Wilkins Co.

GRABDA, E., T. EINSZPORN-ORECKA, C. FELINSKA, AND R. ZBANYSEK. 1974. Experimental methemoglobinemia in trout. Acta Ichthyol. Piscatoria 4(2):43-71.

HAMPSON, B.L. 1977a. Relationship between total ammonia and free ammonia in terrestrial and ocean waters. J. Cons. Int. Explor. Mer 37(2):117-122.

———. 1977b. The analysis of ammonia in polluted sea water. Water Res. 11(11):305-308.

HANSSON, I. 1973. A new set of pH-scale and standard buffers for sea water. Deep-Sea Res. 20(5):479-491.

HARRIS, J.C. 1971. Pollution characteristics of channel catfish culture. M.S. thesis, Environ. Health Eng., Univ. Tex., Austin. 94 pp.

HINSMAN, C.B. 1977. Effect of nitrite and nitrate on the larval development of the grass shrimp *Palemonetes pugio* Holthius. M.S. thesis, Inst. Oceanogr., Old Dominion Univ., Norfolk, Va.

KHOO, K.W., C.H. CULBERSON, AND R.G. BATES. 1977. Thermodynamics of the dissociation of ammonium ion in seawater from 5°C to 40°C. J. Sol. Chem. 6(4):281-290.

KIESE, M. 1974. Methemoglobinemia: A comprehensive treatise. CRC Press, Cleveland, Ohio. 259 pp.

KNEPP, G.L., AND G.F. ARKIN. 1972. Ammonia toxicity levels and nitrate tolerance for channel catfish *(Ictalurus punctatus)*. Paper presented at the Annual Meeting Am. Soc. Agric. Eng., Hot Springs, Ark. 6 pp.

KRAMER, CHIN AND MAYO, INC. 1971. Warm Springs National Fish Hatchery, Wasco County, Oregon-Master Plan. 108 pp.

LARMOYEUX, J.D., AND R.G. PIPER. 1973. Effects of water reuse on rainbow trout in hatcheries. Prog. Fish-Cult. 35(1):2-8.

LARMOYEUX, J.D., R.G. PIPER, AND H.H. CHENOWETH. 1973. Evaluation of circular tanks for salmonid production. Prog. Fish-Cult. 35(3):122-131.

LATIMER, W.H., AND J.H. HILDEBRAND. 1951. Reference book of inorganic chemistry. Third edition. The MacMillan Company, New York. 625 pp.

LIAO, P.B., AND R.D. MAYO. 1974. Intensified fish culture combining water reconditioning with pollution abatement. Aquaculture 3(1):61-85.

LLOYD, R., AND L.D. ORR. 1969. The diuretic response of rainbow trout to sublethal concentrations of ammonia. Water Res. 3(5):335-344.

MAETZ, J., P. PAYAN, AND G. DeRENZIS. 1976. Controversial aspects of ionic uptake in freshwater animals. Pages 77-92 *in* P.S. Davies, ed. Prospectives in experimental biology, Vol. 1. Pergamom Press, London. 525 pp.

MANGUM, C., S.U. SILVERTHORN, J.L. HARRIS, D.W. TOWLE, AND A.R. KRALL. 1976. The relationship between blood pH, ammonia excretion and adaptation in low salinity in the blue crab *Callinectes sapidus.* J. Exp. Zool. 195(1):129-136.

MAYER, F.L. JR., AND R.H. KRAMER. 1973. Effects of hatchery water reuse on rainbow trout metabolism. Prog. Fish-Cult. 35(1):9-10.

MIALE, J.B. 1967. Laboratory medicine-hemotology. Third edition. C.V. Mosby Publisher, St. Louis, Mo. 1257 pp.

OLSON, K.R., AND P.O. FROMM. 1971. Excretion of urea by two teleost exposed to different concentrations of ambient ammonia. Comp. Biochem. Physiol. 40A (4):999-1007.

ORION RESEARCH. 1976. Instruction Manual — Nitrogen Oxide Electrode, Model 95-46. Cambridge, Mass. 28 pp.

PAGE, J.W., AND J.W. ANDREWS. 1974. Chemical composition of effluent from high density culture of channel catfish. Water Air Soil Poll. 3(3):365-367.

PECOR, C.H. 1979. Experimental intensive culture of tiger muskellunge in a water reuse system. Prog. Fish-Cult 41(2):103-108.

PERRONE, S.J., AND T.L. MEADE. 1977. Protective effect of chloride on nitrite toxicity to coho salmon *(Oncorhynchus kisutch).* J. Res. Board Can. 34(4):486-492.

REICHENBACH-KLINKE, H.H. 1967. Investigation on the influence of the ammonia content on the fish organism. Arch. für Fischereiwiss. 17(2):122-132. (In English translation.)

RUANE, R.J., T.-Y. J. CHU, AND V.E. VANDERGRIFF. 1977. Characterization and treatment of waste discharged from high-density catfish cultures. Water Res. 11(9):789-800.

RUSSO, R.C., AND R.V. THURSTON. 1977. The acute toxicity of nitrite to fishes. Pages 118-131 *in* R.A. Tubb, ed. Recent advances in fish toxicology. U.S. Environmental Protection Agency, EPA-600/3-77-085. U.S. Government Printing Office, Washington, D.C. 203 pp.

SHARMA, B., AND R.C. AHLERT. 1977. Nitrification and nitrogen removal. Water Res. 11(10):897-925.

SHAW, J. 1960. The absorption of sodium ions by the crayfish *Astacus pallipes* Lereboullet. III. The effects of other cation in the external solution. J. Exp. Biol. 37(3):548-556.

SMART, G.R. 1978. Investigation of the toxic mechanisms of ammonia to fish — Gas exchange in rainbow trout *(Salmo gairdneri)* exposed to acutely lethal concentrations. J. Fish. Biol. 12(1):93-104.

SMITH, C.E., AND R.G. PIPER. 1975. Lesions associated with chronic exposure to ammonia. Pages 497-514 *in* W.E. Ribelin and H. Migaki, eds. The pathology of fishes. Univ. Wis. Press, Madison. 1004 pp.

SMITH, C.E., AND R.C. RUSSO. 1975. Nitrite-induced methemoglobinemia in rainbow trout. Prog. Fish-Cult. 37(3):150-152.

SMITH, C.E., AND W.G. WILLIAMS. 1974. Experimental nitrite toxicity in rainbow trout and chinook salmon. Trans. Am. Fish. Soc. 103(2):389-390.

SOLORZANO, L. 1969. Determination of ammonia in natural waters by the phenolhypochlorite method. Limnol. Oceanogr. 14(5):799-801.

SOUSA, R.J., AND T.L. MEADE. 1977. The influence of ammonia on the oxygen delivery system of the coho salmon hemoglobin. Comp. Biochem. Physiol. 58A(1):23-28.

SPEECE, R.E., AND R. OROSCO. 1970. Design of U-tube aeration systems. ASCE J. Sanit. Eng. Div. 96(SA3):715-725.

STANDARD METHODS FOR THE EXAMINATION OF WATER AND WASTEWATER. 1976. Fourteenth edition. American Public Health Association, American Water Works Association, and Water Pollution Control Federation, Washington, D.C. 1193 pp.

TABATA, K. 1962. Toxicity of ammonia to aquatic animals with reference to the effects of pH and carbon dioxide. Bull. Tokai Reg. Fish. Res. Lab. 34:67-74.

TRAMA, F.B. 1954. The acute toxicity of some common salts of sodium, potassium and calcium to the common bluegill (Lepomis macrochirus Rafinesque). Proc. Acad. Nat. Sci. Phila. 106:185-205.

TUMMAVUORI, J., AND P. LUMME. 1968. Protolysis of nitrous acid in aqueous sodium nitrate and sodium nitrite at different temperatures. Acta Chem. Scand. 22(6):2003-2011.

WEDEMEYER, G.A., AND W.T. YASUTAKE. 1978. Prevention and treatment of nitrite toxicity in juvenile steelhead trout (Salmo gairdneri). J. Fish. Res. Board Can. 35(6):822-827.

WESTERS, H., AND K.M. PRATT. 1977. Rational design of hatcheries for intensive salmonid culture, based on metabolic characteristics. Prog. Fish-Cult. 39(4):157-165.

WESTIN, D.T. 1974. Nitrate and nitrite toxicity to salmonoid fishes. Prog. Fish-Cult. 36(2):86-89.

WHITFIELD, M. 1974. The hydrolysis of ammonia ions in seawater — A theoretical study. J. Mar. Biol. Assoc. U.K. 54(3):565-580.

WHITFIELD, M. 1978. The hydrolysis of ammonium ions in seawater — Experimental confirmation of predicted constants at one atmosphere pressure. J. Mar. Biol. Assoc. U.K. 58(3):781-787.

WICKINS, J.F. 1976. The tolerance of warm-water prawns to recirculated water. Aquaculture 9(1):19-37.

WUHRMANN, K., AND H. WOKER. 1948. Experimentelle Untersuchungen Uber de ammoniak und Blausaurevergiftung. Schweiz. Z. Hydrol. 11(1.2):210-244.

Bio-Engineering Symposium for Fish Culture (FCS Publ. 1):48-52

Possible Effects of Phototactic Behavior on Initial Feeding of Walleye Larvae

Luciano Corazza and John G. Nickum

New York Cooperative Fishery Research Unit
Fernow Hall, Cornell University
Ithaca, New York 14853

ABSTRACT

During the course of experiments on the culture of walleye *(Stizostedion vitreum vitreum)* larvae, it was observed that they were attracted to the white surface of the experimental unit. This attraction appeared to override normal feeding responses when the larvae were presented with either live or artificial food.

Energetic considerations suggest that uniform distribution of food and larvae should increase feeding efficiencies. Investigations were undertaken to determine the effects of backgrounds of various colors on the dispersion of walleye larvae in experimental rearing units. Previous observations had indicated that feeding was disrupted when the larvae were not distributed uniformly. The experiments were conducted on walleye larvae hatched at 10 °C and maintained at this temperature for five days. Experiments were carried out when the larvae were swimming freely but still had some yolk sac. Although the larvae had not begun to feed, their reaction to light and background coloration was similar to slightly older larvae that had started to feed. Containers of four colors (white, yellow, green, and gray) and two textures (smooth and rough) were used in the experiments.

Fish were dispersed most evenly in gray containers. It appears that this color creates a monotone environment and thereby improves the distribution of the larvae. The results indicated that the color of rearing units and the uniformity of lighting may influence results in walleye culture, but also raised questions concerning the relative roles of vision and other senses in the detection of food particles. The importance of a specific environmental factor, its potential effects on behavior, and subsequently the success of fish culturing systems were established.

Introduction

Initial feeding of larvae is a critical point in the intensive culture of walleye *(Stizostedion vitreum vitreum)* and many other fishes for which culture systems have not been perfected. Survival through larval stages is often insignificant and unreliable (Nickum 1978). Since extensive culture techniques for walleye have been unpredictable and have been demonstrated to be impractical where large numbers of fingerlings or larger fish are needed, there is a need to develop intensive culture systems.

In early feeding trials, we noticed that walleye larvae were strongly attracted to the sides of the white containers in which they were held and did not attempt to feed on either live or artificial diets. Additional observations of larvae in unevenly lighted containers indicated that the attraction to light-colored or brightly lighted areas was sufficient to block response to stimuli other than direct physical contact. Early life stages of walleye under natural lake conditions are characterized by a positive phototaxis, which is abruptly reversed when the animal is approximately 5 weeks old (Houde 1968). Our observations under cultural conditions (Fig. 1) may therefore be related to the typical behavior of larval walleyes.

Our study was conducted under intensive culture conditions to determine the effects of selected background colors and uniformity of lighting on the distribution and feeding behavior of larval walleyes. Some energetic considerations that may relate to the observed behavior and success in culturing larval fishes are also presented. The considerations were developed by utilizing some of the strategies relative to prey concentration and availability that have been documented by several authors (Ivlev 1961; Emlen 1966, 1968; Cody 1974; Stein 1977). Feeding efficiency seems to be a function not only of availability, but also of intrinsic behavioral characteristics of the species (such as

FIGURE 1. — Positive phototaxis of larval walleye in response to a bright background.

phototaxis); therefore, intensive culture systems and techniques should consider all such factors.

Materials and Methods

Walleye eggs, obtained from the Oneida Lake Hatchery at Constantia, New York, were incubated at 10 °C. Larvae hatched after 18 days (average size 7.3 mm) and were maintained in holding tanks at 10 °C for 5 days. Experiments were carried out when the larvae were swimming freely, but still retained some portion of their yolk sac, and had not yet begun to feed; however, their reaction to light and background coloration was similar to slightly older larvae that had begun to feed.

The internal surfaces of translucent fiberglass containers (15 cm × 30 cm) were painted with four different colors:

white, yellow, green, and gray. The amount of light reflected by the walls of the container at wave length ranging from 400 μm to 700 μm (Fig. 2) (the range of the instrument) was measured by means of a scanning spectrometer. Each color had two different textures, smooth and rough. The rough texture was obtained by mixing fine sand with the paint. A grid of 55 squares was drawn on the bottom of each container.

Two containers of the same color but different texture were placed on a platform under a light source of four 60-watt light bulbs symmetrically located. A camera was located over the structure. The containers were then filled with 2 liters of 10 °C water and 100 larvae introduced to each. After 10 minutes, at which point the larvae had assumed a steady pattern of dispersion, a picture was taken. This procedure was repeated 4 times for each set of containers. For each trial new larvae were introduced.

The pictures were enlarged to a 5" × 7" format and larvae in each square of the grid were counted. An index of dispersion (Table 1)

$$I_d = \frac{S^2}{\bar{x}}$$

was utilized in a 2 × 4 factorial analysis of variance to examine differences between treatments.

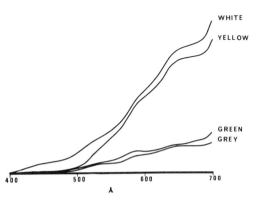

FIGURE 2. — Scanning spectrographs of light reflected from walleyes of fish-rearing troughs of four different colors. (The horizontal axis indicates wave length, in mm, of reflected light. Units of vertical axis are in an arbitrary scale indicating the relative intensity of light reflected.)

TABLE 1. — Index of dispersion values for walleye fry in rearing units of different colors and textures.

Color:	Gray		Green		Yellow		White	
Replicate	Smooth	Rough	Smooth	Rough	Smooth	Rough	Smooth	Rough
1	1.529	1.937	3.261	2.079	3.431	1.774	2.161	2.344
2	1.795	1.388	2.833	2.528	2.752	2.731	2.894	2.707
3	2.099	1.916	3.118	1.978	2.670	2.609	1.794	2.711
4	1.627	1.672	2.228	2.589	2.079	2.711	2.018	2.222

Results

The analyses indicated that a significant difference ($P < 0.005$) (Table 2), attributable to color, existed among the treatments. The effects of textures, as well as texture × color, were not significant ($P > 0.25$) (Table 2).

By partitioning the treatment sum of squares, it was established that the distribution of larvae in the gray troughs was significantly different from those in troughs of other colors ($P < 0.001$) (Table 2). In other words, the fish were more evenly distributed in the gray troughs than they were in the troughs of other colors. The distribution of larvae in the green units was not significantly different from that of the white and yellow troughs ($P > 0.25$) (Table 2).

TABLE 2. — Analysis of variance table comparing the dispersion of walleye larvae in troughs of selected colors and textures.

Sources	df	S.S.	M.S.	F
Total	31	8.454		
Treatments	7	4.736	0.676	4.36**
Color	3	3.783	1.261	8.14**
gray vs. all	1	3.501	3.501	22.60***
green vs. yellow + white	1	0.033	0.033	0.22
yellow vs. white	1	0.119	0.119	0.77
Texture	1	0.179	0.179	1.15
Color × texture	3	0.774	0.258	1.66
Error	24	0.372	0.155	

**$P < 0..01$
***$P < 0.005$

Discussion

The efficiency with which larval fish forage for their diet may have substantial influence on their growth and survival. Factors that influence their ability to obtain an adequate diet with minimal foraging effort can therefore be of considerable importance in rearing larval fishes. Availability of the diet is a function of the distribution and concentration of the food particles. If we assume uniform distribution of the prey, even distribution of the predator should reduce pursuit time and minimize competition. In other terms, the distance between fish and food particles is minimized, and the efficiency of feeding increased.

By modifying the feeding efficiency model proposed by Emlen (1966), we have obtained the following equation:

$$E = \frac{\alpha C_1 - KD}{t + qD}$$

where

E = feeding efficiency or net caloric intake per unit time

α = coefficient of utilization (ratio between caloric content of food introduced and food consumed)

C_1 = caloric content of the ration

K = caloric expenditure per unit distance

D = distance between two food particles

q = reciprocal of speed with which predator moves

t = time expenditure on capture and consumption

50

Given the small size of the food particles (on the order of 500 μm) and the flow-through nature of the intensive culture system, it was difficult to control the distribution of the food particles; however, a uniform distribution was assumed.

The only other parameter that could be controlled was the time expenditure on capture. By achieving a uniform distribution of the larvae in an environment with a finite supply of uniformly distributed food particles, the distance between the predator and prey should be minized. As a result, time expenditure for capture and consumption of food particles can be minimized and feeding efficiency increased.

The results indicated that color and the intensity — or a combination of these two characteristics — of light reflected from the walls of the container could be the most important factors affecting the distribution and therefore the feeding efficiency of walleye larvae (Fig. 3). The fact that only gray, and not green (chosen as an alternate neutral color), fostered a uniform distribution contradicted the hypothesis that light intensity alone triggered photic response. The phototaxis elicited by the green color on the other hand could be explained by the absorption characteristic of the walleye retina. The absorption of the cone pigment is maximal in the green range (560 μm) (Ali et al. 1977). Therefore, the green background may appear very bright to larval walleyes. The scanning spectrographs (Fig. 2) indicated little difference in intensity of light reflected in the green range between the gray and green troughs. This could suggest that light intensity and color might have a synergistic effect on the phototactic response of walleye larvae.

Conclusion

Under the conditions of our laboratory it appears that light conditions are of critical importance in the food search of walleye larvae. The implications of these results on the development of a culturing system are clear. More studies are necessary, however, to determine the mechanism of location and recognition of the food particle. For example, preliminary scanning electron microscopy on newly hatched larvae have brought to light a well-developed olfactory apparatus that could be of crucial importance on their behavior.

Acknowledgment

This project is part of a larger project on intensive walleye culture conducted at Cornell University by the New York Cooperative Fishery Research Unit. It was sponsored by the U.S. Fish and Wildlife Service, the New York State Department of Environmental Conservation, and the New York Sea Grant Institute under a grant from the Office of Sea Grant, National Oceanic and Atmospheric Administration (NOAA), U.S. Department of Commerce. The U.S. Government (including the Sea Grant Office) is authorized to produce and distribute reprints for governmental purposes notwithstanding any copyright notation appearing hereon.

References

ALI, M.A., R.A. RYDER, AND M. ANCTIL. 1977. Photoreceptors and visual pigments as related to behavioral responses and preferred habitats of perches (*Perca* spp.) and pikeperches (*Stizostedion* spp.). J. Fish. Res. Board Can. 34:1475-1480.

CODY, M.L. 1974. Optimization in ecology. Science 183:1156-1164.

EMLEN, J.M. 1966. The role of time and energy in food preference. Am. Nat. 100:611-617.

———. 1968. Optimal choice in animals. Am. Nat. 102:385-389.

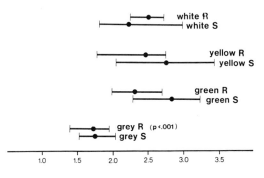

FIGURE 3. — Mean and range of dispersion index $I = \dfrac{S^2}{\bar{x}}$ of larval walleye in troughs of different colors and textures (R = rough and S = smooth).

HOUDE, E.D. 1968. The relation of water currents and zooplankton abundance to distribution of larval walleyes, *Stizostedion vitreum vitreum,* in Oneida Lake, New York. Ph.D. thesis, Cornell Univ., Ithaca, New York. 164 pp.

IVLEV, V.S. 1961. Experimental ecology of feeding of fishes. Yale Univ. Press, New Haven. 302 pp.

NICKUM, J.G. 1978. Intensive culture of walleyes: The state of the art. Pages 187-194 *in* Selected Coolwater Fishes of North America. Am. Fish. Soc. Spec. Publ. 11.

STEIN, R.A. 1977. Selective predation, optimal foraging, and the predator-prey interaction between fish and crayfish. Ecology 58:1237-1253.

Bio-Engineering Symposium for Fish Culture (FCS Publ. 1):53-62

SESSION III   WATER CONDITIONING FOR FISH REARING

Management of Dissolved Oxygen and Nitrogen in Fish Hatchery Waters

RICHARD E. SPEECE

Drexel University
Philadelphia, Pennsylvania 19104

ABSTRACT

Dissolved oxygen is a key component in fish production and excess dissolved nitrogen can result in mortality. Conventional methods of aeration with their limitations, energy consumptions, and costs are reviewed. Typical trout production costs for labor and feed are $1.60/lb ($3.52/kg). Amortization of the hatchery facility adds another $3.50/lb ($7.70/kg). In view of the fact that it requires only one-third of a pound of oxygen (0.73 kg), approximately 2¢ worth of oxygen at $120 per ton to produce a pound of trout, commercial oxygen is attractive and the alternatives for supplying commercial oxygen are presented. The oxygen absorption system must be efficient and economical. It will also strip dissolved nitrogen simultaneously. Alternative methods for efficient absorption of commercial oxygen are presented.

Introduction

Dissolved oxygen (D.O.) is just as basic a necessity in fish production as feed, and excess dissolved nitrogen (D.N.) can be as lethal as "ich." However, the management of these two gases does not generally receive the same priority in hatchery design and operation as do feed and "ich." Hatchery siting and design have favored going to more protected source waters, especially for supply to eggs and fingerlings. This results in greater use of well supplies, which characteristically have little or no dissolved oxygen and dissolved nitrogen levels that are often in excess of 120%. The aeration system, therefore, must simultaneously add dissolved oxygen and strip dissolved nitrogen.

The actual weights of oxygen and nitrogen involved are relatively small. For a trout hatchery producing 100,000 lb/yr (45,359 kg/yr), the peak feeding rate may be about 3.5 times the average, e.g. 1,400 lb (636 kg) of feed per day. This would require a peak water flow of about 5,400 gal/min (20,250 l/min) to keep the ammonia nitrogen from increasing more than 0.5 mg/l. Fish metabolism would consume

a peak of 300 lb/d (136 kg/d) of D.O., resulting in a D.O. decrease of 4.5 mg/l. If the inlet D.N. was 120%, then 220 lb (100 kg) of nitrogen gas would have to be stripped per day to bring the D.N. down to 102%.

The situation of a fish hatchery that greatly magnifies the difficulty of dissolving or stripping such relatively small amounts of oxygen and nitrogen is that the transfer must occur when the concentrations of these gases are so close to saturation. The rate of gas transfer is reduced proportionately as the concentration of dissolved gas approaches saturation according to the gas transfer equation.

Hatchery production can be limited by D.O. and attempts to eliminate this oxygen limitation of hatchery production have resulted in a multitude of aerator designs. It is easy to design an aerator that has a high unit weight of oxygen transfer per kWh of input energy, but the unit capital costs may be prohibitive. Likewise, it is easy to design an aerator with very low unit capital costs, but the unit weight of oxygen transfer may be prohibitively low. Thus it is readily observed that design of an effective aeration system for fish production has to satisfy many constraints:

1. Capital cost ($ per ton/day of oxygen absorbed)
2. Energy consumption (lb O_2 absorbed/kWh consumed)
3. Discharge D.O. (4-8 mg/l, or 40-80% of saturation)
4. Dissolved nitrogen supersaturation (<102% for salmonids)
5. Minimal maintenance
6. Must accommodate the raceway or pond configuration.

With this combination of constraints, most conventional aeration systems are inappropriate.

Dissolving oxygen into hatchery water is not cheap — even though the air is free — in those cases where hydraulic head is not available and the water must be pumped. The electrical energy cost is approximately $30-$50/t (ton) ($33-$55/Mg [tonne or metric ton]) of oxygen dissolved. Even so, it is rather difficult to achieve D.O. values in excess of 6 mg/l, which is considered minimal for trout production. From the author's limited experience, it is not uncommon for fish production to be D.O. limited.

In spite of the relatively high cost of oxygen, the most economical alternative for increasing fish production may well be an increase in oxygenation capacity when compared with a major capital expansion of raceways, assuming the water supply is available.

The political reality in the consideration of state or federal fish hatchery economics is that capital costs are preferred over operating costs. The federal or state government can cut yearly operating budgets, but not completed capital expenditures. There are certain trade-offs during the design phase between capital and operating costs. For hatchery water aeration, a judicious increase in capital expenditure for a small on-site oxygen-production plant or liquid oxygen storage facility can significantly reduce yearly operating costs while eliminating D.O. limitations of fish production.

Objective

The objective of this paper is to address the costs associated with oxygen transfer and nitrogen stripping in fish hatcheries by some of the conventional aeration systems where gravity head is not available and pumped head is required. No consideration will be given to those hatchery situations having ample water supply with naturally occurring head or sufficiently low fish-loading density as to result in no D.O. or D.N. problems. The main thrust of the paper is toward hatcheries that must maximize fish production with a limited water supply and a restricted hydraulic head.

Conventional aerator designs will be reviewed. Energy consumption and operating costs will be considered. The advantages, costs, and absorber design requirements of commercial oxygen for management of D.O. and D.N. in hatchery waters will then be considered. Finally suggested design criteria for use of commercial oxygen in hatchery water management will be given.

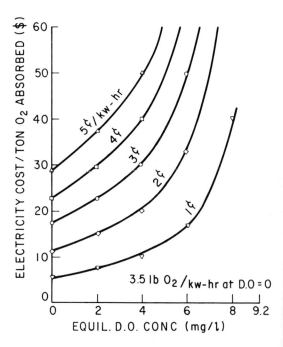

FIGURE 1. — Electricity costs for oxygen transfer by conventional surface aeration.

The electrical operating cost for conventional aeration systems is a function of the discharge D.O. required and the cost of electricity per kW. This relationship is shown graphically in Figure 1. Since the minimum desired D.O. in a trout hatchery is 6 mg/l, the aerator must produce a discharge D.O. preferably in excess of 8 mg/l. At 0.03 kWh, the operating cost for electricity alone would be $60/t ($66/Mg) of oxygen added, or $270/mo for peak production at a 100,000-lb/yr (45,359-kg/yr) trout hatchery. The gas transfer equation shown below indicates a reduction in the rate of gas transfer as the actual D.O. in the discharge increases.

$$dc/dt = K_2 (C_{Sat} - C_{Act})$$

where

dc/dt = rate of gas transfer
K_2 = reaeration coefficient
C_{Sat} = saturation D.O.
C_{Act} = actual D.O.

Discharge D.O.

Conventional aeration systems are progressively more inefficient as the requirements for higher discharge D.O. levels increase. Most off-the-shelf conventional mechanical surface aerators are designed for aeration of wastewater requiring a discharge D.O. of only 2 mg/l. In trout hatchery applications, such aerators would operate at 1/2 to 1/5 of the standard rated efficiency.

Weir Aeration

A common situation in a fish hatchery is that where a low head of 3 in to 3 ft (0.076-0.914 m) exists for free fall over a weir. Maximal utilization of this low head for aeration is desired. Minimal aeration occurs when the flow "clings" to a vertical or inclined surface. The development of algae or a mossy growth on a weir encourages the water to "cling" to the face and thus not aerate as efficiently as with free fall. At low flows with the 2-in (55-m) stop planks used in fish hatcheries, the flow tends to "cling" to the face and thus not aerate as efficiently.

In such cases, D.O. increase would be enhanced by blocking off part of the weir to increase the unit flow per length of weir, assuring free fall. Budenko (1979) found the optimal flow rate for aeration with broad created weirs (e.g., 2-in planks) to be 192 gal/min/ft [2,384 l/(min·m)]. For sharp created weirs the optimal flow rate was 384 gal/min/ft [4,769 l/(min·m)]. For these optimal flow rates for most efficient aeration, the optimal downstream water depth below the weir was 2.6 ft (0.79 m). This depth is normally much greater than exists in raceway hatcheries. Budenko (1979) also found the most effective drops over weirs to be from 2.0 ft to 2.6 ft (0.61 m to 0.79 m). Again, this is normally a much greater drop than is available in raceway hatcheries.

The British Water Pollution Research Laboratory (Gameson et al. 1958) studied aeration at weirs and developed an equation to describe the results. The equation is shown in Figure 2 below the nomograph developed by this author. It can be shown by this nomograph that a free fall from a weir height of 8.25 ft (2.5 m) would be required to raise the D.O. from 6 mg/l to 8 mg/l if the hatchery elevation was 2000 ft (610 m) and the water temperature was 15°C. It can be shown from the nomograph in Figure 3 that 0.4 lb (0.18 kg) of oxygen would be transferred per hp-hr and the electrical cost would be about $60/t ($66/Mg). From this it can be shown that weirs are relatively inefficient aerators. This in no way is meant to discourage the utilization of weir aeration in hatcheries. It is better than no aeration at all.

Cascade and Trickle Aeration

A series of weirs (or cascades) having a given head differential is more efficient in aeration than a single weir of the same total head. Therefore, it becomes a matter of how to "break up the fall" most effectively. Some hatcheries use "Christmas tree" aerators, wood slats placed in layers about 2 in (50 mm) apart, splash plates, and other configurations.

Chemical engineers utilize a packing material called Pall rings to achieve effi-

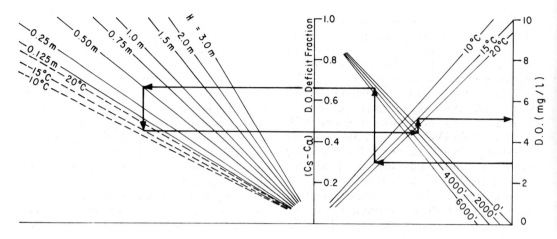

FIGURE 2. — Nomograph for determination of D.O. increase in free fall over weir. (Water Pollution Research Laboratory, England.)

$$R_T = \frac{C_s - C_{up}}{C_s - C_{down}} = 1 + 0.38\,(1.6)\,h\,(1-0.11\,h)\,(1 + 0.046T)$$

Instructions: 1. Enter C_{up} horizontally to intercept Water Temperature.

2. Then vertically to intercept Elevation.
3. Then horizontally to intercept Fall Height-H.
4. Then vertically to intercept Water Temperature.
5. Then horizontally to intercept Elevation.
6. Then vertically to intercept Temperature.
7. Then horizontally to intercept C_{down}.

cient gas transfer. Pall rings have been evaluated extensively at the Dworshak National Fish Hatchery by Owsley (1979) for nitrogen gas stripping and oxygenation. He found 1 1/2-in (38-mm) Pall rings to be best, based on performance and cost. In small quantities of less than 100 ft³ (2.8 m³), the cost of 1 1/2 in (38-mm) Pall rings is about $10/ft³ ($353/m³) F.O.B. Akron, Ohio. Owsley used a depth of 5 ft (1.52 m) and a hydraulic loading rate of about 230 gal/min/ft² [9,274 l/(min·m²)]. Under the above conditions, nitrogen gas supersaturation could be reduced from 130% to nearly 100%. D.O. was simultaneously raised to near saturation.

Commercial Oxygen On-Site Production

The case for use of commercial oxygen in the management of hatchery water D.O. and D.N. must be given a fair hearing and not be dismissed without rational consideration. A ballpark estimate for the cost of constructing a 100,000-lb/yr (45,359-kg/yr) trout hatchery is $5 million (Thoesen 1979). This is no small investment, and thorough planning must insure maximum return in the form of fish production. For approximately 1% added investment, on-site oxygen production could be provided that may possibly allow doubling of the fish production capacity without the need for additional water supply.

British Oxygen Corporation markets a Pulsed Swing Absorption (P.S.A.) oxygen

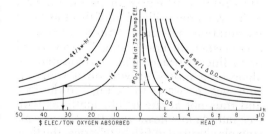

FIGURE 3. — Cost of oxygen for pumped head and electricity cost as indicated.

production system that produces 1000 lb/d (454 kg/d) of 90% pure oxygen. This is the smallest size marketed. The cost was quoted at $60,000 in 1978. It operates automatically, negating the need for attention or skilled help. This capacity would meet the peak oxygen needs of a 300,000-lb/yr (136,078-kg/yr) hatchery. It would require about 20 hp (15 kW) to operate the system and its use would probably result in reduced yearly operating costs when compared to conventional aeration systems requiring pumped head.

D.O. levels could be maintained at saturation so that fish production would never be oxygen limited. At such high D.O. levels, it is speculated that ammonia concentrations well above 0.5 mg/l could be tolerated without adverse effect. This would permit increased fish production from a given water supply. Dissolved nitrogen gas levels could easily be reduced to 80% to 90% of saturation — far below the 102% limit — by the nitrogen stripping action that would occur within the oxygen absorption system. The rate of oxygen demand in a fish hatchery is reasonably uniform 24 hours per day, with a temporary increase at feeding time. This uniform oxygen demand would match reasonably well with the uniform oxygen production rate, negating the need for oxygen storage.

A major issue would be the reliability of the oxygen production system. Power failure or mechanical failure of the oxygen production system would require a back-up supply of liquid oxygen. Liquid oxygen is available in 400-lb (182-kg) cylinders at a cost of about $90/cylinder. About 2%/d is lost because of vaporization when not in use. The geographic proximity of the hatchery to the oxygen supplier would determine the desirable quantity of back-up liquid oxygen supply.

The cost of oxygen must be placed in perspective with the total cost of producing trout. It requires 0.33 lb (0.15 kg) of oxygen (Speece 1973) to produce one lb (454 g) of trout. Even with the most expensive form of oxygen the small 400-lb (182-kg) liquid oxygen cylinders at $450/t ($495/Mg) would add only $0.07/lb ($.15/kg) to pro-

duction costs. This compares with estimated feed and labor costs of $1.60/lb ($.52/kg) plus an additional $3.50/lb ($7.70/kg) for amortization of the hatchery facility ($5 million for 100,000 lb/yr [45,359 kg/yr] at 7% interest). Even if the 1000-lb/d (454-kg/d) PSA oxygen production plant (with 300% of required peak oxygen production capacity) was used only 6 mo/yr at a facility producing 100,000 lb/yr (45,359 kg/yr) with amortization at 7% over 10 years and electricity at $0.03/kWh, it would add only $0.10/lb ($0.22/kg) to production costs.

Commercial Liquid Oxygen

An alternative commercial oxygen source is bulk oxygen trucked to the site in liquid form. Bulk oxygen costs approximately $100/t ($110/Mg) within a 100-mile radius off the oxygen supplier. An on-site cryogenic oxygen storage vessel would be required. A 1500-gal (5.62-m^3) cryogenic oxygen tank costs about $250 to $350/mo to lease, but can be purchased for $20,000 (Minnesota Valley Engineering 1979). It will hold 13,500 lb (6,136 kg) of oxygen and would need to be filled about 3 times/yr for a 100,000-lb/yr (45,359-kg/yr) trout hatchery. Amortization of the tank at 7% for 20 years would be $1880/yr and the oxygen would be $1660/yr for a total of $0.035/lb ($0.077/kg) of trout produced.

In these calculations it is assumed that the water supply enters the hatchery with 6 mg/l of D.O. If a well or spring containing no D.O. and 120% D.N. saturation served as the supply, it is recommended that the entire flow be pumped through a packed bed of Pall rings 3 ft (0.9 m) deep and 25 ft^2 (2.3 m^2) in horizontal cross section. This would raise the D.O. from 0 to 5 mg/l and reduce the D.N. from 120% to 110% saturation. Subsequently, the water supply would pass through the commercial oxygen absorption chamber where the D.O. would be raised from 5 mg/l to as high as 12 mg/l and the D.N. would be reduced to less than 100%. Negligible loss of D.O. would occur in the raceways, even though it represents about 120% of saturation.

Assuming the water supply flow was 5500 gal/min (20,625 l/min), it would require an additional 12 t (10.909 Mg) of oxygen per year to raise the D.O. from the gravity aerator discharge of 5 mg/l to 6 mg/l assumed in the above case. This would raise the oxygen cost to $2680/yr and result in a total storage amortization plus oxygen cost of $0.047/lb ($0.10/kg) of trout produced. This would appear to be a very reasonable cost increment in exchange for elimination of any oxygen limitations and associated stress on the fish.

Commercial Oxygen — Absorption Systems

An integral part of the utilization of commercial oxygen in hatchery water management is the absorption system. Unlike air, commercial oxygen has a significant value and therefore requires an absorption system that achieves at least 90% absorption efficiency. Even though commercial oxygen is about 5 times as soluble as oxygen in air, commercial oxygen is still very difficult to dissolve and requires special attention in the design of the absorption systems.

Five different oxygen absorption systems will be described that are capable of achieving at least 90% oxygen absorption efficiency with reasonable capital costs and energy consumption ratios: (1) enclosed packed column, (2) U-Tube, (3) downflow bubble contact aeration, (4) recycled diffused oxygenation, and (5) rotating packed column.

Enclosed Packed Column

Owsley has used an enclosed column packed with Pall rings for stripping of D.N. and absorption of oxygen from air. It can also be used for efficient absorption of commercial oxygen if it is enclosed and the bottom is submerged, as shown in Figure 4. Commercial oxygen would be injected at the bottom and flow up countercurrent to the water. With a constant side-stream flow rate, the required oxygen flow rate resulting in 90% oxygen absorption at the desired discharge D.O. could be set

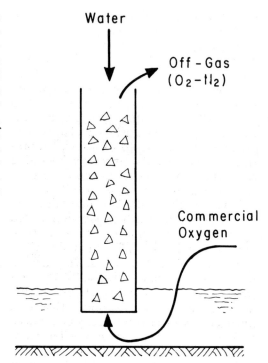

FIGURE 4. — Packed column modified for efficient absorption of commercial oxygen.

manually or a feed-back loop controlled by a D.O. probe in the discharge could regulate the oxygen flow to achieve 90% absorption efficiency. Such a system would be capable of producing in excess of 40 mg/l of D.O. in the discharge. Therefore, one mode of operation would be to pump a side stream of the main flow through the enclosed, packed column and reintroduce this high D.O. side stream back into the main flow through a pipe across the channel bottom to prevent oxygen loss to the atmosphere. For example, if the D.O. in the channel was 6 mg/l and it was desired to raise it to 10 mg/l, then 1/8 of the flow would be withdrawn in the side stream and raised to 38 mg/l and reintroduced through the diffuser back into the channel.

Oxygenation of the side stream results in smaller pumps and smaller enclosed packed columns because only a relatively small fraction of the flow is handled. The packed column would need to be approximately 5 ft (1.52 m) deep. Excellent D.N. stripping characteristics would exist in an enclosed, packed column with such a high oxygen-composition atmosphere.

U-Tube

In U-Tube aeration, water flows down a vertical shaft, under a baffle and back to the surface. Gas bubbles injected at the inlet are swept along with the water. The advantage of U-Tube aeration is that hydrostatic head pressurizes the gas bubbles and thus enhances gas transfer. The U-Tube can be adopted for efficient absorption of commercial oxygen by capture and recycle of the oxygen-rich off-gas, as shown in Figure 5. The design of a U-Tube for 90% absorption of commercial oxygen would incorporate a depth of approximately 40 ft (12.2 m) and a water velocity of 6 ft/s (1.8 m/s). The discharge D.O. would be approximately 40 mg/l if the inlet D.O. was 6 mg/l. The head loss across the system would be 3.8 ft (1.2 m). Absorption costs (operation and amortization) would be about $8/t ($8.80/Mg) of oxygen. D.N. stripping would occur, but less efficiently than with the enclosed packed column. Control of oxygen absorption efficiency and discharge D.O. would be similar to that suggested for the enclosed packed column.

Downflow Bubble Contact Aeration

The principle of DBCA is that bubbles are hydraulically trapped in a flow of water. The DBCA consists of a cone-shaped chamber that facilitates an inlet water velocity at the top of about 6 ft/s (1.8 m/s) and a discharge water velocity at the bottom of less than 0.5 ft/s (0.15 m/s). The bubbles have a nominal buoyant velocity of 1 ft/s (0.3 m/s) and thus are trapped as a swarm in the cone, as shown in Figure 6. The water encounters a very large surface area within the cone and about 40% if the D.O. deficit is satisfied. The DBCA would be about 10 ft (3 m) deep and the discharge D.O. would be approximately 22 mg/l if the inlet D.O. was 6 mg/l. The head loss across the system would be about 3 ft (0.9 m). Here again, control of oxygen absorption efficiency and discharge D.O. would be similar to the procedure suggested for the enclosed packed column.

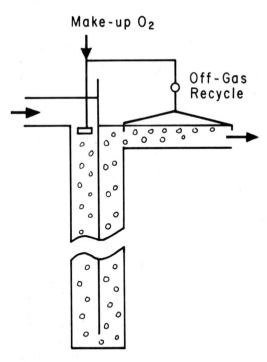

FIGURE 5. — U-Tube with off-gas recycle for efficient absorption of commercial oxygen.

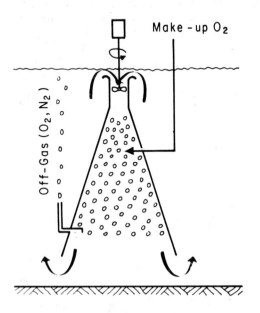

FIGURE 6. — Downflow bubble contact aeration for efficient absorption of commercial oxygen.

Recycled Diffused Oxygenation

The insolubility of oxygen prevents 90% absorption efficiency from being realized by simple injection of oxygen through a bubble diffuser in the bottom of the raceway. A 2-mm bubble requires a water depth of approximately 100 ft (30 m) to achieve 90% absorption efficiency, so it is readily apparent that low absorption efficiency would be realized if oxygen was injected into 1 or 2 ft (0.3 or 0.6 m) of water in a raceway.

However, placement of a hood above the diffuser to collect the oxygen-rich off-gas for subsequent recycle to the diffuser can result in 90% absorption efficiency, as shown in Figure 7. The simplicity of this absorption scheme makes it very attractive as an in-raceway oxygenation scheme to replenish the D.O. to a lower set of raceways. The main attraction of this oxygenation concept is that it requires no pumping of water. It requires only a small blower to force the oxygen-rich gas collected by the hood back through the diffuser in the channel bottom.

Make-up oxygen must be metered into the system continually. Since dissolved nitrogen will be stripped continually, there must be continual wastage of gas from the system. A mass balance on the make-up oxygen flow and the mass of D.O. added to the water must be made to establish oxygen absorption efficiency. There is an inverse relationship between oxygen absorption efficiency and stripping of dissolved nitrogen.

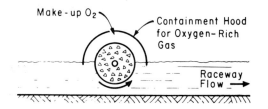

FIGURE 8. — Rotating packed column for efficient absorption of commercial oxygen.

Rotating Packed Column

Another scheme for achieving 90% oxygen absorption efficiency is to provide for rotation of a packed column under a hood to contain the oxygen-rich atmosphere as shown in Figure 8. The axis of rotation would be horizontal and perpendicular to the raceway flow. Half of the packed column would be submerged. Paddles placed on the packed column could cause the water flow to drive the packed column rotation. However, since the oxygenation efficiency of the rotating packed column is proportioned to the rotation speed, it would be desirable to provide a separate mechanical drive. Make-up oxygen would be continually metered to the enclosed space above the hood as discussed above in Recycled Diffused Oxygenation. The same principles would also hold for oxygen absorption efficiency and stripping of dissolved nitrogen.

Summary

Typical aerators used for enhancement of D.O. and stripping of D.N. are reviewed. These conventional aeration systems are relatively inefficient and ineffective in raising trout hatchery waters above the minimum desired D.O. of 6 mg/l. The electrical energy costs are reviewed. The effectiveness of aeration at weirs is discussed and a nomograph presented that predicts the change in D.O. for various conditions.

The exceptional potential of commercial oxygen for eliminating D.O. limitations in fish production and stripping D.N. levels considerably below 100% saturation is discussed. D.O. can be maintained well above 6 mg/l under the most adverse con-

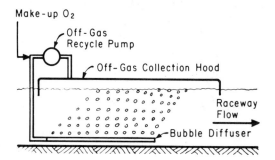

FIGURE 7. — Recycled diffused oxygenation for efficient absorption of commercial oxygen.

ditions with commercial oxygen. Two alternative commercial oxygen sources are discussed, i.e., on-site generation and trucked-in liquid oxygen. The potential for increased fish production with a fixed water supply using commercial oxygen is postulated. The incremental costs of using commercial oxygen in trout production are estimated to be $0.035 to $0.10/lb ($0.077 to $0.22/kg) of trout produced. These costs are compared to $1.60/lb ($3.52/kg) of trout for estimated feed and labor plus $3.50/lb ($7.70/kg) of trout for estimated hatchery amortization. The increase in hatchery production that could result from the use of commercial oxygen in hatchery water quality management could easily result in considerable net savings in trout production.

The use of commercial oxygen would require specially designed oxygen absorption systems to achieve 90% oxygen absorption efficiency. Five different commercial oxygen absorption systems are described and the general design and operational characteristics are noted.

References

BUDENKO, B.M. 1979. Post aeration of wastewater. J. Am. Soc. Civ. Eng. 105 EE2:297.

GAMESON, A.L.H., K.G. VAN DYKE, AND C.G. OGDEN. 1958. The effect of temperature on aeration at weirs. Water Water Eng. 62:489-492.

MINNESOTA VALLEY ENGINEERING. 1980. Personal communication.

OWSLEY, D.E. Packed columns for nitrogen gas removal. Bio-Engineering Symposium for Fish Culture. Traverse City, Michigan, October 16, 1979.

SPEECE, R.E. 1973. Trout metabolism characteristics and the rational design of nitrification facilities for water reuse in hatcheries. Trans. Am. Fish. Soc. 102(2):323-334.

THOESEN, R.W. 1979. U.S. Fish Wildl. Serv. Personal communication.

QUESTIONER: Are there any dangers from supersaturation of oxygen?

RICHARD SPEECE: I am not a fish man, but at one time I worked with Keen Buss and we raised trout in 22-ppm dissolved oxygen for 18 months.

KEEN BUSS: We ran 165 percent D.O. for over a month and had no problems at all. We raised it many times up to 22 ppm for rather short intervals, but oxygen has never been a problem.

W.L. HUGES-GAMES: We have had up to 500% oxygen saturation in some of our ponds in Israel and we have had a very bad gas bubble disease problem. I doubt if many of you would have up to 500% saturation. Our ponds are in a desert area.

E.A. HUISMAN: Dr. Speece, could you enlarge a little more on oxygenation with pure oxygen? For us it is all important. We have a recirculation type of fish rearing in the Netherlands and we have to add at least 50% oxygen as pure oxygen. We also have to aerate. The price you mentioned is about half the price we can buy oxygen for in Europe. Still, we find it to be economical. I do not know of any system that will dissolve oxygen very well in water. The latest thing that has been built is from the Linde Company in Germany. They use big tanks and they have oxygen under pressures of some 10-50 atmospheres at the water surface. They produce some 50 milligrams of oxygen per litre of water and then add it to the conventional fish-rearing unit. This is a pretty expensive method. We must have a high concentration of oxygen, say about 8 ppm to 9 ppm in the water so that the fish will grow fast. We have a fish concentration of 1:10. These are salmonids. We are pending a contract with the AGA Company of Sweden to develop a disperser of oxygen. There seems to be none. You mentioned 90% dissolution of oxygen in water. Is there anything that can do it? We have come up to 85%. The AGA people think that this is excellent. However I have not told them how we do it. It is very simple. Could you enlarge a bit on this? Could you give us, for example, the price of power in the U.S.? In Europe, there are great differences. In Sweden, you have one price and down in the Continent the price is double. In Germany it is even higher. The dispute over atomic energy will result in higher prices for oxygen because it is directly bound to the price for power.

SPEECE: On Friday, I was speaking to two representatives of AGA from Stockholm. This same issue came up concerning the cost of oxygen in Sweden. It was a couple of times higher than the cost quoted by the other countries. The Linde system is a very energy-intensive system. It requires a lot of energy. It also requires a large amount of hardware. When you look at an oxygen system, you have to say, are they trying to sell oxygen or are they trying to sell the system? There are systems available that will efficiently do this in a modification of the packed column that was enclosed with counter-current oxygen flow. There is another system that is referred to as the U-tube. Another system is the down-bubble contact aeration. These are able to achieve that absorption efficiency. There is a trade-off here in that you are trying to do two things. You want to strip dissolved nitrogen and you want to add dissolved oxygen. If you want to strip considerable amounts of dissolved oxygen, you will have to settle for less dissolved oxygen absorption. If you reduce dissolved nitrogen from 105%, then I believe that you could easily meet the 90% absorption of oxygen and end up with dissolved nitrogen between 90% and 100%.

HUISMAN: Dr. Speece, I would like to say that I agree with you on your comments regarding oxygen. When you were talking about dissolution efficiency of 90%, was that dissolution efficiency or utilization efficiency?

SPEECE: I was referring to that as absorption efficiency. For every 100 pounds of oxygen going into the system you are dissolving 90 pounds, or 90%.

HUISMAN: That would be in the stream itself. In a deep pond you are getting 90% utilized oxygen. Are you taking into account any of the off-gases?

SPEECE: That would be essentially utilization as well.

Bio-Engineering Symposium for Fish Culture (FCS Publ. 1):63-70

Performance Ratings
for Submerged Nitrification Biofilters
Development of a Design Calculation Procedure

W. JOHN HESS

U.S. Army Engineer District
123 Union Street
Walla Walla, Washington 99362

ABSTRACT

Biofilters have capability of removing toxic ammonia from fishpond effluent so it can be safely reused. Such reuse can save on water, energy, and costs, and can muliply the number of sites usable and available for hatcheries. But before the cost and effectiveness of reuse development can be calculated, a method of rating biofilters is needed.

To find a rating method that compares biofilters above their many differences depends on developing an expression with value independent of known variables; and that requires an analysis.

Biofilters are special vessels that present bacteria-coated surfaces of their submerged media packing to all parts of the stream flowing through it. By analysis and by tests of many differing biofilters, an expression for a correlation number NXC, held to be theoretically constant for such biofilters, has been developed and its value determined. Separate investigators verified the expression and its value. Having the NXC value enables comparing biofilters with each other, with all those filters tested for development of the NXC expression, and with untried filter arrangements like those still in planning stages, all on the basis of their nitrifying effectiveness.

The ability to compare biofilters gives the NXC concept usefulness. It can be used to design new biofilters to meet any desired fish load, to evaluate the performance of operating biofilters, and to develop equal-effectiveness cost comparisons of even greatly differing media without pilot testing. Examples of its uses are presented.

In a situation of growing scarcity of sites for fish hatcheries, designers and planners are reexamining the resources actually necessary for any given fish production goal. Part of the analysis focuses on cost of development, but mostly the concern is for the land and water needed, especially water.

Hatchery Sites Improved by Biofilters

Specifically hatchery planners look for water of good temperature and little change, and water with appreciable mineral content, for it is now recognized that fish numbers and quality can drop sharply in water of too low a hardness. Sites with good water are limited, and biofilters can help. When properly applied, biofilters rejuvenate pond-effluent streams and return them for reuse ten, even twenty, times with good results. By transforming the toxic ammonia into harmless nitrate, biofilters make used pond water safe to return to the fish for reusing. Such reuse saves water and energy. It also saves on cost of filtering, disinfecting, degassing, and mineralizing. Reuse can also cut site requirements if it enables bringing fish to size in one year instead of two. With reuse as a tool, sites having just a fraction of the flows usually sought would become desirable and generally available opportunities for new hatcheries.

Need for Sizing Biofilters

Before setting production claims for any reuse sites, however, there is need for a method of determining the size and configuration of reuse biofilters that will meet a given production capacity goal. That in turn requires an analysis of the structure and workings of a biofilter.

Features of Biofilters

Structurally, submerged nitrification biofilters are special vessels packed with a porous and washable bed of solid media to which bacteria attach and then ingest the ammonia from pond effluent flowing by. Biofilters do their work by mixing all of the water over bacteria-covered media surfaces. They work more efficiently when there is more surface and when the surfaces "mix" more readily with the water, so bacteria can reach more of it.

Design Problem

But what kind of media? And in how big a bed? Bed and media dimensions are factors the designer controls; so to match them to the fish load is the design problem. The three areas of the design problem — estimating ammonia production by the fish, allowing for effects of the reuse system and its multiple passes on ammonia concentration, and sizing biofilters to remove the ammonia — are all treated in a biofilters manual — Hess (1976). This paper discusses the third area, biofilters sizing.

Starting Point

The approach to establishing a filter-sizing procedure is based on comparing the performance of many different existing biofilters. Two of the filters compared are shown in Figure 1: one-foot depth of 5/16-inch granules on the left; and six-foot depth of 3 1/2-inch rings on the right. The one on the left is six times more effective than the one on the right because of its greater surface and smaller voids. The key to the sizing procedure is the normal vigor level of nitrifying bacteria — or better, of one bacterium. The usual level of activity of a bacterium is the one filter-effectiveness parameter that carries the same value over from one biofilter to another independently of differing flows, loads, and filter dimensions, thus establishing a reference that permits comparison. A designer may select almost any size and proportion of vessel dimensions and almost any solid media from such things as sand, gravel, beads, and plastic shapes or structures for the bacteria to attach to in flowing water; but the activity of the bacteria population determines the filter effectiveness.

Method

So, to find the expression that compares biofilters, simply take account of all variables that affect the total biofilter effec-

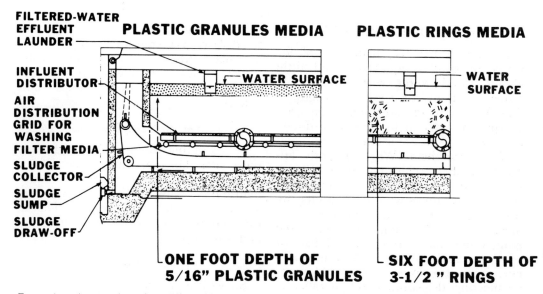

SUBMERGED NITRIFYING BIOFILTERS
(UPFLOW)

FILTERED-WATER EFFLUENT LAUNDER

PLASTIC GRANULES MEDIA

PLASTIC RINGS MEDIA

INFLUENT DISTRIBUTOR

WATER SURFACE

WATER SURFACE

AIR DISTRIBUTION GRID FOR WASHING FILTER MEDIA

SLUDGE COLLECTOR

SLUDGE SUMP

SLUDGE DRAW-OFF

ONE FOOT DEPTH OF 5/16" PLASTIC GRANULES

SIX FOOT DEPTH OF 3-1/2 " RINGS

FIGURE 1. — A comparison of two different biofilters. The one at the left is six times more effective.

tiveness (as distinct from one bacterium's effectiveness). The expression that emerges will be an estimate of a bacterium's effectiveness, with a value that is theoretically always the same at a given temperature in any nitrifying biofilter. Then find that value; and there is the tool for direct comparisons.

The first step in the derivation was to list all the factors that affect biofilter effectiveness. But that is impossible. There are too many. Yes, but it turns out that the most significant ones are also the measurable ones:

1. miC = media-in Concentration, ppm (factors in ACTUAL eXsert, NX)
2. moC = media-out Concentration,
1.
3. Q = Flow rate, gpm
4. Tc = Temperature correction, f (°F)
5. BD = Bed depth, ft (factors in EXPECTED eXsert, NEX)
6. BA = Bed area (plan), ft²
7. VF = Voids fraction
8. SPS = Specific surface, ft²/ft³

The second step in the derivation was to bunch the factors into two clusters: eXsert (or effectiveness) actual, and eXsert expected, label them NX and NEX respectively, and then set their ratio equal to NXC.

NXC Concept For Design

The resulting dimensions-removal comparison number is NXC, Number for eXsert Correlation, and the present best estimate of its value is 4.8, as obtained from tests on actual systems with different loads and dimensions. Finding an expression and a value for NXC means that biofilters can be compared directly on their effectiveness so that designers can size them to match fish loads, and hatcherymen can evaluate the effectiveness of existing biofilters. So NXC is useful, let's see how it works. Although symbols are tedious, they can help at this point.

Defining Expression

Here is the defining expression for the dimensions-removal correlation number called NXC:

$$NXC = NX/NEX$$
NXC Defining Expression

NXC = Number for eXsert Correlation
NX = Number for actual eXsert, a function of ammonia removal.
NEX = Number for Expected eXsert, a function of bed-media dimensions.

This is the brief form of the defining equation, and details of its factors NX and NEX must be developed. But it shows that NXC is the balancing number between actual eXsert NX and the expectation for eXsert NEX. The numerator is a function of the amount of ammonia removed. The denominator quantifies the assistance that the media bed gives to the bacteria by its surfaces and by the way it mixes them with the stream. It says that if the bed is improved to make NEX bigger, then its actual performance will improve to make NX bigger also, and so keep NXC the same.

Factors in Definition

Now let's elaborate the right-side factors.

The defining expression for *Actual effectiveness NX*, Number for eXsert is:

$$NX = \frac{1000\,(miC\text{-}moC)Q}{miC\,Tc}$$

NX = Number for eXsert
miC = media-in Concentration of ammonia ppm
moC = media-out Concentration of ammonia ppm
$miC\text{-}moC$ = eXsert, or removal of ammonia concentration
Tc = Temperature correction (Haug, McCarty 1972)
Q = flow rate in gpm

The equation says that at any given temperature, NX is directly proportional to the eXsert fraction *(miC-moC)/miC,* and to the flow rate. It also says, for instance, that if flow rate Q increases, the eXsert effectiveness will decrease to keep NX the same; and when that happens the concentrations of ammonia will climb.

Expected effectiveness

The other factor, *NEX,* Number for Expected eXsert, answers, "What do dimensions of the bed and media have to do with their nitrification effectiveness?" It was derived by supposing nitrification to be strengthened when there are more media surface, more detention time, and more contact opportunity. Surface, time, and opportunity were then each expressed as proportional to certain bed-media dimensions as follows:

a. For surface for bacteria attachment:

$$NEX \sim MS$$

MS	= Media surface, ft^2
	= $BA\ BD\ SPS$
BA	= Bed area in plan, ft^2
BD	= Bed depth, ft

So that $NEX \sim BA\ BD\ SPS$

SPS	= Specific surface of media, ft^2/ft^3

b. For time of fluid detention:

$$NEX \sim t$$

t	= liquid detention time
	= $BA\ BD\ VF/Q$
VF	= Voids fraction
Q	= flow rate in gpm

So that $NEX \sim BA\ BD\ VF/Q$

c. The idea of opportunity for liquid-to-bacteria contact, sometimes called "scrub factor," seems to say that more portions of the liquid are "touched" by the biomass surfaces when the stream flows closer to them and faster:

$$NEX \sim V/S$$

v	= velocity
	= BD/t (from above) = $BA\ BD\ VF/Q$
then $v \sim$	$BD\ Q/(BA\ BD\ VF)$
$NEX \sim$	= $Q/(BA\ VF)$
s	= average distance across passages.
	= bed volume (empty) / bed surfaces
	$\sim BA\ BD\ VF /(BA\ BD\ SPS)$

$$NEX \sim 1/s \qquad s = VF/SPS$$

So that $NEX \sim v,\ 1/s$

$$= \frac{Q}{BA\ VF} \quad \frac{SPS}{VF}$$

The surface, time, and opportunity factors, when put together on the basis that each is proportional to its effect on expected removal effectiveness, develop the number *NEX* for any submerged nitrifying biofilter:

$$NEX = \frac{BA\ BD\ SPS \quad BA\ BD\ VF}{Q}$$

$$\frac{Q}{BA\ VF} \quad \frac{SPS}{VF}$$

from which obtain the defining expression:

$$NEX = \frac{BA\ BD^d \quad SPS^h}{VF}$$

$$d = 2$$
$$h = 2$$

Meanings

This expression for *NEX* lists bed-geometry factors in the relationship that matches their influence on ammonia removal. Voids and high flows are a hindrance; plan area, depth, and specific surface each help. These factors have seemed to be in similar relationships in operating and testing experiences, although it turns

NXC TESTS-VALUES SUMMARY

BIOFILTER	MIC	BD	BA	SPS	VF	Q	NX	NEX	NXC
MEDIA	MEDIA IN AMMONIA CONCEN- TRATION PPM	BED DEPTH FT	BED AREA SF	SPECIFIC SURFACE OF MEDIA SF/CF	VOIDS FRACTION	FLOW RATE GPM	NUMBER FOR EXSERT NO.	NUMBER FOR EX- PECTED EXSERT NO.	NUMBER FOR EXSERT CORRE- LATION NO.
3-1/2 " PLASTIC RINGS	.564	6	13.2	27	.97	42	3,853.08	796	4.841
1-1/2 " PLASTIC RINGS	.9	4	13.2	40	.96	42.6	4,057	796.6	5.093
3/8" STYROFOAM BEADS	.378	3	40	123	.35	80	57,200	12,939.6	4.42
3/8" STYROFOAM BEADS	.288	2	2.7	123	.35	5.4	3,062	630.6	4.855
3/8" STYROFOAM BEADS	.291	1	1.35	123	.35	1.5	852.9	181.2	4.707
3/8" STYROFOAM BEADS	.309	.5	1.35	123	.35	1.5	483.75	104.01	4.65

$$NXC = NX/NEX$$
$$NEX = BD^{0.8} \times BA \times SPS^{0.8}/VF$$
$$NX \approx 1000\, Q(MIC-MOC)/MIC \times TC$$

AVERAGE NXC = 4.761
say 4.8

FIGURE 2. — Notice the great variations between significant parameters, yet the same effectiveness.

out that the exponents as evaluated by tests are nearer 0.8 than 2:

$$d = 0.8$$
$$h = 0.8$$

The specific-surface being in the numerator, for instance, agrees with observed performance. Three-eighths-inch beads, with their greater specific surface, have shown more removal than higher-void, less-specific surface media, even when shallower. The smaller exponent $h = 0.8$, perhaps results from the greater loss of surface among small media from increased contact points and resulting blanking of surface area.

The bed depth also appears to only the 0.8 power in actual tests, probably because it is attenuated by decreases in ammonia concentration as the flow progresses through the media bed. At any rate, 0.8 is the value for the exponents that brought all

NXC values very close. So the NXC evaluating expression that is best substantiated is:

$$NXC = \frac{1000\,(miC - moC)Q}{miC\ Tc} \left(= NX \right) \times$$
$$\frac{VF}{BD^{0.8}\ BA\ SPS^{0.8}} \left(= \frac{1}{NEX} \right)$$

Dependability of the Design Concept

Steady-State for Testing

Actually, ammonia-removal effectiveness depends on the vigor of the bacterial population. Their vigor, in turn, depends on all the variables in the NXC expression, plus the history of operation preceding the taking of measurements. If the history has been lengthy enough, like one month or more, and free of bacteria-inhibiting events like drastic changes in temperature, ammonia concentration, pH, or changes in toxins, or flow rate, then the bacteria can have reached a relatively steady-state condition. Bacterial vigor can be affected by inadequate dissolved minerals also (G.W. Klontz and Khet Van Lai 1979).

Clustered Values

Even though many things can impact the performance of biofilters, a number of different biofilters with very different parameters have shown their dimensions-performance-correlation, or *NXC,* values clustering closely. See the NXC Tests-Values Summary (Fig. 2).

The table shows great variation within significant parameters as from 0.5 to 6 ft of depth, 1.35 to 40 SF of area, 0.35 to 0.97 voids fraction, and 1.5 to 80 gpm flow rate, yet the values of *NXC* are similar and cluster around

$$NXC = 4.8 \, (\text{est}). \qquad \textit{NXC value}$$

Reconfirmation

NXC was later evaluated again at very nearly 4.8 by Paul E. Cooley (1978) while testing a new kind of filter bed for Dr. A.T. Wallace. His pilot biofilter used two feet of

buoyant, 5/16-inch, plastic granules. In a subsequent work, he showed that for beds of very small granules such as sand the specific surface should appear as part of the exponent of natural log *e.* (Cooley 1979).

Uses of The Design Concept

Having a definite and dependable value for *NXC* establishes a direct path of calculations from biofilter effectiveness needed to the dimensions of media and bed that will provide it. And the clustering of *NXC* values, calculated from multiple tests of differing biofilters, leads to a measure of confidence in sizing of media beds.

New Designs

To size a new media bed, use the brief form of the defining equation transposed to calculate the expected effectiveness *required* for the given load; then select bed and media dimensions for a matching *designed* value of *NEX*:

$$rNEX = NX/(\text{NXC} = 4.8)$$
$$\textit{"Required" NEX}$$

Once the *"Required" NEX* value (*RNEX*) is established through the desired effectiveness numbers in *NX,* then select any

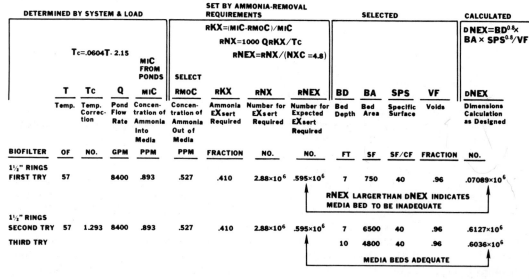

SAMPLE DESIGN CALCULATIONS USING NXC=4.8

DETERMINED BY SYSTEM & LOAD				SET BY AMMONIA-REMOVAL REQUIREMENTS					SELECTED				CALCULATED
					RKX=(MIC-RMOC)/MIC								DNEX=BD$^{0.8}$× BA × SPS$^{0.8}$/VF
					RNX=1000 QrKX/Tc								
	Tc=.0604T- 2.15				RNEX=rNX/(NXC =4.8)								
			MIC FROM PONDS	SELECT									
	T	Tc	Q	MIC	RMOC	RKX	RNX	RNEX	BD	BA	SPS	VF	DNEX
	Temp.	Temp. Correction	Pond Flow Rate	Concentration of Ammonia Into Media	Concentration of Ammonia Out of Media	Ammonia EXsert Required	Number for EXsert Required	Number for Expected EXsert Required	Bed Depth	Bed Area	Specific Surface	Voids	Dimensions Calculation as Designed
BIOFILTER	OF	NO.	GPM	PPM	PPM	FRACTION	NO.	NO.	FT	SF	SF/CF	FRACTION	NO.
1¼" RINGS FIRST TRY	57		8400	.893	.527	.410	2.88×10^6	.595×10^6	7	750	40	.96	.07089×10^6
								RNEX LARGER THAN DNEX INDICATES MEDIA BED TO BE INADEQUATE					
1¼" RINGS SECOND TRY	57	1.293	8400	.893	.527	.410	2.88×10^6	.595×10^6	7	6500	40	.96	.6127×10^6
THIRD TRY									10	4800	40	.96	.6036×10^6
										MEDIA BEDS ADEQUATE			

FIGURE 3. — These sample calculations show the sizing process.

BIOFILTERS WITH EQUAL NITRIFICATION EXPECTATION

	FILTER MEDIA					FILTER BEDS				AMOUNT OF MEDIA	
KIND OF MEDIA	AVERAGE DIMEN-SION (inches)	COST $/CY	$/TON	SPS SPECIFIC SURFACE SF/CF	VF VOIDS FRAC-TION	NO. OF BEDS NO.	BA BED AREA SF	BD BED DEPTH FT	BED VOL-UME CY	MEDIA WEIGHT TONS	MEDIA COST $(1975)
1 RINGS	5/8	675		105	.92	8	7500	6	1667		1,125,000
2 RINGS	1-1/2	148.5		40	.96	8	7500	16.4	4600		745,000
3 RINGS	3-1/2	94.5		27	.97	8	7500	24.6	6833		645,000
4 PLASTIC ROD ENDS	5/16	286.9		200	.27	3	2812	1.939	202		92,777
5 STYRO-FOAM BEADS	1/4	48.6		183	.3	3	2812	2.85	297		14,432
6 NO. 4 ANTHRA-CITE	5/16		50	220	.2	3	2812	1.43	149	120	9,069
7 PEA GRAVEL	1/4	20		183	.3	3	2812	2.3	241		4,820
8 NO.20 SILICA SAND	.05		6.25	5070	.27	2	1875	.15	10.4	13.9	87.5

$$NEX = \frac{BA\ BD^{0.8} SPS^{0.8}}{VF} \quad \text{set } NEX = 1.4 \times 10^6 \quad \text{then } BD = \left(\frac{1.4 \times 10^6 \times VF}{BA \times SPS^{0.8}}\right)^{\frac{1}{0.8}}$$

FIGURE 4. — The table shows the results of comparing different media through the *NEX* relationship.

combination of bed-media dimensions that develops that required value of *NEX*. The other expression is the one that uses dimensions to obtain the *"Designed" NEX* (*DNEX*). Any number of selected sets of dimensions can be explored for adequacy by comparing their *DNEX* value obtained through the following equation, with the required value determined above.

$$DNEX = BD^{0.8}\ BA\ SPS^{0.8}/VF$$

"Design" *NEX*

Sample calculations in Figure 3 show the sizing procedure.

Evaluation of Existing Biofilters

To judge an existing biofilter quickly, evaluate NXC under operating conditions:

use $NXC = \dfrac{(miC - moC)Q}{miC\ Tc} \times$

$\dfrac{VF}{BD^{0.8}\ BA\ SPS^{0.8}}$ NXC equation

Steps:
1. Sample influent and effluent.
2. Determine flow rate.
3. Measure bed and media dimensions.
4. Evaluate *NXC*.

If *NXC* for that bed is higher than 4.8, then it is performing better than the average of all filters used to establish the 4.8 estimate. If the *NXC* value is lower than 4.8, then it is doing less well, and there is a chance its performance could be improved. Possibly better cleaning of the bed, or other measures that would reduce short-circuiting and dead spots, could improve its performance or possibly a new media type is indicated.

69

Media Comparisons

For help in selecting media, compare different ones through the *NEX,* or expected-eXsert, expression. The table Beds With Equal Nitrification Expectation (Fig. 4) shows the results. It was developed by setting an arbitrary value on *NEX,* and adjusting bed dimensions until each of several media types was arranged to produce that same *NEX* value, and then pricing the quantity indicated. The prices shown are 1975 level, but they reveal contrasts in size and costs.

The NEX comparison procedure offers a quick interesting exploration of a new media without pilot testing. However, don't select a media type on cost or effectiveness alone. Washability, for instance, is more important than both of them. Without it the filter will sooner or later become useless.

Conclusion and Recommendation

The NXC concept offers greater confidence and agility in biofilter design. I hope you find it useful and that you go ahead and try it on the biofilters you have and on the ones you want to plan for.

References

COOLEY, P.E. 1978. Nitrification of fish hatchery reuse water utilizing low-density polyethylene beads as a fixed-film media type. Masters thesis, Dep. Civ. Eng., Univ. Idaho, Moscow, Idaho.

———. 1979. Summary of biofilter testing. Report prepared for U.S. Army Engineer District, Walla Walla, Washington.

HAUG, R.T., AND P.L. McCARTY. 1972. Nitrification with submerged filters. Water Pollut. Control Fed. 44(11): 2086.

HESS, W.J. 1976. Aids to evaluation and design sizing of submerged nitrification biofilters in fish-rearing water. Design Manual for U.S. Army Engineer District, Walla Walla, Washington.

KLONTZ, G.W., AND V.L. KHIET. 1979. Evaluation of certain chemical and physical factors influencing the performance of biofilters in fish rearing systems. U.S. Army Engineer District, Walla Walla, Washington.

Bio-Engineering Symposium for Fish Culture (FCS Publ. 1):71-82

Nitrogen Gas Removal Using Packed Columns

DAVID E. OWSLEY

U.S. Fish and Wildlife Service
Dworshak National Fish Hatchery
Ahsahka, Idaho 85320

ABSTRACT

Supersaturation of nitrogen gas continues to be a problem in hatchery water supplies. It usually goes undetected unless frequent monitoring is being conducted. Some hatcheries are aware of the fact that nitrogen gas may be a problem but do not have the equipment to monitor for nitrogen gas. A simple and effective method is available through the Weiss saturometer. If nitrogen gas can be determined to be a problem, the next step is to eliminate or alleviate it in the water supply. There have been various methods used to try to reduce supersaturation of nitrogen gas. These methods range from simple splash plates to elaborate Swedish degassers and costs vary accordingly. A degasser has been developed at Dworshak National Fish Hatchery that is simple in design and economical. It is a very efficient degasser as compared to other methods and has a broad scope of use. This new degasser is called the "packed column." It can be used with any given constant flow over a broad range. The packed column is capable of reducing nitrogen gas from over 130 percent saturation down to or near 100 percent saturation. Not only is the packed column effective at removing excess nitrogen gas but it is also a highly efficient means of aerating water.

Nitrogen supersaturation should not exceed 103 percent for salmonid eggs, fry, and fingerlings, or 105 percent for fish above fingerling size. In the best interest of fish health, nitrogen gas should be kept at or below 100 percent saturation. The packed column cannot guarantee nitrogen gas levels of 100 percent, but to date it is the most economical and efficient degasser available to the fish culturist.

Introduction

Dworshak National Fish Hatchery is located at the confluence of the main stem of the Clearwater River and the North Fork of the Clearwater River in Idaho. The hatchery supply water is received from the North Fork, which has been blocked by Dworshak Dam, located upstream approximately 0.2 kilometers from the hatchery.

Adult steelhead trout return to the hatchery and are spawned in early spring. Incubator water must be heated at this time because of low temperatures in the river (4-7°C).

These two sources are the main causes of nitrogen gas supersaturation at the hatchery. A high incidence of white-spot (coagulated yolk) and early mortality has plagued the hatchery since the beginning of the first adult return.

Early methods of combating nitrogen gas included splash plates, spray nozzles, perforated plates, perforated buckets, and mechanical aerators. None of these methods proved to be very effective except for the mechanical aerators.

Methods

In the spring of 1976, extensive testing of the Swedish degasser (Dennison and Marchyshyn 1973) was conducted. The Swedish degasser proved to be effective but minor problems prompted the hatchery to seek a better method for removing nitrogen gas. Through a joint effort by the Corps of Engineers and hatchery personnel, a structure similar to one used at the Wells Dam hatchery in Washington was designed and tested at Dworshak. The pagoda-shaped structure worked effectively in removing nitrogen gas when the porous substrate was added to the baskets (Fig. 1). After the pagoda was tested, it was apparent that a similar structure was needed to handle smaller flows.

Results

In the spring of 1977, the first packed column was tested with very favorable

FIGURE 1. — Pagoda, nitrogen gas stripping tower. In this pilot study the structure was 3.66 meters in height and flows up to 4542 L/min were tested. For results on the pagoda tests, refer to Table 1.

FIGURE 2. — Packed column degasser. For results on the packed column tests, refer to Table 2.

results. A packed column consists of a length of pipe filled with a porous substrate (Figures 2, 3). Water flows vertically through the pipe and free-falls into a receiving vessel. Pipe diameters tested ranged from 5.08 cm up to 61 cm, and flows ranged from a low of less than 40 L/min up to 568 L/min. Media tested included plastic Koch rings with the following diameters: 1.59 cm, 2.54 cm, 3.81 cm, 5.08 cm, and 8.89 cm. In addition to the rings, Koch flexipac (waffle media) and various sizes of gravel and other miscellaneous materials were tested. The depth of media was varied from 0.8 meter up to 1.5 meters for test purposes.

Water for testing the various methods for removing nitrogen gas was either heated or was supersaturated from the water supply source. Nitrogen gas levels tested ranged from 104 to 140 percent saturation. A Model ES-3 Weiss

saturometer was used to determine the nitrogen gas content of the water. Dissolved oxygen was measured by the modified Winkler method. Barometric pressure was measured using an aneroid barometer. Temperature was measured using a standard mercury thermometer. Flows were determined by V-notched weirs.

For the nitrogen gas test to be standardized, the following procedure was conducted for each reading. A gauge reading (normally zero) is taken prior to immersing the saturometer into the test vessel. The saturometer is submerged between 15 cm to 30 cm in depth in the test vessel. After the saturometer is submerged, the relief valve is closed and the instrument is pumped vigorously for 30 seconds. This pumping action removes air bubbles from the membrane tubing. After 15 minutes have elapsed, another vigorous pumping is

FIGURE 3. — Packed columns and PVC substrate. Media size ranges from 1.59-cm diameter up to 8.89-cm diameter.

required. At the end of a 30-minute period, a gauge reading is taken along with a temperature, barometric pressure, and dissolved oxygen measurement. These readings are then used with the following formula to determine the nitrogen gas content of the water (Weiss 1970).

$$N_2\% = \frac{(P_{atm} + \Delta P) - \left[\dfrac{O_2}{BO_2}(0.532) \right] - P_{H_2O}}{(P_{atm} - P_{H_2O})(0.7902)}$$

where

$N_2\%$ = percent saturation of nitrogen gas

P_{atm} = atmospheric pressure in mm Hg (barometer reading)

ΔP = saturometer reading (final ± initial)

O_2 = dissolved oxygen in mg/l

0.532 = constant

BO_2 = Bunsen solubility coefficient

P_{H_2O} = partial pressure of water

0.7902 = constant

Because these tests were being conducted for design purposes, no fish were used in any of the tests.

The main water treatment facility at Dworshak is a large concrete structure with a total volume of over 1 million liters. Eight surface aerators, platform mounted and supported on steel bridges, are used to aerate and degas all incoming water to the hatchery. Past records indicate that nitrogen gas is removed if the proper number of aerators are operating (Wold 1973). The problem arising with this system is that the water is pumped and sometimes heated after aeration and prior to reaching the fish-rearing facilities.

Systems I, II, and III were all designed to employ aspirators for aeration when on reuse. Aspirators function by introducing atmospheric air into the water under a pressurized system. Tests indicate that aspirators can increase nitrogen gas supersaturation by 2 or 3 percent. The point to be made here is that if aspirators do in fact create nitrogen gas supersaturation, they should not be used in any system having a potential for a gas problem. That is, aspirators will only contribute to or worsen a nitrogen gas problem.

Aspirators with splash plates (located at the water surface) were tested and a slight improvement was seen in reducing nitrogen gas. This improvement was not significant enough to justify conversion of all the systems and ice became a problem during the winter months.

A number of aspirators were replaced with spray nozzles to compare aeration and nitrogen gas removal. Tests showed spray nozzles as efficient as the aspirators with splash plates. The main disadvantage of the spray nozzles was a freezing problem of the mist created by the nozzles.

For nursery tank rearing, aluminum and stainless steel perforated splash plates were placed at the water inlet area to remove nitrogen gas. These plates reduced nitrogen gas by 0.5 to 1.0 percent. Wood splash plates have been tried in the same manner and are less effective than the perforated plates.

Plastic buckets with numerous holes punched in the bottom were also tried for nursery tank nitrogen gas removal. Although the action of the water appeared good, there was no reduction in nitrogen gas.

A living stream fiberglass tank was mounted above the Heath incubator trays and four Crescent aerators were tested for nitrogen gas removal. The aerators worked well but the unit was limited by the

73

amount of flow it could handle (less than 13 L/min) and its size.

The pagoda was tested for use with large flows and was very effective in removing nitrogen gas (Table 1). Since the time of testing, a more efficient (circular tower) design has been tested and is in use at the Jackson Hole, Wyoming, National Fish Hatchery (Schrable 1978).

The need for a degasser capable of handling smaller flows brought about the development of the packed column. Based on the concept of the pagoda degasser, a clear plastic pipe was filled with porous substrate and tested. The results of the packed column tests were the best of any degasser tested at Dworshak (Table 2). From an extensive literature survey, it appeared that the packed column was the most economical and efficient degasser available to the fish culturist.

The best design based on the performance of a packed column to remove nitrogen gas is a 25.4-cm diameter pipe (PVC or aluminum), 1.52 meters in height and filled to 1.37 meters with 3.81-cm diameter plastic rings. This unit will accommodate a flow range between 379 to 568 L/min and has removed nitrogen gas from over 130 percent down to 100 percent saturation. The packed column is not only effective in removing nitrogen gas but is also a highly efficient means of aerating water. An increase of 10 percent was realized when packed columns were tested against aspirators for dissolved oxygen efficiency (packed columns 95 percent, aspirators 85 percent).

A stipulation to the design of the packed column is that there must be an open interval (5 cm to 10 cm) between the bottom of the column and the receiving vessel's water surface. Submergence of the packed column will reduce its efficiency. Another factor to consider is that the packed columns appear to function better under pressurized flows than gravity flows. This would relate to the concept of the pressure change within the column and the reason for the gas removal.

Presently, there are two systems using packed columns for degassing and aerating at Dworshak. One is a 2840 L/min incubator supply line. Five packed columns, each 25.4 cm in diameter, 1.52 meters in height and filled to 1.37 meters with 3.81-cm-diameter plastic rings, keep nitrogen gas levels 102 percent or less for egg incubation and some nursery rearing.

System II with a total flow of 56,775 L/min has been converted to packed columns and the pagoda for nitrogen gas removal and aeration. Each packed column accommodates a flow of 473 L/min and has the same dimensions as the packed columns used in the incubator supply system. The pagoda was built for test purposes and will be eventually replaced by packed columns.

Plans have been made and material requisitioned to convert all three reuse systems completely from aspirators to packed columns. In addition, a new nursery building will have the benefit of packed columns for nitrogen gas removal and aeration.

Several Federal and State hatcheries in the United States are now employing the packed column for nitrogen gas control with repeated success.

Discussion

Nitrogen as a gas is colorless and odorless and makes up 78 percent of the air by volume. Nitrogen gas is considered to be only slightly soluble in water with oxygen being almost twice as soluble. The solubility of nitrogen is based on Henry's Law which states: "The weight of gas that will dissolve in a given volume of liquid, at constant temperature, is directly proportional to the pressure that the gas exerts above the liquid." What this law states is that the solubility of a gas, in this case nitrogen, is dependent upon temperature and pressure as shown by the general gas equation:

$$PV = nRT$$

where

P = pressure
V = volume
n = constant
R = universal gas content
T = temperature

(Continued on page 79)

TABLE 1. — Pagoda degasser

Date (1977)	Test No.	O_2 mg/l	N_2 %	Sample Place	Q l/min	T °C	Comments
4-10	1	9.8	107.1	IN	4542	24	Five baskets & outlet box w/diamond grating.
		8.2	102.9	OUT	4542	24	
4-10	2	9.8	109.6	IN	3785	26	Decreased flow.
		7.8	102.9	OUT	3785	26	
4-18	3	9.4	115.6	IN	3028	24	Raised outlet box splash plate and decreased the flow.
		7.6	104.9	OUT	3028	24	
4-19	4	9.9	114.4	IN	3028	28	Lowered outlet box splash plate.
		7.4	104.8	OUT	3028	28	
4-20	5	10.2	121.2	IN	3028	27	Added stainless steel perforated plates to each basket and
		7.5	104.7	OUT	3028	27	outlet box splash plate.
4-20	6	10.2	121.2	IN	3028	27	Repeat of morning test.
		7.5	104.6	OUT	3028	27	
4.21	7	11.8	107.4	IN	3028	6	Decreased temperature.
		11.6	103.0	OUT	3028	6	
4.21	8	9.5	120.1	IN	3028	28	Increased temperature.
		7.3	105.5	OUT	3028	28	
4-22	9	12.4	105.5	IN	3028	7	Decreased temperature.
		12.6	100.8	OUT	3028	7	
4-26	10	12.0	107.6	IN	3028	6	Added 150 8.89-cm Norton plastic rings to bottom basket.
		12.0	102.5	OUT	3028	6	
4-26	11	11.0	113.6	IN	3028	19	With media and temperature increase.
		9.7	101.1	OUT	3028	19	
4-26	12	11.1	117.1	IN	3028	18	
		10.0	101.5	OUT	3028	18	
4-27	13	12.0	100.0	OUT	3028	18	
5-02	14	12.2	114.4	IN	3028	17	Baskets 4 and 5 have media.
		11.8	106.9	#1	3028	17	
		11.5	103.5	#2	3028	17	
		11.2	102.6	#3	3028	17	
		10.8	100.0	#4	3028	17	
		10.6	100.0	#5	3028	17	
		10.6	100.0	splash plate	3028	17	
		10.4	100.0	OUT	3028	17	
5-10	15	9.9	107.1	IN	4542	24	Increased the flow.
		8.2	102.9	OUT	4542	24	
5-10	16	9.8	109.6	IN	3785	26	Testing higher flows.
		7.9	102.9	OUT	3785	26	
5-11	17	11.2	112.9	IN	3028	17	
		9.2	102.8	OUT	3028	17	
5-11	18	11.2	112.9	IN	3028	16	
		9.8	101.3	OUT	3028	16	
5-11	19	11.2	112.9	IN	3028	16	
		9.8	101.7	OUT	3028	16	
5-11	20	10.8	115.8	IN	3028	16	Raised outlet box splash plate.
		9.6	100.3	OUT	3028	16	
5-11	21	10.8	115.8	IN	3028	16	
		9.6	101.4	OUT	3028	16	

TABLE 1. — (Continued)

Date (1977)	Test No.	O$_2$ mg/l	N$_2$ %	Sample Place	Q l/min	T °C	Comments
5-31	22	10.2	113.5	IN	3028	17	Basket nos. 3, 4, 5 have media.
		10.0	100.0	OUT	3028	17	
6-01	23	10.3	114.1	IN	3028	16	Removed the top basket; still have 3 baskets with media.
		10.6	105.7	#2	3028	16	
		10.8	102.6	#3	3028	16	
		10.5	100.4	#4	3028	16	
		10.4	100.0	#5	3028	16	
		10.4	100.0	OUT	3028	16	
6-02	24	11.4	109.1	IN	3028	18	Removed one more basket; have 3 baskets with media.
		10.8	102.7	#3	3028	18	
		10.4	100.3	#4	3028	18	
		10.4	100.0	#5	3028	18	
		10.5	100.0	OUT	3028	18	
6-03	25	10.4	111.2	IN	4542	23	Increased flow.
		9.6	106.9	#3	4542	23	
		9.2	101.2	#4	4542	23	
		9.2	100.0	#5	4542	23	
		9.3	100.0	OUT	4542	23	
6-08	26	10.5	114.9	IN	3028	19	Removed another basket; have 2 baskets with media.
		9.1	106.9	#4	3028	19	
		9.1	103.4	#5	3028	19	
		9.0	102.5	OUT	3028	19	
6-09	27	10.5	115.2	IN	3028	19	Two baskets with larger bottom openings and without
		10.0	108.0	#4	3028	19	media.
		9.9	104.3	#5	3028	19	
		9.7	103.1	OUT	3028	19	
6-10	28	10.4	113.2	IN	3028	16	Two baskets with larger bottom openings and media.
		10.2	103.4	#4	3028	16	
		10.0	102.2	#5	3028	16	
		10.1	101.6	OUT	3028	16	
6-22	29	9.7	107.0	IN	3028	16	Three baskets with larger bottom openings, no media,
		9.6	104.1	#4	3028	16	no splash plate in outlet box.
		9.6	102.8	OUT	3028	16	
6-23	30	9.5	116.9	IN	3028	23	Three baskets with larger bottom openings, media and
		8.7	101.0	#4	3028	23	no outlet box splash plate.
		9.1	100.8	OUT	3028	23	

TABLE 2. — Packed column degasser tests

Date (1977)	Test No.	O$_2$ mg/l	N$_2$ %	Sample Place	Q l/min	T °C	Comments
6-28	1	10.1	109.3	IN	379	21	25.4-cm column filled with 1.52 meters of 3.81-cm rings
		9.6	100.0	OUT	379	21	and using gravity flow.
6-28	2	10.2	111.1	IN	189	21	25.4-cm column filled with 1.52 meters of 3.81-cm rings
		8.7	101.4	OUT	189	21	and using gravity flow.
6-29	3	9.8	112.5	IN	379	18	25.4-cm column with .91 meter of 3.81-cm rings and
		9.6	102.4	OUT	379	18	gravity flow.
6-30	4	10.4	108.7	IN	379	19	25.4-cm column with .91 meter of 3.81-cm rings and
		10.0	100.7	OUT	379	19	gravity flow.
6-30	5	9.8	110.2	IN	189	22	25.4-cm column with .91 meter of 3.81-cm rings and
		9.3	100.0	OUT	189	22	gravity flow.

TABLE 2. — (continued)

Date (1977)	Test No.	O_2 mg/l	N_2 %	Sample Place	Q l/min	T °C	Comments
7-04	6	10.4	109.4	IN	379	21	25.4-cm column with 106.68 cm of 3.81-cm rings and
		9.4	100.0	OUT	379	21	gravity flow.
7-04	7	10.4	109.4	IN	379	21	Repeat of No. 6.
		9.4	100.0	OUT	379	21	
11-16	1	10.3	127.4	IN	1136	24	1st test on new Incubator degasser. Two 25.4 cm diameter
		8.4	100.9	OUT	1136	24	columns with 1.37 meters of the 3.81-cm plastic rings. Each column has 568 l/min flow.
11-16	2	10.8	130.1	IN	1136	24	Same as above with three columns being tested.
		8.4	100.7	OUT	1136	24	379 l/min per column.
11-17	3	10.7	120.9	IN	2271	18	All five columns at 454 l/min each.
		9.7	100.0	OUT	2271	18	
11-17	4	10.8	122.9	IN	2271	18	All five columns at 454 l/min each.
		9.7	100.0	OUT	2271	18	
11-18	5	10.8	122.7	IN	2271	18	All five columns at 454 l/min each.
		9.7	100.0	OUT	2271	18	
11-02	1	10.8	130.3	IN	390	18	First test on a 25.4-cm diameter column 1.52 meters in
		8.8	106.7	OUT	390	18	height using 1.22 meters of 8.89-cm rings.
12-02	2	10.6	130.2	IN	390	18	Added 10.16 cm of 3.81-cm rings to top of the column
		8.4	106.6	OUT	390	18	containing 1.22 meters of 8.89-cm rings.
12-06	3	10.8	129.8	IN	390	25	25.4-cm column filled to 1.22 meters of depth with
		7.8	105.9	OUT	390	25	"crushed" 8.89-cm rings.
12-07	4	10.4	130.8	IN	390	17	25.4-cm column with bottom .61 meter filled with
		8.8	106.3	OUT	390	17	8.89-cm rings and .30 meter on top with 3.81-cm rings.
12-07	5	10.4	130.8	IN	390	18	25.4-cm column with bottom filled with .61 meter of
		8.8	104.3	OUT	390	18	8.89-cm rings and top filled with .61 meter of 3.81-cm rings.
12-08	6	10.4	130.8	IN	390	18	25.4-cm column filled with .91 meter of 3.81-cm rings.
		8.8	102.8	OUT	390	18	
12-08	7	10.6	130.7	IN	390	18	25.4-cm column filled 1.37 meter of 3.81-cm rings
		8.5	100.4	OUT	390	18	for a control test.
12-08	8	10.3	130.8	IN	390	18	25.4-cm column filled with 1.22 meter of 5.08-cm rings.
		8.7	104.4	OUT	390	18	
12-09	9	10.4	136.5	IN	390	18	25.4-cm column with 1.07 meters of 2.54-cm rings.
		8.8	103.0	OUT	390	18	
12-09	10	10.4	136.4	IN	390	18	25.4-cm column with 1.22 meters of 2.54-cm rings.
		9.0	102.4	OUT	390	18	
12-09	11	10.4	136.7	IN	390	18	25.4-cm column with 1.37 meters of 2.54-cm rings.
		9.0	101.9	OUT	390	18	
12-09	12	10.4	136.7	IN	390	18	25.4-cm column with 1.37 meters of 3.81-cm rings
		9.0	101.4	OUT	390	18	for a control.
12-09	13	10.4	136.7	IN	390	18	25.4-cm column with 1.37 meters of 5.08-cm rings.
		8.6	104.0	OUT	390	18	
12-12	14	11.4	129.3	IN	473	17	25.4-cm column with 1.22 meters of 3.81-cm rings
		8.9	102.7	OUT	473	17	at increased flow.
12-12	15	11.4	129.3	IN	473	17	25.4-cm column with 1.37 meters of 3.81-cm rings
		9.4	102.2	OUT	473	17	at increased flow.
12-12	16	11.4	129.3	IN	473	17	25.4-cm column with 1.37 meters of 5.08-cm media
		9.6	102.3	OUT	473	17	at increased flow.

TABLE 2. — (continued)

Date (1977)	Test No.	O₂ mg/l	N₂ %	Sample Place	Q l/min	T °C	Comments
12-13	17	11.0	114.2	IN	473	10	25.4-cm column with 1.37 meters of 5.08-cm media.
		10.5	102.0	OUT	473	10	
12-14	18	11.0	114.2	IN	473	10	25.4-cm column with 1.37 meters of 5.08-cm media.
		10.4	101.9	OUT	473	10	
6-22	1	11.0	104.3	IN	189	13	15.24-cm column filled with 1.52 meters of 3.81-cm
		11.2	100.0	OUT	189	13	rings and gravity flow.
6-22	2	11.0	104.4	IN	189	13	Repeat of test no. 1.
		11.2	100.3	OUT	189	13	
6-22	3	11.3	106.5	IN	379	13	15.24-cm column filled with 1.37 meters of 3.81-cm
		11.4	102.5	OUT	379	13	rings and gravity flow.
6-23	4	9.4	113.8	IN	379	21	15.24-cm column with 1.37 meters of 3.81-cm rings
		9.3	102.7	OUT	379	21	and gravity flow.
6-27	5	10.2	110.0	IN	379	19	15.24-cm column with 1.37 meters of 3.81-cm rings
		9.1	101.7	OUT	379	19	and gravity flow.
12-08	1	10.0	140.8	IN	250	20	15.24-cm column filled with 1.22 meters of 2.54-cm rings
		9.0	106.9	OUT	250	20	at maximum flow without overflowing the column.
12-09	2	10.6	130.2	IN	303	18	15.24-cm column with .61 meter of 2.54-cm rings
		9.4	112.2	OUT	303	18	at maximum flow.
12-09	3	10.2	130.2	IN	390	18	15.24-cm column with 1.52 meters of 5.08-cm rings.
		8.8	105.9	OUT	390	18	
(1978)							
1-04	4	11.4	129.5	IN	284	17	15.24-cm column with .61 meter of 2.54-cm media.
		10.1	112.0	OUT	284	17	
1-04	5	11.0	129.5	IN	284	18	15.24-cm column with .91 meter of 2.54-cm media.
		10.0	106.7	OUT	284	18	
1-04	6	11.0	129.5	IN	284	18	15.24-cm column with 1.22 meters of 2.54-cm rings.
		9.8	105.4	OUT	284	18	
1-04	7	11.2	129.1	IN	189	18	15.24-cm column with 1.22 meters of 2.54-cm rings.
		9.8	102.5	OUT	189	18	
1-04	8	11.4	129.1	IN	114	18	15.24-cm column with 1.22 meters of 2.54-cm rings.
		9.2	101.0	OUT	114	18	
1-04	9	10.8	129.0	IN	303	19	15.24-cm column with 1.22 meters of 2.54-cm rings.
		9.9	106.8	OUT	303	19	
1-05	10	10.8	133.0	IN	284	18	15.24-cm column with 1.22 meters of 5.08-cm rings.
		9.6	105.2	OUT	284	18	
1-05	11	10.8	133.0	IN	284	18	15.24-cm column with 1.22 meters of 3.81-cm rings.
		9.5	105.2	OUT	284	18	
1-05	12	10.8	133.0	IN	284	18	15.24-cm column with 1.22 meters of 3.81-cm rings
		9.4	110.8	OUT	284	18	and bottom of column submerged.
1-09	1	12.4	120.8	IN	151	13	12.70-cm column with 1.22 meters of 3.81-cm rings.
		11.5	101.3	OUT	151	13	
1-09	2	12.4	120.8	IN	57	13	12.70-cm column with 1.22 meters of 3.81-cm rings.
		11.2	100.3	OUT	57	13	
1-09	3	12.0	120.8	IN	57	13	12.70-cm column with 1.22 meters of 3.81-cm rings
		10.7	101.0	OUT	57	13	and a reduction in the outlet fall from 22.86 cm to 10.16 cm.

TABLE 2. — (continued)

Date (1977)	Test No.	O₂ mg/l	N₂ %	Sample Place	Q l/min	T °C	Comments
1-09	4	11.9	120.8	IN	76	14	12.70-cm column with 1.22 meters of 2.54-cm rings.
		11.0	100.0	OUT	76	14	
1-09	5	11.9	120.8	IN	151	14	12.70-cm column with 1.22 meters of 2.54-cm rings.
		11.1	102.9	OUT	151	14	
1-09	6	12.4	120.8	IN	212	13	12.70-cm column with 1.22 meters of 2.54-cm rings
		11.4	104.5	OUT	212	13	at maximum flow.
(1977)							
12-06	1	10.8	129.8	IN	390	25	10.16-cm column filled to 1.07 meters with flexipac
		8.8	117.9	OUT	390	25	(waffle type) media.
12-08	2	10.0	140.7	IN	132	22	10.16-cm column with .91 meter of 2.54-cm rings
		8.8	112.4	OUT	132	22	at maximum flow without overflowing the column.
12-09	3	10.0	142.4	IN	68	22	10.16-cm column with 1.22 meters of 1.59-cm rings
		8.4	113.7	OUT	68	22	at maximum flow without overflowing the column.
(1978)							
1-04	4	11.0	129.0	IN	114	19	10.16-cm column with 1.22 meters of 2.54-cm rings.
		9.9	106.8	OUT	114	19	
1-09	5	12.4	120.8	IN	151	13	10.16-cm column with 1.22 meters of 2.54-cm rings.
		11.4	110.6	OUT	151	13	
1-09	6	12.1	120.8	IN	151	13	10.16-cm column with 1.22 meters of 3.81-cm rings.
		11.3	103.0	OUT	151	13	
7-19	1	11.8	104.8	IN	379	19	30.48-cm column filled with 76.20 cm of 3.81-cm rings
		9.0	102.5	OUT	379	19	and gravity flow.
7-20	2	10.1	110.3	IN	379	19	30.48-cm column filled with .91 meter of 3.81-cm
		9.1	102.2	OUT	379	19	rings and gravity flow.
7-20	3	9.7	110.5	IN	379	19	Repeat of test no. 2.
		8.7	102.1	OUT	379	19	
7-20	4	9.7	110.5	IN	379	21	30.48-cm column filled with .91 meter of 3.81-cm rings
		8.6	102.3	OUT	379	21	and splash plate at the top of the column, gravity flow.
7-21	5	10.1	110.2	IN	379	19	30.48-cm column with 106.68 cm of 3.81-cm rings
		9.1	102.1	OUT	379	19	and gravity flow.
7-21	6	8.5	110.9	IN	473	27	30.48-cm column with 106.68 cm of 3.81-cm rings
		7.7	102.3	OUT	473	27	and gravity flow.
(1978)							
1-09	1	12.0	120.8	IN	38	13	7.62-cm column with .61 meter of 1.59-cm rings.
		10.8	108.9	OUT	38	13	

(From page 74)
Rearranging the general gas equation, we get:

$$V = \frac{n\,RT}{R}$$

This means to the fish culturist that if the water supply pressure is changed (pumping) or the temperature changed (heating), nitrogen supersaturation could occur. There are several causes of nitrogen super-saturation due to a temperature or pressure change and these are

1. High rise dams — spilling, turbines (pressure)
2. Pumping — air leaks (pressure)
3. Heating — (temperature)
4. Snow melt — (temperature)
5. Subsurface waters — springs, wells (temperature and/or pressure)

6. Waterfalls — (pressure)
7. Gravity intakes — air leaks (pressure)
8. Thermal power plants — (temperature)

A normal nitrogen gas reading is one that does not exceed 100 percent. At 100 percent a liquid is saturated with a particular gas. This means that the water contains all the gas that it can hold under given temperature and pressure. When the reading exceeds 100 percent, the water has become supersaturated with the gas. Supersaturation is an "unstable" condition; and if given enough time at a normal temperature and pressure, this condition will be reduced to saturation.

The supersaturation of nitrogen gas is the state that causes fish health problems. Unlike oxygen, which can be utilized by the fish, nitrogen does not dissolve in the blood stream because the gas is inert (chemically inactive). Nitrogen cannot be metabolized and may at higher levels build up in the blood of fish to the extent that bubbles form in the blood and tissues (Figure 4).

Symptoms of fish pertaining to nitrogen gas supersaturation:
1. One or both eyes extruded from the socket.
2. Hemorrhaging in the eye area.
3. Bubbles in the fins.
4. Bubbles in the skin tissue.
5. Bubbles in the roof of the mouth — which could relate to untypical feeding patterns.
6. Sounding — fish seeking the lowest depth of water.
7. Extended abdomen — yolk sac fry are sometimes so full of gas that they swim belly-up or in an erratic manner near the surface.
8. White-spot — coagulated yolk.
9. Off feed — low growth rate.

Nitrogen gas causes "Gas Bubble Disease" (Rucker 1972). Lethal concentrations depend upon size and species of fish. Smaller fish (fry and fingerlings) and the salmonids appear to be more susceptible. Most losses from gas supersaturation can be directly attributed to the secondary invasion of bacteria and parasites.

Probably the main concern about

FISH HEALTH PROBLEMS ASSOCIATED WITH NITROGEN GAS ·GAS BUBBLE DISEASE·

EXTENDED YOLK SAC
WHITE SPOT - COAGULATED YOLK
POP-EYE & HEMORRHAGING EYES
BUBBLES & BLISTERS

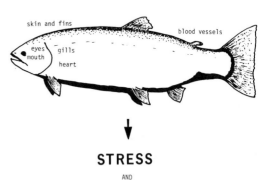

STRESS

AND
SECONDARY INVASION BY
PARASITES AND BACTERIA CAN LEAD TO DEATH

FIGURE 4. — Fish health diagram.

nitrogen supersaturation is that it causes a "stress" on the fish. The fish are already subjected to more stresses than needed in a hatchery situation. If any of these stresses can be eliminated, they should be done so without question.

It was mentioned earlier that nitrogen gas supersaturation is an "unstable" condition. Because of this unstable condition, it is possible to bring it near a stable condition (down to or near saturation). To do this, either the temperature or the pressure must be changed. Normally, temperature is not a very flexible parameter when rearing fish, so a change in pressure must be made. This change in pressure is the concept behind the packed column. Swiftly moving rivers can naturally reduce nitrogen gas through aeration and turbulence. This turbulence and aeration is what is taking place within the packed column.

An example of how to change pressure is, first, to take a bucket of water (at room temperature) and supersaturate it with nitrogen gas. Take a measurement at this time for nitrogen gas. Wait several hours for the water to come to equilibrium with the atmosphere pressure in the room and take another reading. The water will be saturated (100 percent) or less. Because the fish culturist does not have several hours to let water come to equilibrium with the atmosphere, he must rely upon mechanical or physical means to remove excess nitrogen gas.

Conclusion

The packed column is a tool by which the fish culturist can attain a better hatchery environment and a more successful rearing program.

Acknowledgments

This study could not have been conducted without the support and help of the Dworshak staff and lab personnel; namely, Mr. Wayne Olson, Hatchery Manager; Joe Lientz, Hatchery Biologist; and Mrs. Susan Espinosa, Lab Technician.

A special thanks is given to Mr. George L. Williams, whose participation proved invaluable in this study.

The support of the U.S. Army Corps of Engineers — especially Mr. Morris C. Croker and Mr. W. John Hess — is gratefully acknowledged.

References

DENNISON, B.A., AND M.J. MARCHYSHYN. 1973. A device for alleviating supersaturation of gases in hatchery water supplies. Prog. Fish-Cult. 35(1):55-58.

OWSLEY, D.E. Nitrogen gas removal at Dworshak National Fish Hatchery. (Unpublished.) 15 pp. (1977.)

RUCKER, R.R. 1972. Gas bubble disease of salmonids: A critical review. U.S. Fish Wildl. Serv. Tech. Pap. No. 58. 9 pp.

SHRABLE, J. B. Removal of nitrogen gas from water at Jackson National Fish Hatchery. (Unpublished.) 16 pp. (1978.)

WEISS, R.F. 1970. The solubility of nitrogen, oxygen and argon in water and seawater. Deep-Sea Res. 17:721-735.

WOLD, EINAR. 1973. Surface agitators as a means to reduce nitrogen gas in a hatchery water supply. Prog. Fish-Cult. 35(3):143-146.

VICTOR SEGARICH: Mr. Owsley, is there any relationship between the diameter of the column and the flow? You were talking about 10-inch columns. How low can you go with a 10-inch column?

DAVID OWSLEY: In a previous report that I have published I have data for various size diameter columns for various flows. The column that I am suggesting here is just one example. The size of the column will depend upon the flow. You can go anywhere from a 2-inch to a 24-inch column. At Jackson Hole, there is even a 5-foot diameter packed column.

SEGARICH: Mr. Owsley, how close can you get with a Weiss saturometer? We have been talking about significant figures in the 103-105% range. There are considerable variations. I was wondering how your measurements compared with VanSlyke determinations?

OWSLEY: There are variations in nitrogen as reported. When you get a Weiss saturometer, you receive a manual with an altitude and approximate barometric pressure reading. When you are using this particular method, 104%, 105%, or 106% might not mean too much. When you get a good reading from your barometer at your station, a good temperature reading, and good dissolved oxygen data, you can obtain readings to two decimal places. I think that one decimal place is actually good.

QUESTIONER: Mr. Owsley, I have been using cold-current-packed columns for about six years now. Have you calculated it out in terms of mass transfer coefficients as standardly employed by chemical engineers?

OWSLEY: No, I have not.

DALLAS E. WEAVER: It might be useful to analyze some of the data in terms of conventional engineering technology. Back to the old problem — two engineers do not talk to each other. Chemical engineers have been working with this gas-liquid transport problem for a long time. They are rather good at it when you look at the petrochemical processes.

DOUGLAS JESTER: Mr. Owsley, I would like to address a comment to you. In late 1976 or 1977, there was a comparison of the Weiss saturometer and the standard gas chromatograph methods published in the *Transactions of the American Fisheries Society*. The comparison was done by the standard T-test statistical methodology. The entire data were presented in the paper. If you examine the data, they do not meet the assumptions of a T-test application of the appropriate nonparametric test. This indicates that the Weiss saturometer is significantly different from the standard gas chromatograph. The difference is primarily in the range of 100% to 110% saturation. I would suggest a great deal of caution in using the numbers interchangeably.

SEGARICH: In the instance I was talking about, the comparison between the Weiss saturometer and the VanSlyke apparatus was made on-site with the same water.

QUESTIONER: Dave, could you tell me what pressure you are using for the water coming into the packed columns?

OWSLEY: The pressure ranges from 5 to 25 PSI. The higher pressures worked better as far as the efficiency of the column itself is concerned. This was without airflow going in.

QUESTIONER: What was the maximum super-saturation of nitrogen you could live with in rearing smaller fish? What sizes were the most vulnerable?

OWSLEY: I reported 103% as the maximum limit for fingerlings and yolk-sac fry. We do not like to see anything over 105% for salmonids. That may seem low to some people and some of the literature never stated it that low. The fish shown on my slides were from 105% saturation.

QUESTIONER: There are figures in the literature for adults that are as high as 110%. Are these logical?

OWSLEY: I have seen these figures. The problem is that damage can occur and not be visible. The damage was visible at 105%. At 103%, there was damage but it was not visible. I would recommend for salmonids, not over 105% for adults and 103% or less for fingerlings and yolk sac fry.

Bio-Engineering Symposium for Fish Culture (FCS Publ. 1):83-89

An Overview of the Various Techniques to Control Infectious Diseases in Water Supplies and in Water Reuse Aquacultural Systems

HARRY K. DUPREE

U.S. Fish and Wildlife Service
Fish Farming Experimental Station
Post Office Box 860
Stuttgart, Arkansas 72160

Introduction

Production and profitable aquacultural systems must provide for good water quality and disease control. Problems arise at all levels of production, but the problems intensify exponentially as the size of the system and level of production increase. The difficulties encountered are now magnified as the quality of ground and surface water available for fish production is less and the discharge regulations for effluents are more stringent. These factors encourage the utilization of closed or semiclosed systems where water is recirculated and only small amounts of make-up water are needed. Under these conditions the risk of infectious diseases and the difficulty of their control greatly increase.

Bacteria, viruses, and other single-celled organisms may be present in the incoming water supply or in the water to be reused, i.e. culture water. Some of these organisms are highly infectious and may cause weight loss and mortality in the production system, while others may actually be beneficial in providing supplemental food or assisting in waste nitrification. It may be desirable and economical to control only the harmful organisms. Such a technique is termed "disinfection." In contrast, the destruction of all organisms, both harmful and beneficial, is termed "sterilization."

Disinfection and/or sterilization can be accomplished by several methods including ultraviolet radiation, ozone, chlorine, oxidizing agents, surface active chemicals, filtration, heat, and time. An overview of these various methods for disinfecting and sterilizing water in fish culture systems follows.

Methods and Techniques for Disinfecting and Sterilizing Water

Ultraviolet Radiation

The observation that sunlight could destroy some bacteria was reported by Dawns and Blunt (1878). Later, Ward (1893) determined that the germicidal component of sunlight was contained in the ultraviolet spectrum. Ultraviolet radiation includes the light wavelengths from 150 Å to 4000 Å. The killing effect of ultraviolet radiation on bacteria, viruses, and other small organisms is a function of the wavelength. Koller (1965) reported that the most effective wavelength is 2600 Å. The effectiveness decreases rapidly on either side of the peak and at 3200 Å and 7000 Å it drops to 0.4 and 0.002% respectively. Hoffman (1974) reported that 2537 Å is effective and practical for killing certain fish pathogens, and is non-toxic to fish.

The mechanism by which microorganisms are killed or rendered harmless by UV is not completely known. Some evidence has been presented that the UV energy interacts with or adversely affects the nucleic acids in the cells.

Koller (1965) reported that the UV dosage required to kill all but 36.8% of the organisms is the mean lethal dose, lethal exposure, or inactivation dose. The lethal dose varies with the age, size, and species of microorganism, and water turbidity, ions in the water and water depth. The microorganism survival ratio is a function of the product of ultraviolet radiation intensity and exposure time. The same mortality can be obtained with a short exposure time at high intensity as with a long exposure time at low intensity.

Water temperature has little effect on the bactericidal effects of ultraviolet radiation (Koller 1965). UV intensity and exposure time required to kill 90-100% of the various fish pathogens and other organisms have been reported by Kelly (1974) and Hoffman (1974).

It is not possible to predict with certainty the absorption coefficient of UV for a given water sample. This information can only be determined by experimental means. Impurities in water will significantly reduce transmission of UV. Plankton, silt or other particulate matter will obviously hinder tranmission and UV is ineffective in waters of high turbidity (Honn and Chavin 1976). Effective exposure is apparently related to the size and transparency of the organism to UV (Hoffman 1974). Normandeau (1968) reported that as little as 3620 microwatts sec/cm^2 killed *Aeromonas salmonicida*, which is 0.5-1.0 μm wide. Vlasenko (1969) reported that the 800 μm diameter *Ichthyophthirius multifiliis* trophozoites required 1.7×10^6 microwatts sec/cm^2 of UV radiation for kill.

Burrows and Combs (1968) in treatment of fresh water used a sand filter to remove opaque particles larger than 15 μm and treated the filtered hatchery fresh water with UV. In the raw water some trout were killed by *Columnaris* and the rest died from *Ceratomyxa shasta*. Those fish in the treated water remained healthy. Similar results were reported by Sanders et al. (1972) and Bedell (1971). A very good review on the disinfection of water by UV and whirling disease of trout has been published by Hoffman (1974).

Ozone

Ozone has been used as a purifying agent since 1782. Ozone has recently attracted attention in intensive cultural systems for fishes since an inexpensive disease control agent is needed (Roselund 1974; Conrad et al. 1975). Honn and Chavin (1976) proposed that ozone could be used as an oxidative supplement in water reuse systems for fishes.

Ozone is the triatomic molecule of oxygen formed when the diatomic oxygen molecule is sufficiently excited to decompose into atomic oxygen. Collisions of these atomic units cause the formation of ozone (O_3). Klein et al. (1975) and Koller (1965) reported that ozone formation is accomplished by passing air or oxygen through a high voltage corona, or by subjecting it to ultraviolet radiation with a wavelength range between 1000-2000 Å.

Effectiveness of ozone as a disinfectant is a function of contact time and dosage. Pavoni et al. (1975) indicated that virus and bacterial survival, as a function of time, is a sigmoid curve.

Ozone dispersion in the liquid brings about the required contact between the ozone and target organism. In the simplest form, the ozone is introduced in small bubbles with an aerator stone into the bottom of a column of water. The necessary time of contact and ozone concentration vary with the target organism and water quality. The scientific literature suggests that 1-5 minute contact time and 0.56-1.0 mg O_3/l are sufficient to kill most pathogens in aquacultural systems. However, performed on-site tests are the only positive method to determine the required ozone concentration and contact time. It should be noted that the analytical methods for ozone quantitation (Schechter 1973; APHA et al. 1975) are not exact.

Several factors affect the concentration of ozone. Organic matter exhibits an ozone demand and thus water containing high levels of organic materials must be treated with a larger quantity of ozone. That color, odor, and turbidity in water can be removed by ozone indicates they have an ozone demand. Bailey (1975) discussed the reaction sites of ozone on organic matter and showed that inorganic chemicals may exhibit an ozone demand. Kjos et al. (1975) were able to reduce ferrous iron in fresh water from 9.54 mg/l to 0.07 mg/l with ozone. Spotte (1970) reported that the use of ozone in closed culture systems must be done with care because some desirable salts may be oxidized to insoluble forms thus removing them from the system.

Colberg and Ling (1978) found that concentrations of ozone in aqueous systems were indirectly related to pH. However, when the enteric red mouth pathogen was tested at pH 6.0-9.3, and mixed bacterial-protozoan populations at pH 8.2-9.3, the difference in pH had no apparent effect on rate of kill. Consequently, the bacterial activity of ozone appears to be due to a mechanism more specific than random oxidation.

Ozone is toxic to aquatic animals. Good toxicity data is not available except for scattered points. However, it appears that any measurable ozone in the culture system is detrimental and should be avoided. Arthur and Mount (1975) reported that 0.2-0.3 ppm ozone was toxic to fathead minnows but this may be high in view of the following references. Roselund (1974) reported gill epithelial damage and death to rainbow trout exposed to an estimated 0.010-0.060 mg O_3/l. In other chronic toxicity testing Wedemeyer et al. (1978) observed no recognizable differences between salmon exposed for three months to 0.023 mg O_3/l and the control. However, at the 0.05 mg O_3/l level, the pathological changes were typical to those following exposure to agents such as chlorine, ammonia, and formalin. Roselund (1974) reported that low level exposure may cause gill adhesions and mortality.

Small amounts of unmeasurable or undetectable ozone appear to be lethal to oyster larvae. De Manche et al. (1975) tested previously ozonated water that had been "aged" and aerated and found all were lethal to oyster larvae. However, filtration through activated charcoal was effective in producing water safe for oyster larvae.

Conrad et al. (1975) showed that ozone destroyed *Flexibacter columnaris* in spring water. Wedemeyer and Nelson (1977) reported that enteric red mouth (ERM) bacterium was more sensitive to ozone than *Aeromonas salmonicida*. An ozone residual of 0.01 mg/l caused complete inactivation of ERM in 30 seconds while 10 minutes was required to inactivate *A. salmonicida*. Colberg and Ling (1978), however, found *A. salmonicida* more sensitive to ozone than ERM. They could not explain the differences in the results and suggested that strain differences, pH differences in the water, or other factors may affect the response to ozone.

In soft (30 mg/l as $CaCO_3$) and hard (120 mg/l) lake water, the studies by Wedemeyer and Nelson (1977) illustrate the difficulty in obtaining stable ozone residuals, and especially the difficulty in disinfecting "hard" waters within short periods.

Wedemeyer, Nelson and Yasutake (1978) reported that the minimum effective therapeutic dose of ozone to control *Saprolegnia* during egg incubation was about 0.03 mg/l, but 0.3 mg/l was toxic to the eggs. At 0.03 mg/l O_3 fry mortalities and abnormalities were not appreciably increased over the control values.

Ozone is an effective disinfectant and offers much potential in aquaculture. However, its acute toxicity is high and the margin of safety is low.

Chlorine

Chlorine is an inexpensive water disinfectant. Chlorine is readily available as the gas, or as sodium or calcium hypochlorite. Good equipment that is relatively inexpensive is available for its application. Analytical methods and techniques for quantifying residual chlorine in water are easily obtained, and no special instrumentation is required.

However, the use of chlorine in aquaculture systems is risky. Most aquatic animals are highly sensitive to chlorine and chloramines (Kelly 1974). Toxicity of chlorine to cultured fish can range as low as 0.1-0.3 ppm.

When chlorine is used, sufficient time must be allowed for the chlorine to kill the target organism, and then for the removal of the residual chlorine and chloramines. While it is relatively easy to kill waterborne pathogens with chlorine, it is more difficult to remove the excess chlorine and the chloramines. Activated carbon can remove both the chlorine and chloramines. Reducing agents such as sodium thiosulfate and ferrous salts will reduce the residual chlorine to an inactive product but they have no effect on the toxic chloramines. Adding reducing agents must be done cautiously since they themselves could be toxic (Coventry et al. 1935). Aeration alone is probably not sufficient to remove the toxic factors associated with chlorine, especially in those waters containing large amounts of organic matter (Bedell 1971). An excellent review of the toxicity of combined chlorine residuals to freshwater fish was given by Zillich (1972).

Wedemeyer and Nelson (1977) reported that in demand-free buffered water 0.01 mg/l chlorine for up to 10 minutes did not significantly affect ERM and *A. salmonicida*. Increasing chlorine residual, up to 0.05 mg/l inactivated ERM in 30 seconds but little change was noted in *A. salmonicida* even after 10 minutes. The amount of chlorine needed to inactivate these two pathogens varied according to water type, but in contrast to ozone, a stable residual could be obtained.

Wedemeyer et al. 1978 reported that infectious hematopoietic necrosis virus (IHNV) appeared to be more intrinsically sensitive to chlorine than infectious pancreatic necrosis virus (IPNV). For IHNV disinfection in hard lake water, 0.5 mg Cl/l for 10 min was sufficient. Increasing it to 1.0 mg Cl/l decreased the contact time to 30 seconds. Soft lake water was easier to disinfect: 0.5 mg/l chlorine level required only a 5-minute contact time.

IPNV was inactivated by a chlorine level of 0.2 mg/l of chlorine in 10 minutes in soft water, but in hard water this concentration essentially had no effect. Increasing the chlorine dose to 0.7 mg/l in the hard water destroyed the IPNV within 2 min.

Bedell (1971) and Sanders et al. (1972) eliminated *Ceratomyxa shasta* from water supplies of rainbow trout using a practical application of chlorination followed by dechlorination with activated charcoal.

Oxidizing Agents

Several oxidizing agents are effective in killing fish pathogens. These include the halogen compounds such as the various iodine and bromine formulations, hydrogen peroxide, and potassium permanganate. These materials are generally used in disinfecting and sterilizing nets and other equipment, troughs, raceways, and buildings. This use of chemicals would be more appropriately covered in another overview.

Surface Active Materials

Researchers have shown that viruses can be removed in varying degrees by carbon absorption (Cookson 1967, 1970; Sproul et al. 1967). It is suggestive that activated carbon is not an effective absorber of virus. *Echerichia coli* in the amount of only about 18% was apparently absorbed on the surface of the carbon. Of particular interest was the observation that when water containing organic matter was pumped through a virus-saturated filter, the organic matter would replace the virus, thus releasing the virus into the effluent.

Studies by Fina and Lambert (1972) and Taylor et al. (1970) suggest that quaternary ammonium anion exchange resins combined with triodials form stable insoluble complexes that have remarkable antibacterial properties. It appears to the investigators that pathogen contact with the resin beads is necessary; the iodine is released on demand. This technique appears to lend itself to high volume systems. Since iodine is released only on demand, residual materials are no problem. The susceptibility of the test organisms points to broad spectrum capabilities. While there were no fish pathogens included in the study, it did include *Salmonella typhimurium, Streptococcus faecalis, S. aureus,* and *Pseudomonas aeruginosa* thus suggesting that the system would eliminate fish pathogens. The organism cells were not filtered from the water but emerged from the resin column in a nonviable form.

Filtration

Numerous filter types exist. Many of these will filter out organisms and particles in commercial water quantities. Where large disease organisms (mostly parasites) are present, the filtration technique is suggested. It is not very effective on smaller organisms such as virus and bacteria.

In a modified filter technique, Combs (1968) reported that salmon poisoning fluke *(Nanophyetus salmincola)* cercariae could be killed by electrocution. The study showed that 60-cycle alternating current at a voltage gradient of 230 volts per inch and an exposure time of one second was sufficient. Power demand of the parallel plates was affected by the quantity of water treated, conductivity of the water, voltage gradient, and area of the plates.

Heat

Water can be sterilized or disinfected by heating it to a sufficiently high temperature to kill the pathogenic organism. Examples are common in food processing and preservation. In aquaculture, the primary disadvantage is the energy required to heat the water and then to cool it before it flows into or through the culture system. However, there are several promising observations reported in which the water was warmed to less than the sterilization temperature. In these cases disease was either prevented, or the signs eliminated.

Amend (1970) reported that mortalities in sockeye salmon from hematopoietic necrosis virus (IHN) could be prevented if the water temperature was raised to at least 18°C within the first 24 hours after infection of the fish and if the fish were maintained at this temperature for 4-6 days. The disease did not recur after the elevated temperature treatment, but the fish would still contact the disease if they were reinfected.

Plumb (1974) reported that water temperature had an effect on the mortality rate of virus-infected channel catfish. CCV-infected channel catfish held in water above 25°C developed clinical signs, followed by rapid mortality. By reducing the water temperature to 20-25°C the clinical response is irregular but some mortality occurs. Further reducing the temperature to 15°C eliminates the clinical signs and mortality.

Time

Survival time of pathogens in water supplies that are left untreated is also important in disease control. Wedemeyer and Nelson (1977) reported that *A. salmonicida* died out in soft lake water in two days at 20°C, but lived for about two weeks in "hard" (120 mg/l as $CaCO_3$) lake water. The enteric red mouth survived well in both hard and soft lake water. The author suggests that soft hatchery water supplies where furunculosis is enzootic might not need to be treated if the reservoir of infection is sufficiently removed from the hatchery intake to allow a 1-2 day detention time. The same would not be true for ERM.

Wedemeyer et al. (1978) and Tu et al. (1975) reported that IHNV and IPNV of salmonoids survived up to several weeks in lake water. Survival of both viruses in untreated hard or soft lake water was so prolonged that sufficient die-off for disease control is unlikely unless the reservoir, such as a salmon spawn ground, is quite distant from the hatchery intake.

Comparison of Methods

Ultraviolet radiation is an effective tool for disinfecting water in aquacultural systems provided that turbidities are not excessive. Filtration equipment is readily available for removing particulate matter in water. UV does not produce toxic residues, the equipment for its production is readily available and the system is simply installed, operated, and maintained. Electricity is a continuing expense. In the laboratory and small aquacultural operations, ultraviolet radiation is probably the preferred technique for disinfecting culture waters.

Ozone is highly effective as a disinfectant. Ozone kills bacteria and virus quickly and does not leave harmful residues such as the chloramines produced by chlorine. However, ozone itself is toxic to aquacultural organisms and must be removed by filtration through activated carbon or by a similar technique. Ozonators are somewhat inefficient in converting electrical energy to ozone. Ozone does have, however, an affinity for organic and some inorganic materials in water and thus the dosage must allow for this absorption. Analytical techniques are limited and time consuming to perform when compared to those used on chlorine. In the large scale aquacultural operation and where water quality permits, ozone should be considered as a disinfection method.

Chlorine/dechlorination is an effective disinfectant. The chemical quickly destroys most bacteria but considerable contact time is required to kill virus. Chlorine combines with organic materials to form chloramines. Both chlorine and the amines are extremely toxic to fish and are difficult to remove except by carbon absorption. Equipment for applying chlorine and removing the residue is readily available. Because of the difficulty in preventing equipment failures, chlorine should be used only as the last resort.

The remaining techniques including filtration and surface active chemicals for purifying water may be preferred in some specific applications. It is not possible in this overview to discuss and evaluate all these options.

References

AMEND, D.F. 1970. Control of infectious hematopoietic necrosis virus disease by elevating the water temperature. J. Fish. Res. Board Can. 27(2):265-270.

APHA (AMERICAN PUBLIC HEALTH ASSOCIATION), AMERICAN WATER WORKS ASSOCIATION, AND WATER POLLUTION CONTROL ASSOCIATION. 1975. Standard methods for the examination of water and wastewater. 14th edition. Washington, D.C. 874 pp.

ARTHUR, J.W., AND D.I. MOUNT. 1975. Toxicity of a disinfected effluent to aquatic life. Pages 778-786 *in* First international symposium on ozone in water and wastewater treatment. International Ozone Institute, Waterbury, Conn.

BAILEY, P.S. 1975. Reactivity of ozone with various organic functional groups important to water purification. Pages 101-119 *in* First international symposium on ozone in water and wastewater treatment. International Ozone Institute, Waterbury, Conn.

BEDELL, G.W. 1971. Eradicating *Ceratomyxa shasta* from infected water by chlorination and ultraviolet irradiation. Prog. Fish-Cult. 33(1):51-54.

BURROWS, R.E., AND B.D. COMBS. 1968. Controlled environments for salmon propagation. Prog. Fish-Cult. 30(3):123-136.

COLBERG, P.J., AND A.J. LING. 1978. Effect of ozonation on microbial fish pathogens, ammonia, nitrate, nitrite, and BOD in simulated reuse hatchery water. J. Fish. Res. Board Can. 35(10):1290-1296.

COMBS, B.D. 1968. An electrical grid for controlling trematode cercariae in hatchery water supplies. Prog. Fish-Cult. 30(2):67-75.

CONRAD, J.F., R.A. HODT, AND T.D. KREPS. 1975. Ozone disinfection of flowing water. Prog. Fish-Cult. 37(3):134-136.

COOKSON, J.T., JR. 1967. Absorption of viruses on activated carbon-absorption of *Echerichia coli* bacteriophage T_4 on activated carbon as a diffusion-limited process. Environ. Sci. Tech. 1(2):157-160.

———. 1970. Removal of submicron particles in packed beds. Environ. Sci. Tech. 4(2):128-134.

COVENTRY, F.L., V.E. SHELFORD, AND L.F. MILLER. 1935. The conditioning of a chloramine treated water supply for biological purposes. Ecology 16:60-66.

DAWNS, A., AND T.P. BLUNT. 1878. Research on the effects of light upon bacteria and other organisms. Proc. R. Soc. London 26:488.

DEMANCHE, J.M., P.L. DONAGHAY, W.P. BREESE, AND L.F. SMALL. 1975. Residual toxicity of ozonated seawater to oyster larvae. Publ. No. ORESUT-003, Sea Grant Coll. Progr. Oregon St. Univ., Corvallis.

FINA, L.R., AND J.L. LAMBERT. 1972. A broad-spectrum water disinfectant that releases germicide on demand. Proc. Second World Congress, International Water Resources Assoc., New Delhi, India 2:53-59.

GIESE, A.C. 1967. Effects of radiation upon protozoa. Pages 267-356 *in* Research in protozoology, 2. Pergamon Press, New York.

HOFFMAN, G.L. 1974. Disinfection of contaminated water by ultraviolet radiation, with emphasis on whirling disease *(Myxosoma cerebralis)* and its effect on fish. Trans. Am. Fish. Soc. 103(3):541-550.

HONN, K.V., AND W. CHAVIN. 1976. Utilizing of ozone treatment in the maintenance of water quality in a closed marine system. Mar. Biol. 34:201-209.

KELLEY, C.B. 1974. The toxicity of chlorinated waste effluents to fish and considerations of alternative processes for the disinfection of waste effluents. Va. St. Water Control Board.

KJOS, D.J., R.R. FURGASON, AND L.L. EDWARDS. 1975. Ozone treatment of potable water to remove iron and manganese: Preliminary pilot results and economic evaluation. Pages 194-203 *in* First international symposim for water and wastewater treatment. International Ozone Institute, Waterbury, Conn.

KLEIN, M.J., R.I. BRABETS, AND L.C. KINNEY. 1975. Generation of ozone. Pages 1-9 *in* First international symposium for water and wastewater treatment. International Ozone Institute, Waterbury, Conn.

KOLLER, L.R. 1965. Ultraviolet radiation. Second edition. John Wiley, New York. 312 pp.

NORMANDEAU, D.A. 1968. Progress report, Proj. F-14-R-3. State of New Hampshire. (mimeo.)

PAVONI, J.L., M.E. TITLEBAUM, H.T. SPENCER, AND M.R.FLEISHMAN. 1975. Ozonation as a viral disinfection technique in wastewater treatment systems. Pages 340-366 in First international symposium on ozone for water and wastewater treatment. International Ozone Institute, Waterbury, Conn.

PLUMB, J.A. 1974. Channel catfish virus. The Catfish Farmer and World Aquaculture News 6(3):40-42.

ROSELUND, B. 1974. Disinfection of hatchery effluent by ozonation and the effect of ozonated water on rainbow trout. Pages 59-65 in W. Blogoslawski and R. Rice, eds. Aquatic applications of ozone. International Ozone Institute, Inc., Syracuse, N.Y.

SANDERS, J.E., J.L. FRYER, D.A. LEITH, AND K.D. MOORE. 1972. Control of the infectious protozoan Ceratomyxa shasta by treating of hatchery water supplies. Prog. Fish-Cult. 34(1):13-17.

SCHECTER, H. 1973. Spectrophotometric method for determination of ozone in aqueous solutions. Water Res. 7:729-739.

SPOTTE, STEPHEN. 1970. Fish and invertebrate culture: Waste management in closed systems. Wiley-Interscience, New York.

SPROUL, O.J. AND B.M. SOMENDU. 1975. Poliovirus inactination with Ozone in water. Pages 288-295 in First international symposium for water and wastewater treatment. International Ozone Institute, Waterbury, Conn.

TAYLOR, S.L., L.R. FINA, AND J.L. LAMBERT, 1970. New water disinfectant: An insoluble quaternary ammonium resin-triiodid combination that releases bactericide on demand. Appl. Microbiol. 20(5):720-722.

TU, K.C., R.S. SPENDLOVE, AND R.W. GOEDE. 1975. Effect of temperature on the survival and growth of infectious pancreatic necrosis virus. Infect. Immunol. 11: 1409-1412.

VLASENKO, M.I. 1969. Ultraviolet rays as a method for the control of diseases of fish eggs and young fishes. Probl. Ichthyol. 9:697-705.

WARD, H.M. 1893. The action of light on bacteria. Proc. R. Soc. London 54:472.

WEDEMEYER, G.A., AND N.C. NELSON. 1977. Survival of two bacterial fish pathogens (Aeromonas salmonicida and the Enteric Redmouth Bacterium) in ozonated, chlorinated, and untreated waters. J. Fish. Res. Board Can. 34(3):429-434.

WEDEMEYER, G.A., N.C. NELSON, AND C.A. SMITH. 1978. Survival of the salmonoid viruses infectious hematopoietic necrosis (IHNV) and infectious pancreatic necrosis (IPNV) in ozonated, chlorinated, and untreated waters. J. Fish. Res. Board Can. 35(6):875-879.

WEDEMEYER, G.A., N.C. NELSON, AND W.T. YASUTAKE. 1978. Potentials and limits for the use of ozone as a fish disease control agent. Presented at the International Ozone Institute, Second Workshop of Marine and Freshwater Applications of Ozone, Orlando. (mimeo.)

ZILLICH, JOHN A. 1972. Toxicity of combined chlorine residuals to freshwater fish. J. Water Pollut. Control Fed. 44(2):212-220.

Bio-Engineering Symposium for Fish Culture (FCS Publ. 1):90-91

Recent Advances in Evaluating Biofilter Performance

GEORGE W. KLONTZ

Fishery Resources
University of Idaho
Moscow, Idaho 83843

Introduction

During the past ten or twelve years, there has been an increasing interest in raising fish in recycled water systems. The two major reasons for this interest are to make the available water go farther and to make the available water more suitable for raising fish, especially with respect to temperature.

Conceptually, hatcheries using recycled water systems consist of two biological systems, each operating independently of the other. That is, the fish in the system must be raised as though they were in an open water system and the biofilter must operate as though there were no fish in the system. To be sure, each provides a service for the other but each must be independent of the other for life support.

To some, that last statement might sound somewhat paradoxical. How can the fish not be dependent upon the biofilter, when the task of the biofilter is to remove the nitrogenous waste products that are injurious to the fish above certain levels? Also, how is the biofilter going to function optimally without the fish providing the nitrogenous waste products for it to utilize? As recently as six months ago, I asked the same questions, but the answers were not easy to come by. I do not admit or even aspire to having the questions answered, but I think we have come a long way.

Methods

Our approach to defining the biological requirements of a functioning biofilter system centered around our setting up closed, fish-free biofiltration systems and simulating the fish-derived inputs.

The first of many inputs measured was NH_3-N oxidation to NO_3-N. The biofiltra-tion systems, four of them, were "fed" NH_4Cl daily at levels equivalent to multiples of 0.012 mg/l NH_3. This base level was chosen because, according to several studies, it is the level above which fish growth would be adversely affected. During the ensuing weeks the daily levels of dissolved oxygen, hardness (total and calcium), alkalinity (carbonate), pH, carbon dioxide, ammonia (both NH_3 and NH_4^+), nitrite, nitrate, and temperature were determined according to standard methods using Hach chemicals.

After nearly 16 months of isolating and testing various chemical components, we have established that optimum biofilter performance at 11-15°C is dependent upon pH, hardness (magnesium and calcium), dissolved oxygen, inorganic carbon, and rate of water flow through the medium. The recommended levels are

pH	7.5-8.4
D.O.	>80% saturated
Hardness	
Mg^{++}	>80 mg/l
CA^{++}	>100 mg/l
Alkalinity	>50 mg/l as HCO_3^-
Water flow	1-2 gpm/ft² of medium

The required conditions can be maintained by adding oyster shell (for CA^{++} and CO_3^{--}), $NaHCO_3$ (for inorganic carbon), and $MgCl_2$ (for MG^{++}) to the system. In addition to the inorganic chemicals required for optimal bacterial nitrification in a biofilter, the system should be supplied with an outside nitrogen source, thus, not relying on the fish-derived NH_3-N. We have used NH_4Cl in an aqueous drip over an 8-hour period daily at a level to maintain 0.02-0.03 mg/l NH_3 (calculated) in the entire system. The NH_4Cl should be added to the intake of the biofilter and, if there is 100% nitrification, the effluent from the biofilter should be 0.00 mg/l NH_3. This has been the case in our studies.

The daily feeding of the biofilter with NH_4Cl is analogous to feeding fat and protein to fish. The nitrifying bacteria require nitrogen as an energy source and the fish use organic carbon as an energy source. If either is deprived of its respective energy source, growth virtually ceases.

Conclusion

According to recent unpublished data from Dr. Meade's group at the University of Rhode Island, the chemical requirements we have determined for optimum biofilter performance are compatible with the environmental requirements for fish, particularly salmonids. The next step taken was to connect the two biological systems — the fish and the biofilter — and monitor the effects throughout the presmolt rearing period.

Early this past November [1978] juvenile steelhead at 50/lb were put into two steady-state biofilter systems. One system was fed NH_4Cl and the other was not — supposedly. However, as it turned out, a mix-up in communication occurred and both systems were fed NH_4Cl. For two weeks the fish showed no evidence of being adversely affected by the chemically manipulated systems. However, because of the lack of an untreated control. the test was discontinued to be reinitiated at a later date.

Bio-Engineering Symposium for Fish Culture (FCS Publ. 1):92-96

Preliminary Results and Design Criteria from an On-Line Zeolite Ammonia Removal Filter in a Semi-Closed Recirculating System[1]

WILLIAM BRUIN, JOHN W. NIGHTINGALE, AND LAURA MUMAW

The Seattle Aquarium
Pier 59, Waterfront Park
Seattle, Washington 98101

DENNIS LEONARD

The Anaconda Company
660 Bannock Street
Denver, Colorado 50204

ABSTRACT

A water reuse system for a salmon-rearing facility at The Seattle Aquarium is described. The system utilizes clinoptilolite, a zeolite mineral, for ammonia removal. Water quality parameters were observed during the rearing of several lots of salmon. Quantitative assessments of ammonia produced by the fish, filtered from the water, and recovered from the clinoptilolite filter show good agreement. The ammonia levels in the rearing water were effectively maintained at values acceptable for good fish growth.

The Seattle Aquarium, a municipally operated facility, has reared chum, coho, and chinook fingerlings for educational display purposes since opening in 1977. The small scale (125,000 released per year) rear-release program takes place in a fish ladder that winds around the outside of The Aquarium building and empties into Elliott Bay on the downtown Seattle waterfront. A 9500-gallon (Table 1) L-shaped section at the head of the ladder, termed the raceway, can be isolated and filled with fresh water to rear salmon fingerlings, or with seawater pumped from Puget Sound. Salmon in the rear-release program are obtained from the Washington State Department of Fisheries hatcheries shortly before smolting. At the appropriate time, they are exposed to an organic imprinting agent (morpholine), introduced to salt water, and released down the ladder to sea. The facility was designed in the mid-1970's by Kramer, Chin & Mayo, Inc.

Fresh water for the raceway rearing facility is obtained from the city water system, passed through a sand filter to remove particulates, and subsequently dechlorinated by passing through activated carbon filters. The limited amount

TABLE 1. — Flow rates in raceway water system.

Raceway capacity:	9500 gal (35,950 l)
Overflow rate:	13 gpm — 12-h turnover (49.21 l/min)
Total recirculating flow rate:	120 gpm — 1-h turnover (454 l/min)
Clinoptilolite filter flow rate:	50 gpm — 3-h turnover (189.25 l/min)

of available fresh water necessitated the design and use of a recycling water system. Filters containing clinoptilolite, a zeolite mineral shown to be highly efficient in ammonia removal (Williams 1973) were incorporated in the system (Table 2) to remove the metabolic nitrogenous wastes. Clinoptilolite functions as an ion-exchange medium, showing a high affinity for ammonium ions. As fresh water containing ammonia waste passes over the clinoptilolite, ammonium ions are preferentially bound to the medium, displacing sodium ions. When saturated with ammonia, clinoptilolite can be recharged using a sodium chloride solution in which the con-

---

[1] The Seattle Aquarium Technical Report Number 7.

TABLE 2. — Clinoptilolite filter characteristics (two in parallel).

Surface area:	9.8 ft^2
Bed depth:	42 in
Bed volume:	34.3 ft^3, 256 gal (969 l)
Clinoptilolite weight:	848 kg

centration of sodium ions greatly exceeds that of the ammonium ions being displaced. This ammonia removal system was considered particularly suitable for The Seattle Aquarium because of its location on a saltwater site. With seawater readily available, recharging the clinoptilolite would involve minimal preparation and expense. Major advantages of such a physico-chemical ammonia-removal filter over a biological filter include its relative independence from factors such as temperature and the presence of antibacterial agents, and its ability to function at maximum effectiveness almost immediately.

The raceway water system (Fig. 1) consists of 90% recirculating water flow with 10% fresh 'new' water input. The recirculating water is first passed through pressure sand filters. Approximately 45% of this water is subsequently passed through filters containing the clinoptilolite before returning to the raceway. Normal seawater from Puget Sound is used to recharge the clinoptilolite filters when necessary. Flow meters (Eagle Eye Portable Meter, Model 77C) were installed in June 1979 at points shown in Figure 1. Measured flow rates are shown in Table 1. Flow rates through the clinoptilolite filters are within the 12-15 bed volumes/hour recommended by the supplier.

Various chemical parameters are monitored routinely. Dissolved oxygen (YSI Model 57 DO meter), salinity, temperature (YSI-SCT Model 33 meter), pH (Orion Model 407A pH meter), and ammonia (Stickland and Parsons 1972) are monitored in the raceway, the clinoptilolite influent, and the clinoptilolite effluent lines. If these parameters reach potentially dangerous levels, steps are taken to correct them. A value of 0.005

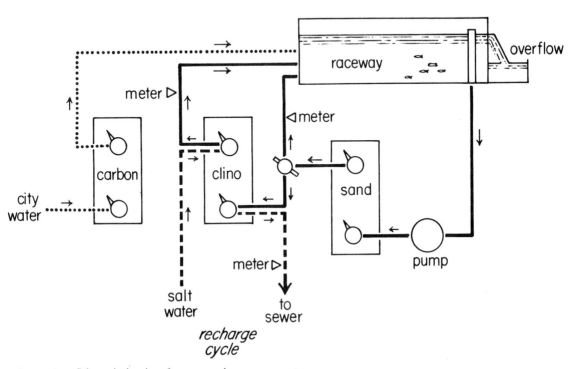

FIGURE 1. — Schematic drawing of raceway and water reuse system.

mg/l un-ionized ammonia (the form toxic to fish) is taken as the maximum allowable level (Spotte 1970; Rice 1977).

In 1978 a total of 32,000 chinook and coho fingerlings were reared in the raceway. Quantitative aspects of ammonia removal for that season are difficult to interpret because operating procedures were still being refined and standardized. Nevertheless, the clinoptilolite filters worked efficiently to keep the raceway ammonia at acceptable levels. The origin (species) of clinoptilolite installed in the filters was not known.

By the beginning of the 1979 season, procedures for operating the water reuse system were well defined. A new clinoptilolite species, designated Anaconda-Becker 1010, was placed in the filters. From February 13 to April 11, 1979 — 57 days — 13,000 coho fingerlings (127 kg initial total weight) were reared in the raceway. Water quality parameters were monitored as described previously for two weeks while the clino filters were kept off line (Table 3). Water quality parameters for the succeeding 43 days (Table 4) with the clinoptilolite filters in use differed primarily in increased temperature, concomitant decrease in oxygen, lower pH, and reduction in ammonia. The latter two appear to be a function of clinoptilolite filter operation. The greatest un-ionized ammonia value (0.001 mg/l) was well below our chosen limit. Clinoptilolite filters were recharged every 14 days. After a 3-h saltwater recharge (salinity 28°/oo, pH 7.8), the first few bed volumes of fresh water run through the filter had an unusually high pH and visible turbidity, apparently due to very fine suspended matter. This phenomenon had not been observed with the different clinoptilolite species used in the 1978 season. A 30-min

TABLE 4—Water quality for coho (February 28 to April 11, 1979) with clinoptilolite filters.

	T°C	pH	NH$_4$-N (mg/l)	DO (mg/l)
Mean	10.0	6.8	0.23	8.7
Range	8.0-11.2	6.5-7.2	0.10-0.75	6.4-11.5
S.D.	1.7	0.2	0.12	1.4

13,000 coho fingerlings; 127 kg initial weight.

freshwater rinse after the saltwater backwash was therefore instituted as a part of the standard backwash procedure.

A more quantitative study of clinoptilolite filter performance and dynamics was undertaken with the subsequent rearing of a group of 30,000 chinook fingerlings (118 kg initial weight). Salmon were fed 5.5% of their body weight of Oregon Moist Pellets (OMP-II, Moore Clarke, Co.) daily (2.2 kg at 0500, 1100, and 1700 hours). Over the eighteen days they were held in fresh water, the fish mass increased to 186 kg. These are loads of 2.0 lb/gpm and 3.1 lb/gpm respectively (flow-regulated density measures). Expressed as a static volume, the figures are 0.2 lb/ft^3 and 0.31 lb/ft^3.

Flow rates as well as water quality parameters were monitored daily during the experimental period of 14 days (June 20 to July 3, 1979) (Table 5). The seasonally high temperatures and lower dissolved oxygen levels are apparent. Clinoptilolite filters were recharged once a week. Un-ionized ammonia values in the raceway did not exceed 0.002 mg/l. Ammonia determinations were made at 2-h intervals for the first 48 hours. When it became apparent that values were not fluctuating greatly, sampling frequency was decreased to 4-h intervals. Figure 2 shows the daily variation in raceway ammonia levels, averaged from seven days. Values were lowest early in the morning, rising shortly after the first feeding (0500).

Performance of the clinoptilolite filters was assessed by comparing average daily values of ammonia in the clinoptilolite influent and effluent lines (Fig. 3). During the course of both weeks, the influent ammonia level gradually rose and leveled off at 0.45 mg/l, while the effluent ammonia level in the clinoptilolite filter began rising

TABLE 3. — Water quality for coho (February 13-27, 1979) without clinoptilolite filters.

	T°C	pH	NH$_4$-N (mg/l)	DO (mg/l)
Mean	7.9	7.2	0.66	10.0
Range	7.0-8.9	6.9-7.3	0.40-0.90	9.2-11.2
S.D.	0.5	0.1	0.16	0.5

13,000 coho fingerlings; 127 kg initial weight.

TABLE 5. — Water quality for chinook (June 20 to July 3, 1979) with clinoptilolite filters.

	T °C	pH	NH₄-N (mg/l)	DO (mg/l)
Mean	17.9	6.7	0.43	5.4
Range	17.0-19.8	6.5-7.0	0.21-1.0	3.4-6.9
S.D.	0.6	0.1	0.11	0.6

30,000 chinook fingerlings; 118 kg initial weight. (0.9 kg fish/gpm flow); 186 kg final weight (1.4 kg fish/gpm flow).

TABLE 6. — Ammonia production and removal (June 20-26, 1979).

	NH₄-N (g)
Produced from food (f = 0.0203)	980
Removed by clinoptilolite (790 g)	1,000
Discharged in overflow (210 g)	
Recovered in clinoptilolite backwash	910

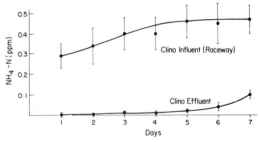

FIGURE 3. — Ammonia concentrations in clinoptilolite filter influent and effluent.

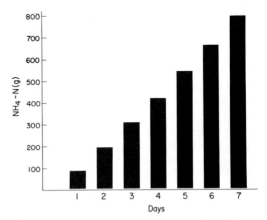

FIGURE 4. — Ammonia accumulation on clinoptilolite.

about the fifth day (to 0.1 mg/l). Recharging occurred on the seventh day. By using the difference between the clino influent and effluent ammonia levels, and multiplying by the flow rate through the clinoptilolite filters, the theoretical amount of ammonia accumulating in the filters can be calculated (Fig. 4). At the end of the seventh day of operation, 790 g of ammonia had been filtered out of the raceway water.

The ammonia recovered in the salt water recharging solution is shown in Figure 5. This curve is reproducible and has been obtained many times. A large peak is seen in the first 5 minutes; the bulk of the ammonia appears to be removed within the first 90 minutes. This curve is also plotted as a function of bed volumes of

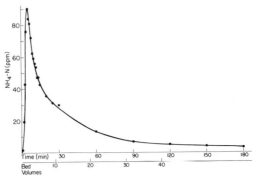

FIGURE 5. — Ammonia recovered in salt water recharge solution.

backwash water (Fig. 5). Most ammonia is removed in the first 40 bed volumes (10,300 gal; 39,400 l). The amount of ammonia recovered in the recharging solution can be calculated by integration of the area under the curve.

A balance sheet of ammonia production and removal in the raceway system for one week is presented in Table 6. Similar

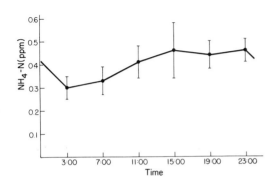

FIGURE 2. — Diurnal variation in raceway ammonia levels.

results were obtained the following week. The factor used to convert food consumed into ammonia produced by the fish was 0.0203 lb (kg) ammonia nitrogen produced per lb (kg) of food fed (OMP) (Liao 1976); Speece 1973). Based on these figures, the capacity of the clinoptilolite filters was 1.1 g ammonia nitrogen per kg of clinoptilolite with a breakthrough value of 0.1 mg/l ammonia (the breakthrough value is the level of ammonia in the clinoptilolite filter effluent just prior to recharging). Functional capacity of the clinoptilolite filter is undoubtedly higher since raceway levels of un-ionized ammonia were maintained below the desired level (0.002 mg/l) and clinoptilolite filters were still operating at 80% efficiency at this breakthrough value.

Further experimentation on the operating characteristics of the clinoptilolite filters will be conducted during the 1980 rearing season. A predicted biological program developed from engineering biological criteria will be tested. A comparison of actual values for growth, oxygen consumption, ammonia production, and ammonia removal will be compared with the prediction model. A revision of the prediction criteria is anticipated.

References

LIAO, P.E. 1976. Personal communication.

RICE, L. 1977. Salmon culture. Pages 165-217 *in* C. Dyckman and S. Garrod, eds. Small streams and salmonid: A handbook for water quality studies. UNESCO, Paris.

SPEECE, R.R. 1973. Trout metabolism characteristics and the rational design of nitrification facilities for water reuse in hatcheries. Trans. Am. Fish. Soc. 102(2):323-334.

SPOTTE, S.H. 1970. Fish and invertebrate culture. Wiley-Interscience, New York. 145 pp.

STRICKLAND, J.D.H., AND T.R. PARSONS. 1972. A practical handbook of seawater analysis. Fish. Res. Board Can. Bull. 167. 310 pp.

WILLIAMS, W.G. 1973. Evaluation of clinoptilolite for ammonia removal. Tech. Repr. No. 33. Kramer, Chin and Mayo, Inc., Seattle, Washington. 2 pp.

Bio-Engineering Symposium for Fish Culture (FCS Publ. 1):97-103

Determination of Operating Parameter Values for Water Reuse Aquaculture

WILLIAM E. MANCI[1] AND JOHN T. QUIGLEY

University of Wisconsin-Extension
Madison, Wisconsin 53706

ABSTRACT

The paper presents the results of a series of flow-through bioassays that were conducted to determine the optimum values of key water quality parameters for the rearing of yellow perch (*Perca flavescens*) in a recirculating water aquaculture system. Specific parameters include ammonia gas, nitrite, and chloride ion concentration, evaluated at certain pH and temperature conditions set for maximum growth rate. The components of a bioassay apparatus are briefly described and the results of bioassay are related to operating system performance.

Introduction

The importance of considering total water chemistry when one designs a recirculating water rearing system has become increasingly evident. Published data for aquaculture of salmon indicate that un-ionized ammonia (NH_3-N) should not exceed 0.0075 mg/l (Meade 1977) and that calcium and chloride ion concentrations equivalent to at least 25 mg/l $CaCl_2$ should be assured (Wedemeyer and Yasutake 1978) as a means of mitigating the toxic effects of nitrite ion for steelhead trout culture. Other observations suggest that an interrelationship may exist between the chronic effect of low levels of un-ionized ammonia and nitrite ion concentration, and pH value (Wedemeyer 1977).

The first objective of engineering studies in support of the University of Wisconsin Aquaculture Project was to conduct a series of bioassays as outlined below in order to determine the relative importance of these factors and others for the intensive culture of yellow perch (*Perca flavescens*). A further objective of these bioassays was to determine specific operating values or limits for key water quality parameters in a recirculating water system for rearing perch.

Other general objectives of these studies included the determination of improved operating modes and practices to achieve and maintain control of key operating parameters during an entire rearing cycle for recirculating water treatment aquaculture systems. A further study undertaken cooperatively with the systems analysis group was intended to determine carrying capacity of the present upflow and trickling biological filtration treatment processes.

Methods and Materials

A 120-day toxicity experiment was conducted using a modified DeFoe bioassay apparatus (DeFoe 1975). Sample populations, each containing twelve two-inch perch fingerlings, were exposed to various levels of ammonia, nitrite, and chloride in factorial combination. A high and low level of concentration was chosen for each chemical. Duplicating each of the eight possible combinations required sixteen sample populations and aquaria in all.

During the course of this experiment, chemical concentrations in the tanks were monitored and all mortalities recorded. Growth was monitored as a cumulative indicator of progress. Dead fish were examined on a gross microscopic level in an attempt to observe any abnormalities and to determine cause of death. Autopsies were performed on three fish from each sample population after exposure for the

[1]Present address: Aquaculture Laboratory, 6080 McKee Road, Madison, Wisconsin 53711.

TABLE 1. — Bioassay test combinations.

Aquarium	NH₃[a]	NO₂-[b]	Cl[c]
1, 2	High	Low	High
3, 4	High	High	High
5, 6	Low	Low	High
7, 8	Low	High	High
9, 10	High	Low	Low
11, 12	High	High	Low
13, 14	Low	Low	Low
15, 16	Low	High	Low

[a]High = 2.0 mg/l NH₃-N total;
Low = 0.5 mg/l NH₃-N total.
[b]High = 0.2 mg/l NO₂-N;
Low = 0.1 mg/l NO₂-N.
[c]High = 60.0 mg/l Cl⁻;
Low = 0.0 mg/l Cl⁻.

removal of gills for histological purposes. Gill tissue samples were examined for damage as a possible indicator of chronic ammonia toxicity (Smart 1976).

Control Parameters

The choice of ammonia and nitrite values was based upon two preliminary bioassays on yellow perch fingerlings. As a result of these bioassays, ammonia levels were chosen to be 0.0075 mg/l NH₃-N gas or 0.5 mg/l NH₃-N total and 0.030 mg/l NH₃-N gas or 2.0 mg/l NH₃-N total. Likewise, nitrite levels were chosen to be 0.1 mg/l NO₂-N and 0.2 mg/l NO₂-N. Chloride additions were maintained at 0.0 mg/l Cl and 60 mg/l Cl. Table 1 shows the various chemical combinations and the associated aquaria.

Sample populations of fish were housed in all-glass aquaria, each with an effective volume of 19 litres. Each aquarium received 200 ml of solution every 1½ minutes, resulting in a volumetric flow-through or turnover rate of 10 turnovers/day. Water was supplied from the city system and filtered through activated carbon and a physical filter to remove chlorine, various organics, and rust before entering the system head tank reservoir. Water hardness was measured at 290 mg/l (as $CaCO_3$) (USGS 1974). Water temperature was maintained at 20 ± 0.5 °C. Dissolved oxygen was kept above 6 mg/l and pH held to 7.6 ± 0.1 unit by con-

tinuous adjustment with sulphuric acid in the system head-tank reservoir. Dissolved oxygen and pH measurements were taken with a Yellow Springs model 54A meter and an Orion model 901 meter, respectively. Fish were exposed to a 14-h light photoperiod and fed to satiation twice daily with Oregon Moist Pellets (Bioproducts, Inc.). Generally, the feedings required about 6 g per aquarium and were conducted at 0800 and 1600 hours. Tank cleaning proceeded one-half hour after every morning feeding. Sources of ammonia, nitrite, and chloride were ammonium sulfate, sodium or potassium nitrite, and sodium chloride, respectively. In an effort to suppress bioconversion of ammonia to nitrite, 4 ml of N-Serve 24E (Dow Chemical) were added to the head tank reservoir every 2 days.

Description of Bioassay Apparatus

The apparatus in Figure 1 is a modified version of a system first described by DeFoe (1975). (See also the photo, Fig. 2.) The operating principle behind the apparatus is as follows.

Two plexiglass metering chambers, each consisting of four sections of equal volume (500 ml) and one section of greater volume (1000 ml), begin filling with water initially. As the first section fills it overflows into the second, and so on. When the last section (largest section) fills, it starts a siphon and drains itself into a shut-off chamber. Once this chamber is full it actuates a switch (normally closed) to deenergize two solenoid valves and shut off incoming water to the metering chambers. The shut-off chamber then empties itself through a venturi to generate a vacuum, which is used to prime siphons in the eight remaining metering chamber sections (smaller sections). To prevent inflow water from immediately refilling the metering chambers before they have emptied, a time delay relay is placed between the shut-off chamber switch and the solenoids.

Each metering chamber section empties into an erlenmeyer, which is suspended by a toxicant injector lever arm. As the erlenmeyer fills with metered water, the

TANK BIOASSAY APPARATUS

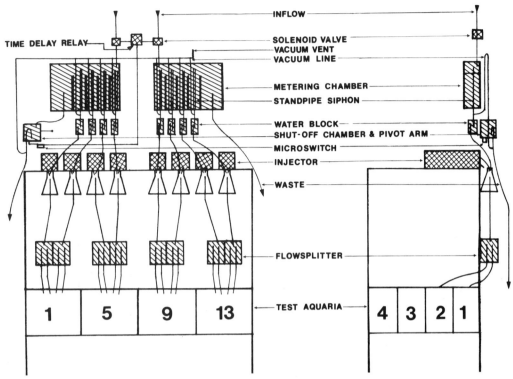

FIGURE 1. — Bioassay apparatus.

FIGURE 2. — Photo of tank bioassay apparatus.

toxicant test combination is injected into the water. Once filled, the contents of the erlenmeyers are siphoned into flow-splitting chambers. The flow splitter divides the sample in half and delivers it to duplicate tanks.

Each tank received 200 ml per cycle. The time delay relay was set so that the whole process repeated every 1½ minutes, resulting in a turnover rate of ten turnovers/day per tank as noted above. Because of solubility problems in the injector syringe, chloride salts were added at the flow splitter for those tanks that required it.

Water Quality Analysis

Samples of 80 ml were taken from each of the sixteen test aquaria at least one a week. Temperature and pH readings were taken immediately. Samples were then passed through a 0.45 μm, 47-mm Acropor membrane filter (Gelman Instruments) to remove suspended solids. Two 25-ml aliquots of each sample were analyzed for ammonia and nitrite following the procedure of Standard Methods for Water and Wastewater (1975) and of the U.S. Environmental Protection Agency (1974), respectively. All spectrophotometric measurements were taken on a Bausch and Lomb Spectronic 88 spectrophotometer.

Histological Methodology

Immediately following final weighing, measuring, and autopsy, gill structures of three fish from each of the sixteen aquaria were removed and fixed in Bouin's fluid for later histological manipulation. Table 2 outlines the steps that preceded embedding in paraffin. Gill tissues were sectioned at 16 μm and stained with acid alum hematoxylin and eosin following the procedure outlined by Quay (1970). Sections were then examined and compared for differences in structure. Since examination and comparison tended to be subjective, slides were given special code numbers and inspected in random order to minimize any bias that would otherwise be present during this procedure.

TABLE 2. — Tissue embedding procedure.

Date	Treatment
5/16-5/18	Bouin's fluid
5/18-5/21	50% EtOH
5/21-5/22	3 changes 70% EtOH
5/22-5/25	3% HCl in 70% EtOH (decalcification)
5/25-5/28	70% EtOH
5/28	1 hour; 95% EtOH
5/28	2 hours; 2/3 100% EtOH + 1/3 xylene
5/28	2 hours; 1/3 100% EtOH + 2/3 xylene
5/28-5/29	2 changes xylene
5/29-5/30	1/4 xylene + 3/4 paraffin 57 °C
5/30-5/31	paraffin 57 °C

Results

The data show little or no difference with respect to growth or mortality between the eight different treatment groups. Tables 3, 4, and 5 show the comparisons between treatment groups in terms of percent weight growth and mortality. Any differences between mortality figures are misleading since all but three mortalities

TABLE 3. — Ammonia treatment comparison.

Tank	Growth % by weight	High level treatment[*]	Fish remaining (of 12)
1	252	+ – +	9
2	276	+ – +	9
3	231	+ + +	9
4	287	+ + +	10
9	292	+ – –	9
10	249	+ – –	11
11	306	+ + –	12
12	250	+ + –	9
x̄% growth =	268		(Total) 78

Tank	Growth % by weight	Low level treatment[*]	Fish remaining (of 12)
5	210	– – +	11
6	213	– – +	9
7	209	– + +	16
8	262	– + +	11
13	273	– – –	11
14	244	– – –	10
15	289	– + –	11
16	272	– + –	11
x̄% growth =	247		(Total) 90

[*]Pluses and minuses in this column refer to ammonia, nitrite and chloride, in that order.

100

TABLE 4. — Nitrite treatment comparison.

Tank	Growth % by weight	High level treatment[*]	Fish remaining (of 12)
3	231	+ + +	9
4	287	+ + +	10
7	209	− + +	16
8	262	− + +	11
11	306	+ + −	12
12	250	+ + −	9
15	289	− + −	11
16	272	− + −	11
x̄ % growth =	263		(Total) 89

Tank	Growth % by weight	Low level treatment[*]	Fish remaining (of 12)
1	252	+ − +	9
2	276	+ − +	9
5	210	− − +	11
6	213	− − +	9
9	292	+ − −	9
10	249	+ − −	11
13	273	− − −	11
14	244	− − −	10
x̄ % growth =	251		(Total) 79

[*]Pluses and minuses in this column refer to ammonia, nitrite, and chloride, in that order.

TABLE 5. — Chloride treatment comparison.

Tank	Growth % by weight	High level treatment[*]	Fish remaining (of 12)
1	252	+ − +	9
2	276	+ − +	9
3	231	+ + +	9
4	287	+ + +	10
5	210	− − +	11
6	213	− − +	9
7	209	− + +	16
8	262	− + +	11
x̄ % growth =	242		(Total) 84

Tank	Growth % by weight	Low level treatment[*]	Fish remaining (of 12)
9	292	+ − −	9
10	249	+ − −	11
11	306	+ + −	12
12	250	+ + −	9
13	273	− − −	11
14	244	− − −	10
15	289	− + −	11
16	272	− + −	11
x̄ % growth =	272		(Total) 84

[*]Pluses and minuses in this column refer to ammonia, nitrite, and chloride, in that order.

can be attributed to fish jump-out during the first two weeks of the experiment. Whether or not these fish would have eventually died in the tanks is uncertain since mortalities were spread fairly evenly throughout the sixteen tanks.

Ammonia, pH, and chloride levels were maintained at design levels throughout the experiment. However, nitrite levels tended to fluctuate drastically because of problems with bioconversion to nitrate. Table 6 shows the mean (x̄) and the standard deviation (σ) for ammonia, pH, and nitrite for each aquarium based on the weekly water analysis.

The most important result was that even at high levels of ammonia and nitrite and low levels of chloride, 88% of the fish survived the experiment. In fact, 88% survived overall while only three of the original 192 fish actually died in the tanks. The remainder were "jump-outs." What made this even more impressive was the fact that the fish survived peaks of 3.4 mg/l

NH_3-N (total) at pH 7.6 and 0.75 mg/l NO_2-N, and had substantial growth through the 120-day exposure as shown in Table 6. These results also supported findings by Wedemeyer and Yasutake (1978) that adequate levels of calcium and chloride can serve to mitigate the toxic effects of nitrite.

Examination of gill sections revealed significant gill damage to those fish exposed to high levels of ammonia and nitrite. Damage was characterized by hematomas, hemorrhage, and hypertrophy of the secondary lamellae of the gills. The extent of damage appeared to be directly related to the degree of exposure. Those fish exposed to high levels of ammonia and nitrite, in the absence of chlorides, had far more extensive damage (30% of the gill tissue) than those exposed to low levels in the presence of chlorides (5% of the gill tissue). Gills from other treatment groups showed moderate degrees of damage (5-20% of the gill

TABLE 6. — Weekly water analysis ($\bar{x} \pm \sigma$; mg/l).

Tank	NH$_3$ Total	NO$_2$	pH	Peak Total NH$_3$	Peak NO$_2$
1	1.8 ± 0.5	0.174 ± 0.150	7.62 ± 0.17	2.8	0.558
2	1.9 ± 0.4	0.164 ± 0.142	7.61 ± 0.19	2.8	0.576
3	2.1 ± 0.4	0.173 ± 0.101	7.64 ± 0.19	3.1	0.445
4	2.1 ± 0.4	0.210 ± 0.130	7.63 ± 0.17	3.1	0.555
5	0.52 ± 0.26	0.091 ± 0.041	7.64 ± 0.18	1.2	0.178
6	0.62 ± 0.26	0.070 ± 0.033	7.64 ± 0.18	1.3	0.122
7	0.58 ± 0.28	0.128 ± 0.067	7.58 ± 0.21	1.4	0.222
8	0.52 ± 0.26	0.145 ± 0.068	7.64 ± 0.18	1.3	0.233
9	1.9 ± 0.4	0.150 ± 0.155	7.64 ± 0.19	2.6	0.686
10	2.1 ± 0.3	0.127 ± 0.129	7.64 ± 0.16	2.6	0.618
11	2.2 ± 0.4	0.210 ± 0.123	7.64 ± 0.18	3.1	0.678
12	2.1 ± 0.4	0.224 ± 0.154	7.62 ± 0.21	3.4	0.755
13	0.54 ± 0.26	0.091 ± 0.049	7.65 ± 0.18	1.1	0.231
14	0.66 ± 0.50	0.083 ± 0.039	7.68 ± 0.20	1.2	0.192
15	0.52 ± 0.22	0.152 ± 0.063	7.67 ± 0.18	1.0	0.172
16	0.57 ± 0.24	0.141 ± 0.070	7.66 ± 0.18	1.1	0.245

tissue). In general, it appeared that chlorides mitigate the gill damage induced by the effects of NH$_3$ and NO$_2$. However, this statement was based on a trend rather than conclusive evidence.

Discussion

As results indicate, this experiment was not without its problems. Bioconversion of nitrite to nitrate and to a much lesser extent conversion of ammonia to nitrite were responsible for the great degree of variability in toxicant levels. It was also possible that the relatively small aquarium size may have amplified the problem by allowing "outside sources" of ammonia and nitrite (i.e., from the fish themselves) to become significant. If larger tanks and smaller fish had been used, larger dilutional capacity may have alleviated this problem. Another problem lay within the chemical nature of nitrite itself. At high concentrations, such as those that existed within the injector syringes, nitrite tended to react with other constituents of the solution to form a gas believed to be nitrous oxide with the result that some nitrite nitrogen was lost to the atmosphere. This was a persistent problem throughout the experiment and further confounded efforts to stabilize nitrite levels.

The growth data from this experiment were difficult to interpret. Because of the length of the experiment, the fish in all aquaria may have reached a size and density (∼22 grams of fish to a liter) where the size of the aquarium was a growth-limiting factor. Unfortunately no growth data were taken midway in the experiment to determine if fish growth rates were segregating according to treatment.

The results from examination of stained gill sections raised an important point. The gill sections revealed varying degrees of damage according to treatment. However, growth did not seem to vary according to treatment. The obvious question was, should ammonia and nitrite be thought of as significant water quality parameters? In terms of this experiment, the answer may lie in the degree and length of time of exposure. Ammonia and nitrite levels were not high enough to kill large numbers of fish, yet were high enough to cause gill damage. It appears that this damage was not severe enough to retard growth at this point. It is possible that the fish were able to adapt physiologically to their polluted environment, as was suggested by Wedemeyer and Yasutake (1978) with steelhead trout. If so, ammonia and nitrite were significant in that they placed the fish under stress, but this stress was not severe enough to hamper growth to any great extent. It was concluded that higher levels of these toxicants would be required to cause greater mortality or to hinder growth.

Although the findings of this study were not conclusive, it is felt that they do represent a significant step forward for future yellow perch aquaculture. By careful control of such key factors as pH, ammonia, and nitrite, fish loadings can be maximized for optimum yield. These results serve as a base for future work on maximum allowable ammonia and nitrite levels and their effects on mortality and growth.

Summary

In order to place these observations into perspective, it may be useful to note that two groups of ca. 450 perch each has now been reared from 8-g fingerlings to 150-g maturity with relatively minor mortality in two recirculating water systems operated with the following parameter values:

Control parameters	Set-Point
Ammonia	0.50 mg/l NH$_3$-N
Nitrite	0.02 mg/l NO$_2$-N
Chloride	ca. 60.00 mg/l
pH	7.4 ± 0.2

In both systems pH control was provided by using an alum drip. Water clarity was better in the upflow biological filter sytsem. The alum flow tended to accumulate within the upflow filter media and solids removal was accomplished by periodically decanting the upper portion of the filter contents. Rearing tank turbidity was much higher in the alternate system, which contained a trickling filter and tube settler.

It has been estimated that pumping costs for water recirculation will run less than one cent per pound of fish in the round for the upflow system. Alum costs should be 1-5¢ per pound.

Recent observation of the eighteen-month-old perch indicate that they are healthy and vigorous, and should be suitable for release.

References

APHA, AWWA, WPCF. 1975. Standard methods for water and wastewater, Nesslerization method. 14th edition. Washington, D.C.

DeFoe, D.L. 1975. Multi-channel toxicant injection sytem for flow-through bioassays. J. Fish. Res. Board Can. 33:544-546.

Meade, T.L. 1977. Nitrite toxicity and other unfavorable parameters contributing to problems in a reuse system. Proceedings of a water reuse meeting, Dworshak National Fish Hatchery, USFWS, Ahsahka, Idaho, March 1977.

Quay, W.B. 1970. Comparative histology: A laboratory manual. California Book Co., Berkeley, California.

Smart, G. 1976. The effects of ammonia exposure on gill structure of the rainbow trout (*Salmo gairdneri*). J. Fish. Biol. 8:471-475.

U.S. Environmental Protection Agency. 1974. Manual of methods of chemical analysis of water and wastes. Washington, D.C.

U.S. Geological Survey. 1974. Report to town of Fitchburg Water Utility.

Wedemeyer, G.A. 1977. Some alleviating parameters in nitrite toxicity. Proceedings of a water reuse meeting, Dworshak National Fish Hatchery, USFWS, Ahsahka, Idaho, March 1977.

Wedemeyer, G.A., and W.T. Yasutake. 1978. Prevention and treatment of nitrite toxicity in juvenile steelhead trout (*Salmo gairdneri*). J. Fish. Res. Board Can. 35(6):822-827.

Bio-Engineering Symposium for Fish Culture (FCS Publ. 1):104-115

A Closed Vertical Raceway Fish Cultural System Containing Clinoptilolite as an Ammonia Stripper[1]

WILLIAM J. SLONE

Soil Conservation Service, USDA
Las Cruces, New Mexico 88001

DOUGLAS B. JESTER

1804 Hamilton Road, Apt. 21-A
Okemos, Michigan 48864

PAUL R. TURNER

Department of Fishery and Wildlife Sciences
New Mexico State University
Las Cruces, New Mexico 88003

ABSTRACT

An experimental closed vertical raceway or silo fish culture system that contained several physical innovations and an ammonia-stripping zeolite filter was tested for 15 months. The physical innovations were self-cleaning spiral-flow patterns, flushable settleable-solids sumps, a tortuous-path bacterial digestion chamber, an aerator above the biofilter, and a positive-head siphon operated by a float valve to maintain water level. Their specific values are discussed in detail.

The zeolite material, clinoptilolite, maintained ammonia nitrogen and, indirectly, nitrate nitrogen at concentrations safe for channel catfish and less-tolerant rainbow trout. Dissolved oxygen was the only chemical factor that became critical. It was supplied adequately by addition of a small aerator.

Biomass of channel catfish increased from 19.2 kg/m³ to 112.3 kg/m³ of water in the vertical raceway during the first seven months, with a crude feed conversion rate of 1.7. The experiment was continued for eight more months to evaluate reliability of the new concepts. The system was operated for a significant length of time with a fish biomass load of slightly more than 160 kg/m³.

Introduction

Rapid growth of sport and especially commercial fish culture in recent years has demonstrated a need for efficient high-density-and-biomass systems. A "silo" or vertical raceway system that utilizes recycled water appears to offer technical feasibility for high rates of production with maximum conservation and utilization of water and minimum discharge of effluents.

The Soil Conservation Service, USDA, provided technical assistance for conducting field trials to produce channel catfish, *Ictalurus punctatus* (Rafinesque), commercially in the Pecos River Valley of southeastern New Mexico. These trials revealed that commercial production is feasible but also indicated a need to solve problems caused by limited fluctuating water sources, erratic water temperatures, and evaporation. Vertical raceways provide an opportunity to control these factors but require control of dissolved oxygen, settleable organic solids, which generate biochemical oxygen demand (BOD) and toxic gases, and nitrogenous wastes.

The vertical raceway is a relatively new concept, first used in Pennsylvania (Buss et al. 1970). Experimental "silos" were constructed by the New Mexico Depart-

[1]Based upon the senior author's M.S. thesis (Slone 1975), which was funded in part by Project 4106-520 (Douglas B. Jester, Principal Investigator) and Project 4105-79 (Paul R. Turner, Principal Investigator, of the Agricultural Experiment Station, New Mexico State University).

ment of Game and Fish in 1972 for production of rainbow trout, *Salmo gairdneri* Richardson (Moody and McCleskey 1978). These silos, based on a "flow-through" water-use regime, were successful in terms of labor efficiency and increased production per unit of volume over horizontal raceways, and additional units were built. However, "flow-through" represents a limited increase in production per water-volume and no real saving of water, and does not address the problem of disposal of organic effluents. Therefore, this study was designed to develop a vertical raceway system incorporating methods for controlling water temperature, BOD, toxic gases, dissolved oxygen, and nitrogenous wastes so that maximum production efficiency may be achieved along with maximum conservation and utilization of water. Also, in accordance with a currently popular theme, energy lost through cooling systems of electrical generators could be used to regulate

water temperatures to optimum levels for fish growth by use of thermostatically controlled heat exchangers. Solar heating and cooling are also compatible for regulating water temperature in such a unit.

Channel catfish were used in this experiment because of demand for information regarding commercial catfish culture in the Pecos Valley. However, all physical and chemical parameters in the system except temperature remained within ranges of tolerance of rainbow trout, so that trout culture should require only establishing temperature control. Some marine species also may be adaptable to a closed system of this type, allowing their culture in inland locations.

Quantities and volumes used in this system were based upon judgment of adequacy to produce a workable experimental model. Thus, they are subject to considerable engineering refinement that could be tested best in a larger, flexible pilot model in order to maximize

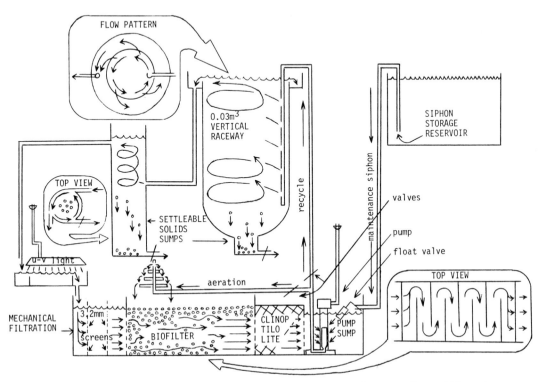

FIGURE 1.—The flow pattern through the biofilter allows near-maximum contact of raceway effluent with bacteria growing on the surfaces of the rock substrate.

benefit/cost ratios. Economic feasibility would be determined from such a unit.

Methods and Materials

The experimental vertical raceway was a translucent fiberglass cylinder. All other components were made of colorless transparent plexiglass, and tubing between components was a mixture of blue and colorless transparent plastic (Fig. 1). Transparent materials were used in the experimental model to afford observation of activity within the components and would be replaced by cheaper opaque and inert materials in a production model. A production raceway might be made of fiberglass, metal coated with epoxy paint, plastic, or other material. However, transparent lower portions of sumps would be advantageous by allowing visual observation of accumulations and flushing of settleable solids. Primary producers tended to accumulate on tube walls but did not impede flow of water significantly. They were removed periodically by use of a long-handled laboratory glassware brush.

The vertical raceway consisted of a 0.03 m^3- or 32-liter-capacity cylinder with counter-clockwise spiral flow and a settleable-solids (silo) sump (Fig. 1). The raceway was stocked with 35 channel catfish, mean length 89 mm and total weight 595 g. Water overflowed from the raceway into a collecting ring and flowed by gravity into a separate (remote) settleable-solids sump, which was intended to collect any feces, uneaten feed, and other organic solids that were not trapped in the silo sump. Both sumps were flushed through valves at the bottom twice daily until experience revealed that one flushing per day was adequate. Supernatant water from the remote sump continued by gravity flow into a distribution header of a shallow pan under an ultraviolet radiation device. After passing under this device, water dropped through a standpipe into a chemical component of a series of filters. Two 3.2-mm-mesh plastic screens functioned as a substrate for an aufwuchs community, became clogged frequently, and were removed from the system. In retrospect, alternation and cleaning of replaceable screens or use of self-cleaning rotating screens would be worthwhile to reduce aufwuchs and their nutrients entering the biofilter. Water entered the biofilter by gravity flow and hydraulic head through slots in the headwall, which separated it from the mechanical filter.

The biofilter or bacterial digestion chamber contained volcanic-ash rock, which was available locally. Weight of this rock is substantially less than that of angular granite, limestone, or river rock, any of which would be acceptable for use in a biofilter. The gas pores in the volcanic rock provide greater angularity and bacterial-attachment surface. Thus, it is preferred over the other materials if cost and availability are similar. A tortuous-path flow pattern through the biofilter (Fig. 1) allowed near-maximum contact of raceway effluent with bacteria growing on the surfaces of the rock substrate. The tortuous path resulted in a flow distance of approximately 2.4 m through a filter 0.6 m long. A riser-pipe aerator with splash collars was placed above the biofilter to provide oxygen for aerobic activity within the filter as well as for introduction into the vertical raceway.

The effluent from the biofilter flowed through slots in the headwall into the 0.005-m^3 or 4.7-l clinoptilolite filter chamber. Clinoptilolite is a zeolite earth material that adsorbs ammonium ions selectively. It is a natural greenish rock mined near Hector, California, and Tucson, Arizona. It was used as a final stripper to remove any NH_4^+ ions that were not oxidized to NO_2 or NO_3 in the biofilter. Clinoptilolite chips are available in graded and ungraded sizes. Ungraded chips that ranged approximately 6.4 mm in diameter to dust were used because they cost less than graded material and dust is eliminated simply by washing. NH_4^+ ions are adsorbed onto the surface of clinoptilolite until the entire surface is occupied. When ammonia-stripping efficiency decreased to a range of 3.0 to 5.0 mg/l breakthrough concentration, clinoptilolite was replaced in the filter chamber so that water never reentered the vertical raceway with an ammonia concentration greater than 5.0 mg/l.

From the clinoptilolite filter, water flowed through slots in the headwall into a pump sump for recycling through the vertical raceway and the biofilter aerator. Water was lifted from the sump by a 0.014-HP evaporative-air-conditioner pump and was divided between the raceway and the biofilter aerator by a "T." Rates of flow to each were controlled by separate valves. The spiral-flow pattern in the raceway was produced by injecting water through perforations in a vertical pipe in a counterclockwise direction to complement the coriolis effect of the earth's rotation. The end of the pipe was capped so that all water was injected horizontally through a single line of perforations. Water entering the raceway was aerated by the aerator above the biofilter and by turbulence in the pump sump. Horizontal injection resulted in a slow upward-spiral flow pattern during normal operation and a downward spiral during flushing of the silo sump. Spiral flows established a self-cleaning system and provided several operational advantages.

The water level in the entire system was maintained by a positive-head siphon that operated when the float valve in the pump sump opened in response to a lowered water level.

Chemical analyses made on various schedules were dissolved oxygen, BOD, total phosphate, ammonia nitrogen, nitrite nitrogen, nitrate nitrogen, total hardness, methyl orange and phenolphthalein alkalinities, and pH. APHA (1965) methods, adapted to the Hach photoelectric colorimeter, were used for all determinations except O_2 and BOD. Dissolved oxygen was analyzed by titration with the standard Alsterberg-azide modification of the Winkler method and five-day BOD was determined with the Hach manometric apparatus. Samples were taken at appropriate points throughout the system.

Ammonia breakthrough experiments were conducted to calibrate the clinoptilolite used and to test the material at higher flow rates than those extant in engineering literature. A gravity-flow two-column series was used. It consisted of columns 44.5 mm diameter packed with 60 linear cm of clinoptilolite, operated at 40 column volumes per hour with an ammonia concentration of 15 mg/l. The ammonia solution was NH_4Cl dissolved in tap water.

Exposed metallic apparatus in both systems was coated with epoxy paint to prevent metallic ions from entering the water in the systems and to prevent corrosion of the apparatus.

The fish were fed daily with Rangen's[2] sinking trout feed at a rate of 3% of fish biomass, as required by a temperature of $26.7 \pm 1.1\,°C$, which was maintained in the system. Water temperature was regulated by room temperature with fine control provided by thermostated aquarium heaters. Additional fishes were maintained in aerated and filtered aquaria for reserve biomass to allow experimental manipulation of loading in the vertical raceway.

Results and Discussion

Several logical innovations were devised by the authors and incorporated into the experimental model, only one of which has appeared in literature concerning high-density experimental or production raceway systems. They consisted of (1) the coriolis-enhanced spiral flow pattern (Leitritz and Lewis 1976) in the vertical raceway; (2) a settleable solids sump at the bottom of the vertical raceway and a remote settleable solids sump in series with the raceway; (3) tortuous path in the biofilter; (4) placement of an aerator above the biofilter; and (5) a positive-head maintenance siphon with a float valve for automatic control of water level in the system. Ammonia stripping (6) with clinoptilolite was the primary experiment designed into the system. However, the other five innovations have demonstrable values, which are discussed in context with the corresponding components of the system. The water reconditioning series was designed as a flexible modular system

---

[2] No endorsement of this or any other product is given or implied.

in which each component may be enlarged, flow patterns modified, additional components "plugged in" (e.g., foam fractionator, carbon bed, diatomaceous earth, or other mechanical filtration units), or existing components could be bypassed. Thus, design of the system allows versatility for both experimental and production purposes.

Vertical Raceway

The system contained 60.2 liters of water when it was operational; 31.8 liters in the vertical raceway, and 28.4 liters in the reconditioning components. The spiral-flow pattern in the vertical raceway was established by attaching the inflow pipe to the side of the silo and injecting water throughout the profile through perforations directed along the wall in a counter-clockwise direction (Fig. 1). The end of the pipe was capped to prevent upwelling from the bottom of the raceway so as not to destroy the upward-spiral flow pattern or the hydraulically dead space below the pipe. Advantages of a spiral flow pattern are lower current velocities for fish to oppose, resulting in uniform distribution of fish throughout the raceway (Leitritz and Lewis 1976), less loss of feed from floating over the lip into the collection ring, uniform distribution of oxygen, less interference with sinking of settleable solids, minimized turbulence, friction at the shearing plane, and upwelling at the top of the hydraulically dead space, and a self-cleaning system. In addition, reduced velocity probably required a smaller volume of water per kg of fish production. Exchange rates ranged from 8.2 to 20.4 minutes per raceway-volume with a mean of 14.3 minutes (or 1.56 l/min to 3.88 l/min, mean of 2.22 l/min). Slower rates of water exchange occurred during the last 11 days of the experiment and are discussed under *Catfish Culture.* Curvature of the bottom of the raceway and presence of the silo sump combined to produce progressively lower flow velocities down to a hydraulically dead space just above and in the sump, resulting in the collection of most of the fecal solids and uneaten feed in the sump. When the sump was flushed, solids were removed from the system and the spiral-flow pattern became oriented downward, still counter-clockwise. Thus, spiral flow results in a self-cleaning system when it is oriented either upward or downward.

A feed ring was used to prevent pellets from floating over the lip of the raceway before they could sink or be eaten. Most feed was eaten as it sank through the raceway because of relatively uniform vertical distribution of the fishes (the objective of a translucent or opaque raceway). Several individuals learned to attack floating feed but they did not appear to take a disproportionate share because their attacks broke the surface tension and caused the pellets to sink sooner than they did when they were undisturbed. Disintegration of feed pellets that reached bottom occurred in approximately eight minutes. However, several fishes would enter the silo sump vertically, head down, and salvage pellets until they disintegrated.

The fish were fed and the silo sump was flushed once each day after flushing twice was found to be unnecessary. In practice, the number of flushings of the silo sump should equal the number of feedings per day.

Remote Settleable Solids Sump

The effluent from the raceway was injected counter-clockwise into the remote settleable-solids sump near the middle of the vertical column (Fig. 1). This created an upward-spiral flow pattern above a hydraulically dead space during operation and a downward-spiral flow when the sump was flushed, imparting the same self-cleaning characteristics that occurred in the vertical raceway. BOD and H_2S problems were eliminated from the system by flushing the silo sump and the remote sump, as noted under *Water Quality* below. Therefore, both settleable-solids sumps were valuable innovations that should be incorporated into production systems. The remote sump should be flushed when visual inspection reveals an accumulation of solids and, in practice, would require flushing less frequently than the silo sump, although a minimum of once daily would be essential.

Ultraviolet Radiation

Ultraviolet radiation has long been known to be an effective antibacterial agent for disease control in intensive fish-culture systems. The ultraviolet unit, consisting of three 15-watt, short-wave length, high-intensity fluorescent tubes with reflectors, was included in the system because crowding of fishes is conducive to epidemics. The unit was located to treat raceway effluent where disease organisms would be concentrated but well ahead of the biofilter where bacteria are essential. Ultraviolet light was used continuously for three months but was deleted when the system was moved from one building to another. After it was moved, the system was operated for 12 months without the ultraviolet light. No disease occurred but the fish appeared to be in excellent condition when they were received from the Dexter (New Mexico) National Fish Hatchery and were treated for 10 days with feed containing 1.0% Terramycin before the raceway was stocked. Therefore, it is not known whether lack of disease was due to ultraviolet radiation, Terramycin, isolation, a combination of these, or absence of pathogenic bacteria when the raceway was stocked (highly unlikely).

Biofilter

The bacterial digestion chamber, utilizing a tortuous-path flow pattern and volcanic-ash rock substrate, developed a dense biotic buildup and resistence to flow. The result was an undesirable flow across the surface. Over time, a dense aufwuchs community formed a mat on the surface of the rocks but was of little or no consequence to efficiency of the filter. Minor redesign to increase the height of baffles, depth of rock, spacing of baffles (increasing length of filter path), and increasing water depth in the preceding chamber to increase the hydraulic head would eliminate surface flow. The tortuous path resulting from baffles caused retention of water in the biofilter for a long period of time, implying more effective treatment

for less cost in space and construction funds.

Bypass Filter

A bypass filter unit consisting of a 10.2-cm-diameter tube containing aquarium filter-floss and clinoptilolite was built to bypass the reconditioning components in case of need for treating disease in the system. Almost any chemical used to treat fish would retard or kill the microflora in a biofilter. This unit was not used because of lack of disease, but any recycle system should provide for this contingency by including alternate components to allow bypassing the biofilter and clinoptilolite chambers temporarily.

Positive-Head Siphon

The positive-head siphon, operated by an evaporative-air-conditioner float valve, maintained a constant water level in the system by replacing water lost to evaporation and providing a means for replacing water rapidly when sumps were flushed. Since these water losses could be replaced manually, either in a small experimental or large production system, the greatest value of the automatic siphon was demonstrated when unanticipated leaks occurred. On several occasions leaks occurred at night and disaster was averted only because the siphon functioned to replace water until the leaks were discovered and repaired.

The float valve in the siphon system could be adapted to any gravity-flow or pressure-operated water source and could function as part of a temperature control system to allow recycling water for trout culture. For example, a thermostatically controlled valve could waste water from a raceway system when a given temperature is reached and a float valve would replace the water wasted with cooler water until the lower control temperature of the thermostat was reached and the wasting-valve closed.

Water Quality

Source water

Tap water used to fill the system contained a total (versenate) hardness concentration of 100 mg/l (as $CaCO_3$). Phenolphthalein alkalinity was 0.0 mg/l and methyl orange alkalinity (= bicarbonates) was 125 mg/l. Therefore, alkalinity consisted entirely of bicarbonates. The pH was 7.8.

Dissolved oxygen

Sampling points for dissolved oxygen were the pump sump to represent water entering the vertical raceway and the remote settleable-solids sump to represent water leaving the vertical raceway. Dissolved oxygen was analyzed daily. Differences in oxygen concentrations between sampling points were minor at any given time but large differences occurred over time. Concentrations ranged from 6.0 to 1.5 mg/l during the time of the experiment. As fish biomass increased, it became obvious that aeration was becoming a limiting factor when O_2 concentrations decreased to 3.0 mg/l with low points near 1.5 mg/l soon after the fish were fed. The only mortality encountered during the experiment resulted from oxygen depletion. Fishes that died were replaced with reserve fishes from the same original stock held in aquaria. Airstones operated by an aquarium aerator were installed in the mechanical filter chamber and in the vertical raceway, and dissolved oxygen increased to the range of 5.0 to 6.0 mg/l or approximately to saturation for the temperature of 27 °C. Thus, an aerator capable of maintaining O_2 concentrations at or near saturation must be an integral part of any high-density-and-biomass fish culture system. The riser-pipe aerator shown in Figure 1 was inadequate to support a heavy fish biomass load but could be replaced by a more efficient riser-pipe with a deflector and perforated collars described by Chessness et al. (1972) or other aeration device.

Biochemical oxygen demand

Five-day BOD was analyzed at seven points in the fish cultural system at the end of the third month of the experiment. Fish biomass increased from 32.1 to 48.1 kg/m³

TABLE 1.—Five-day biochemical oxygen demand as mg/l and day during which maximum BOD occurred at seven points in the vertical raceway system. Fish biomass was 32.1 kg/m³ and 48.1 kg/m³ when the water samples were taken

Sampling points	First analysis		Second analysis	
	BOD	Day	BOD	Day
Raceway surface			53.0	2
Silo sump	240.0	4	135.0	3
Remote sump surface			14.5	2
Remote sump bottom	300.0	2	250.0	5
Ultraviolet unit	12.5	1		
Mechanical filter unit	12.5	1		
Pump sump	12.5	1	0	

during a week between two samples. Analyses were made on water from the silo sump where fecal material and uneaten feed settled and at the surface of the vertical raceway to determine the portion of BOD from the sump that reached the surface. Analyses also were made on water from three other points to determine BOD in the remainder of the system, especially in water entering and leaving the filter components. BOD values of 240 and 135 mg/l were found in the silo sump (Table 1). When 135 mg/l BOD occurred in the sump, the BOD at the surface of the raceway was 53 mg/l or 39% of the BOD, in the sump. BOD values of 300 and 250 mg/l occurred at the bottom of the remote settleable-solids sump. When BOD at the bottom was 250 mg/l, the BOD at the surface was 14.5 mg/l, or only 5.8% of the BOD at the bottom. Large decreases in BOD from bottom to surface in these units demonstrate effectiveness of settleable-solids sumps. However, greater depth of the hydraulically dead space and larger BOD gradient in the remote sump indicate that the silo sump should be lengthened to increase the hydraulically dead space vertically.

BOD values of 12.5 mg/l in the ultraviolet unit, mechanical filtration chamber, and pump sump show that the silo sump and remote sump removed the BOD effectively from the system and that no change occurred in the filter components.

110

Hydrogen sulfide

Concentrations <3.0 to >5.0 mg/l H_2S developed consistently in both the silo and remote sumps within 24 hours after the sumps were flushed. The odor of H_2S was strong when organic effluents were flushed from the sumps but dissipated rapidly as water from the hydraulically dead space followed the solids. H_2S did not occur at the surface in the raceway or the remote sump, indicating that it remained concentrated in solids, and the hydraulically dead water functioned as a buffer to limit or prevent diffusion of gas into flowing water in the system. H_2S was not detected in any other components of the system including effluent from the bacterial digestion and clinoptilolite chambers. Therefore, flushable settleable-solids sumps and hydraulically dead spaces below upward-spiraling flows have a demonstrated value in eliminating H_2S and most of the BOD from a closed vertical raceway system.

Phosphorus

Initially it was anticipated that ortho-, meta-, and total phosphates would be monitored throughout the system. However, total phosphorus usually was 0.0 and the maximum concentration recorded was a trace. Therefore, settleable-solids sumps and biofiltration were completely effective and phosphates were eliminated from the water in the system.

Nitrogen

Nitrogen compounds were the only metabolites that required consideration in this experiment. All other potentially toxic metabolites apparently were removed from the system in the biofilter and by flushing the sumps. At least, no observable stress was manifested in fish behavior, feeding activity, or feed conversion. Much potential dissolved nitrogen is also removed from the system by flushing organic solids. However, the nitrogen sources of urea, liquid fractions of feces, fish slime, and metabolic wastes voided through the gills (NH_4^+ and urea) were in solution or suspension, which required removal by other means.

Ammonia is the most significant toxic nitrogenous metabolite (Burrows 1964; Spotte 1970). Urea, feces, and slime are reduced by biofiltration to produce ammonia, which is added to that produced by fish respiration. A nitrogen-fixing bacterium, Nitrosomonas, oxidizes ammonia to nitrites and Nitrobacter oxidizes nitrites to nitrates (Spotte 1970). The maximum NO_2 concentration was 0.8 mg/l and maximum NO_3 was 60 mg/l in reconditioned water in the pump sump and vertical raceway during the experiment.

Water that was removed by flushing sumps ranged from <5% to >12%/day of the volume of water in the system. The mean volume flushed out and replaced was approximately 6%/day (more than was necessary). This small volume obviously did not represent removal and dilution adequate to prevent accumulation of large concentrations of NO_3 in the system. Therefore, a substantial amount of nitrogen was recycled in the biofilter, with autotrophic organisms utilizing most of the NO_3 rather than allowing it to pass out of the filter in solution to circulate through the system. Any limited substrate would have a maximum capacity to support a nitrogen-fixing flora but the low breakthrough and small increase in concentrations of NO_3 demonstrated that the biofilter did not reach its nitrogen-retaining capacity. Knepp and Arkin (1973) reported that concentrations of $NO_3 \geqslant 400$ mg/l are safe for channel catfish and the maximum of 60 mg/l observed in this system was only 15% of that safe concentration.

[Note: Our statement that autotrophic organisms retain most of the NO_3 in the biofilter summarizes a complex phenomenon. The biofilter developed a small ecosystem containing bacteria, algae, an aufwuchs community, other protozoa, a nematode, a microdrilid, and two families (one species of each) of midges. These organisms undoubtedly established both autochthonous and allochthonous nitrogen cycles in the biofilter and emerging midges removed nitrogen from the

111

system. Adult midges were rarely abundant around the filter but were always present so that emergent biomass was consistent and large over time.]

Knepp and Arkin (1973) also reported that $LC_{50} = 37.5 \pm 1.7$ mg/l and $LC_{100} = 45.7 \pm 6.0$ mg/l of total ammonia ($NH_3 + NH_4^+$) for channel catfish. They recommended that total ammonia be held to <30 mg/l to avoid symptoms of ammonia poisoning that occur when ammonia concentrations approach the LC_{50} value. However, these values were unpublished at the beginning of the experiment and ammonia concentrations that are safe for salmonids in the anticipated pH and temperature ranges were chosen as the required degree of control. Burrows (1964) reported that a lethal effect would occur in rainbow trout at approximately 18 mg/l of ammonia nitrogen under a given set of conditions, or at approximately half the LC_{50} of total ammonia for channel catfish. Because we wanted the closed vertical raceway system to be adaptable to trout culture by controlling temperature, salmonid tolerances were more appropriate objectives for the system.

Ammonia breakthrough rates were lower and pH became more acid than anticipated under heavy fish biomass loads. Therefore, the lethal form of ammonia, un-ionized NH_3, was present in safe concentrations for catfish or salmonids throughout the experiment. It will be shown that ammonia breakthrough over time from a clinoptilolite filter is exponential (ultimately sigmoid) and concentrations would increase rapidly as clinoptilolite becomes saturated with ammonia. Therefore, since maintaining ammonia nitrogen or total ammonia concentrations at ≥5.0 mg/l is feasible with clinoptilolite, and concentration of ammonia increases rapidly as clinoptilolite becomes saturated, it is deemed advisable to accept 5.0 mg/l or other low concentration of ammonia nitrogen as a maximum allowable breakthrough concentration from the clinoptilolite filter chamber. This establishes a large safety factor against sublethal effects for either trout (Burrows 1964) or catfish (Knepp and Arkin 1973).

TABLE 2.—Concentrations of ammonia nitrogen (total ammonia) in mg/l at three points in the vertical raceway system.

Concentration	Remote settleable solids sump	Mechanical filter	Pump sump
Minimum	0.20	0.20	0.10
Maximum	5.00	5.00	5.00
Mean	2.25	1.83	1.66

Clinoptilolite is rechargeable so that only enough to fill the filter chamber twice plus a small amount to replace wasted material are required to replace it as frequently as may be necessary.

At total ammonia levels observed, 0.10 to ≃5.0 mg/l (Table 2), NH_3 or toxic ammonia could have occurred in a maximum concentration of 0.18 mg/l when pH was 7.8 and temperature was approximately 25.6°C. However, by the time total ammonia had increased to 5.0 mg/l, pH had declined to a range of 6.0 to 7.5 and NH_3 ranged from 0 to a maximum concentration of 0.09 mg/l if maximum breakthrough of total ammonia occurred when pH was 7.5. NH_3 concentrations were computed from ammonia nitrogen data with appropriate pH and temperature values shown in Table 3.

The mean concentration of ammonia nitrogen entering the pump sump and vertical raceway was 1.66 mg/l and pH was usually about 7.5. Therefore, the mean concentration of NH_3 was approximately 0.03 mg/l. Ammonia nitrogen rarely exceeded 3.0 mg/l and its NH_3 fraction rarely exceeded 0.05 mg/l as it entered the raceway.

After water left the raceway and remote sump, it entered the mechanical filter

TABLE 3.—Mean concentrations and decreases of ammonia nitrogen in mg/l and percent decrease at three points in the vertical raceway system immediately before and 24 hours after clinoptilolite was replaced.

	Remote settleable solids sump	Mechanical filter	Pump sump
Before	2.11	2.11	2.01
After	1.41	1.44	1.35
Decrease (mg/l)	0.70	0.67	0.66
Decrease (%)	33.20	31.80	32.80

chamber with a mean ammonia nitrogen concentration of 1.83 mg/l along with an abundance of potential ammonia in the forms of urea, liquid feces, and slime. Thus, there was considerable ammonia produced in the biofilter to increase the concentration in solution. With ammonia produced in the filter and a large portion of it almost certainly not oxidized by *Nitrosomonas*, the mean ammonia concentration of 1.66 mg/l in the effluent from the clinoptilolite chamber was 0.17 mg/l less than the concentration entering the biofilter. Therefore, the clinoptilolite adsorbed the equivalent of all ammonia formed but not oxidized in, plus about 10% of that entering, the biofilter.

The mean concentration of 1.66 mg/l ammonia remaining after treatment may be accounted for by progressive reduction of ion exchange surfaces on clinoptilolite by NH_4^+, organic coating of clinoptilolite surfaces, ammonia production in the pump sump, and observed channelization of biofilter effluent over the clinoptilolite surface. Ion "coating" or adsorption was the objective of the clinoptilolite filter and could be controlled simply by replacing the clinoptilolite when ammonia breakthrough reaches the arbitrary maximum concentration of 5.0 mg/l. Otherwise, the clinoptilolite filter may be made slightly more efficient by lowering the slots in the biofilter-clinoptilolite headwall or deepening the clinoptilolite bed so that water cannot channelize over the top of the filter.

Breakthrough

Dean (1971) computed curves showing ammonia breakthrough from 20 column volumes per hour through two columns in series (Fig. 2). Forty column volumes per hour in this experiment gave comparable results with a tendency to be erratic. Apparently, difference in size of clinoptilolite and channelization at the higher flow rate caused the erratic results. The curve shows that breakthrough, once it begins, increases rapidly and daily monitoring of ammonia in the fish culture system is essential.

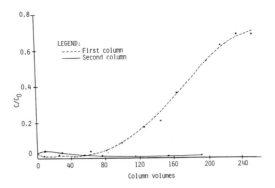

FIGURE 2.—Dean (1971) computed curves showing ammonia breakthrough from 20 column volumes per hour through two columns in series.

Regeneration

Dean (1971) also reported regeneration of clinoptilolite or removal of ammonium ions by washing the material with a caustic solution of CaO and NaCl. However, regeneration may also be accomplished by washing with tap water (pH 7.8-8.4) or by drying the clinoptilolite. In this experiment, regeneration with CaO and NaCl was done only in the two-column series on clinoptilolite loaded with NH_4Cl solution. When 70% of the ammonia in the ammonium chloride solution (15 mg/l ammonia nitrogen) broke through the columns, a standard solution of CaO and NaCl was run through the columns. Ten column volumes of solution was adequate to reduce ammonia breakthrough to 0.0 mg/l. Washing clinoptilolite with tap water was also tested in the two-column series. Again when ammonia breakthrough was 70%, water was run through the columns for 30 minutes at 40 column volumes per hour and ammonia breakthrough was reduced to 0.0 mg/l. Neither of these methods was refined or tested in the vertical raceway system.

Clinoptilolite was replaced as needed in the vertical raceway system. Regeneration was accomplished by washing to remove as much organic coating as possible and then air-drying the clinoptilolite until it was needed for reuse. Ammonia nitrogen was reduced from a mean of 2.1 mg/l to a mean of 1.4 mg/l twenty-four hours after the clinoptilolite was replaced (Table 3). Therefore, approximately 1/3 of the breakthrough ammonia nitrogen was removed. Complete stripping of ammonia nitrogen did not occur with new or regenerated clinoptilolite for reasons discussed under *Nitrogen* above. Therefore, drying was an effective regeneration method. However, regeneration with CaO and NaCl probably would be necessary in prolonged use or with a large volume of clinoptilolite because organic material collects on the clinoptilolite and is not removed by washing and air-drying.

pH

A critical factor in ammonia toxicity to fishes has been shown to be pH (Burrows 1964; Spotte 1970). It varied from 7.8 to 5.7 in this experiment. The un-ionized fraction of ammonia is the toxic form and percentage of NH_3 in total ammonia varies with pH and temperature (Table 4). Temperature was maintained at about 26.7 °C. Therefore, variation in pH was the only factor controlling proportions of NH_3 and NH_4^+ in ammonia nitrogen in the system. At 3.0 mg/l, the highest concentration of ammonia nitrogen that occurred frequently, NH_3 would have been 3.7% or 0.11 mg/l at the most alkaline pH encountered. Since, as noted above, the most alkaline pH and maximum ammonia nitrogen did not occur together, NH_3 concentrations were much lower and were inconsequential to fishes in this experiment. In the pH range of 5.7 to 7.0, NH_3 would range from 0.0 to 0.01 mg/l demonstrating the value of maintaining pH near neutrality. If pH of source water is in a range of values >8.0, buffering should be done to reduce pH to ≤8.0 before fish are stocked.

Catfish Culture

Fish production *per se* was not a major objective of this experiment. Rather, results were noteworthy by demonstrating the potential value of a vertical raceway system with nitrogen stripping and recycled water. Feed and feeding were based on convenience, with only one size of crumbles and one size of pellets used in 15 months. Fish were fed daily part of the

TABLE 4.—Percentages of un-ionized ammonia in ammonia nitrogen (total ammonia) in solution at different temperatures and pH values. Percent NH_3 at 25 °C was computed from a simple regression interpolated between 20 and 30 °C. All data except for 25 °C was reported by Burrows (1964) and Spotte (1970).

pH	Percent NH_3 in ammonia nitrogen when temperature is				
	10 °C	15 °C	20 °C	25 °C	30 °C
7.0	0.3	0.4	0.5	0.6	0.7
7.5	1.1	1.3	1.5	1.7	1.9
7.6	1.4	1.6	1.9	2.3	2.8
7.7	1.8	2.1	2.4	2.7	3.0
7.8	2.3	2.6	3.0	3.5	4.1
7.9	2.9	3.3	3.8	4.4	5.1
8.0	3.6	4.1	4.7	5.4	6.2
8.1	4.5	5.2	6.0	6.9	7.9
8.2	5.7	6.5	7.3	8.1	9.0
8.3	7.1	8.0	9.1	10.4	11.9
8.4	8.9	9.9	11.2	12.9	14.8
8.5	11.1	12.3	13.7	15.3	17.1
8.8	20.3	22.1	24.2	26.6	29.3
9.0	29.1	32.5	35.8	38.9	42.2
9.5	57.6	59.8	62.1	64.5	67.0

time and on alternate days part of the time. Nevertheless, in seven months, catfish biomass increased from 19.2 kg/m³ to 112.3 kg/m³ with a crude feed conversion rate (weight of feed fed/weight of fish produced) of 1.7. At that time, only dissolved oxygen had become critical. Low rates of ammonia breakthrough out of the reconditioning filters and lack of problems with other chemical variables indicated that only aeration was required to achieve a larger biomass load.

Maximum production and carrying capacity of the experimental system were not determined. However, the vertical raceway was loaded with 156.4 kg/m³ fish biomass for 11 days during the final phase. On the 11th day, biomass had increased to 162.0 kg/m³. The system was monitored intensively during the 11 days to detect effects of the large biomass load. Physical and chemical parameters remained within ranges of values which occurred during loadings of ≤64.2 kg/m³ except for flow rate.

The flow rate varied from 12.5 to 45.9 minutes per raceway volume because of a buildup of microflora in tubes, which

114

resulted from increased loads of fecal material and other metabolic products, and daily adjustments. No stress occurred in the fishes. Thus, the water reconditioning system and the spiral flow pattern are effective in maintaining adequate water quality at the slower rates of flow.

Implications

The experiment was continued for 15 months in order to discover and correct any problems that might result from continuous operation over a long period of time. The primary reason was the high rate of failure of large-scale high-density-and-biomass fish production systems based on recycled-water-use regimes and a resultant skepticism toward these systems created in technical field people and commercial growers. No new problems occurred after the first seven months, confirming the reliability of the new concepts and making the vertical raceway system with ammonia stripping and recycled water an efficient system for production of channel catfish, and probably for rainbow trout by including a reliable temperature control system.

Acknowledgments

We thank Dr. Willie P. Isaacs, Professor of Civil Engineering, New Mexico State University, for materials used in the experimental fish cultural system and for suggestions for design and operation of the system. Engineering technicians Tommy Black and Don Richardson fabricated the system.

References

APHA (AMERICAN PUBLIC HEALTH ASSOCIATION). 1965. Standard methods for the examination of water and wastewater. Twelfth edition. American Public Health Association, New York.

BURROWS, R.E. 1964. Effects of accumulated excretory products on hatchery-reared salmonids. Bur. Sport Fish. Wildl. Res. Rep. 66. U.S. Fish Wildl. Serv., Washington, D.C.

BUSS, K., D.R. GRAFF, AND E.R. MILLER. 1970. Trout culture in vertical units. Prog. Fish-Cult. 32(4):187-191.

CHESSNESS, J.L., J.L. STEVENS, AND T.K. HILL. 1972. Gravity flow aerators for raceway culture systms. Agric. Exp. Sta. Res. Rep. 137. Univ. Georgia, Athens.

DEAN, R.B. 1971. Removal of ammonia nitrogen by selective ion exchange. Advanced waste treatment and water use symposium, Dallas, Texas. Pap. 5. U.S. Environ. Prot. Agency, Washington, D.C.

KNEPP, G.L., AND G.F. ARKIN. 1973. Ammonia toxicity levels and nitrate tolerance of channel catfish. Prog. Fish-Cult. 35(4):221-224.

LEITRITZ, E., AND R.C. LEWIS. 1976. Circular ponds. Page 164 in Trout and salmon culture. Calif. Dep. Fish Game, Fish Bull. 164.

MOODY, T.M., AND R.N. McCLESKEY. 1978. Vertical raceways for production of rainbow trout. New Mexico Dep. Game Fish. Bull. 17.

SLONE, W.J. 1975. A new closed vertical raceway fish culture system containing clinoptilolite as an ammonia stripper. M.S. thesis. New Mexico St. Univ., Las Cruces, N.M.

SPOTTE, S.H. 1970. Fish and invertebrate culture: Water management in closed systems. John Wiley and Sons, Inc., New York.

Bio-Engineering Symposium for Fish Culture (FCS Publ. 1):116-127

Management and Cost Implications in Recirculating Water Systems

JAMES F. MUIR

Institute of Aquaculture
University of Stirling
Stirling, Scotland, U.K.

ABSTRACT

Design requirements of recirculated water systems are assessed to develop probable costs for fish production. Effects of possible benefits in reduced risk, disease control, feeding efficiency, and stock management are also determined. Open-flow and recirculated systems are compared on these bases and the areas where advantage is obtained by recirculating water are identified. On an overall basis, the influence of energy costs is great and recycling is unlikely to offer advantages in full ongrowing. Operating costs at similar production rates are higher by 44% and 69% at 3-m pumping head and 20% and 40% at 1-m pumping head at energy costs of 5¢/kWh and 10¢/kWh, respectively, in the model developed. Heating costs, in spite of increased growth rate, are unlikely to be justifiable unless producing a higher-value product.

There is at present a potential for high-value products though the existence of lower-cost alternatives increases the risk of competition. The use of recycled systems for early rearing appears to be feasible, if advantages of low-cost ongrowing or shortened ongrowing time are available.

The greatest improvement in costs of recycled systems is likely to be in pumping costs, representing up to 25% of operating cost, while reduced capital costs, brought about by improved system design, affect only 3-10% of operating cost.

Recirculated water systems have received considerable interest in intensive fish culture, and a number of advantages have been attributed to their use, such as water and heat conservation, environmental control, disease control, stock management, and freedom from normal site limitations. There are at the same time additional outlays in capital and operating cost and level of management to provide for treatment, water movement, and the increased complexity of the system. It is therefore useful to consider how effectively recirculated water systems can be expected to perform and what role they are likely to play in future developments in aquaculture, where constraints in costs and energy input are increasingly apparent.

Details of a number of water reuse designs have appeared in the literature and analyses of systems and comparisons have been provided. (Liao & Mayo 1972; Muir 1978; Wheaton 1977). The basic concepts are illustrated in Figures 1 and 2 while Figure 3 illustrates the ratios of holding volume and treatment volume for given stock weights.

System Design

Design of basic requirements depends, as does that of open-flow systems, on the environmental requirements of stock held, levels of wastes produced, together with the value and operation of the treatment system employed. Relationships for waste production (Liao 1970; Muir 1978; Forster et al. 1977; Wheaton 1977), environmental requirements (Liao 1971; Liao and Mayo 1973; Mayer and Kramer 1973), and integrated design methods (Haskell 1955; Piper 1970; Westers 1970; Liao and Mayo 1972; Speece 1973) have been reported in the literature, factors such as oxygen, ammonia, carbon dioxide, suspended solids, flowrate, and crowding being considered.

More recently information is being developed on long-term effects of these factors; rather than using arbitrary standards based on LC50 results, measurements of

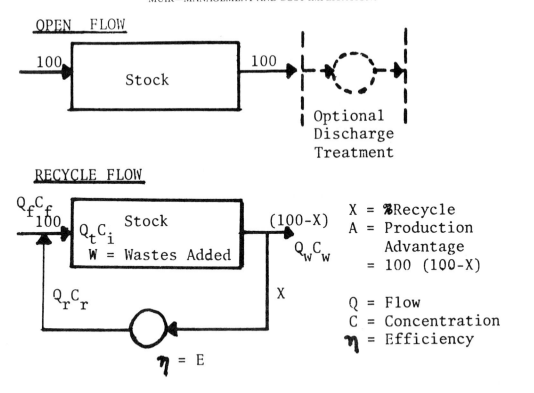

OPEN FLOW

RECYCLE FLOW

X = %Recycle
A = Production
    Advantage
  = 100 (100-X)

Q = Flow
C = Concentration
η = Efficiency

$$C_i = \frac{XW(1-E)}{1+EX-X} \quad : \quad C_w = \frac{W}{1+EX-X} \quad : \quad C_r = \frac{W(1-E)}{1+EX-X}$$

FIGURE 1. — Open and closed systems.

EC50 on growth rates obtained give more appropriate and economically measurable indicators of environmental effects.

The combined effects of several environmental factors have as yet been inadequately determined; it may be useful to consider a multiplicative threshold value, or matrix of values, to express possible combining effects (Muir, ms). Although disease is often associated with suboptimal conditions (Roberts 1979), there is as yet little reported evidence in measurable terms of the additional risk involved.

Typical waste productions are shown in Figure 4, and outlet levels and environmental requirements summarized in Table 1. It is important to note, however, the effect of diurnal activity, feeding regime, and specific dynamic action, from which it is apparent that waste levels may fluctuate considerably (Elliot 1969; Muir 1978; Nava 1977) and that in fully con-

trolled conditions it may be possible to tailor water flow and treatment to these variations, rather than designing for peak

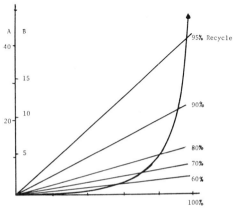

A Advantage in production relative to percent recycle
B Advantage for given percent recycle relative to treatment efficiency.

FIGURE 2. — Production in recycled systems.

117

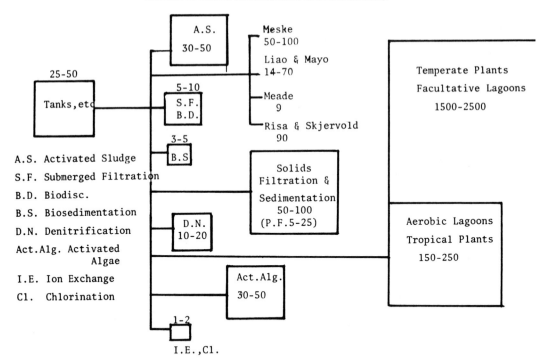

FIGURE 3. — Systems and treatment sizes.

loads. Where limited measurements are made, designs are normally based on averaged waste output figures. Typically, where expected metabolite levels are calculated on the basis of design flowrate and inflow water quality, and are compared with limiting environmental requirements, a system of priorities for treatment design can be developed (Fig. 5).

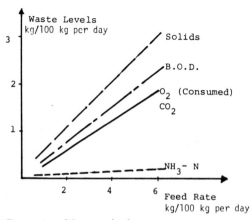

FIGURE 4. — Waste production.

System Performance

The effectiveness and hence design size and cost of individual components depends closely on the design of other components, on overall layout, and on species and husbandry conditions adapted. Using the equations of Figure 1 it is possible to identify the effect of treatment efficiency on overall design in providing for given environmental requirements and degree of recycle.

Information on unit design has until recently derived largely from domestic water and waste treatment practice, where nature and quantity of treated parameters can differ markedly from those encountered in fish culture. Designs can be evaluated on the basis of dependence on flow, concentration, or other factors. Thus biological filtration is dependent both on flow and concentration of substrate, in addition to temperature and presence of inhibitors, and as such is affected by design parameters in a different way from a physico-chemical treatment such as ion-exchange. It is notable that most designed

TABLE 1. — Outlet levels and environmental requirements.

Water quality factor	Typical production from fish tank	Typical acceptable levels for culture	Typical acceptable levels for discharge
Ammonia (NH_3)	0.5 mg/L (NH_3-N)	0.1 mg/L	1-5 mg/L
Nitrites (NO_2)	Negligible	0.1 mg/L NO_2-N	1-3 mg/L
Nitrates (NO_3)	1.5 mg/L NO_3-N	100 mg/L NO_3-N	1-5 mg/L
Phosphate	0.1 mg/L PO_4-P	Not defined	1-10 mg/L
Oxygen	2-5 mg/L	More than 5 mg/L	<5 mg/L
pH	Reduced (more acid)	6-8	6-8
Biological oxygen demand (BOD)	5.5 mg/L	(Not defined) Less than 20 mg/L	10-30 mg/L
Suspended solids	7 mg/L	(Not defined) Less than 20 mg/L	10-30 mg/L
Bacteria or other pathogen levels	Variable	Minimize disease agents	10-30 mg/L

recycle systems are considerably more conservative than loadings indicate.

Comparison with Open Systems

The construction of a recycled water system will involve greater costs than an open system of equivalent capacity. In Figure 3 it can be seen that on volumetric terms solids removal, biological treatment, and fine filtration can represent respectively an additional 20-200% of the basic holding volume. A pro-rata increase in capital cost is likely to be involved, and media, chemicals, and control systems are likely to add further costs.

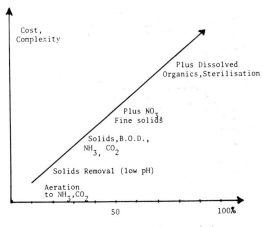

FIGURE 5. — Degree of recycle and system design.

Work by Collins et al. (1975, 1976), Levine and Meade (1976) have indicated the problem of disease treatment in biological systems. The need to maintain multiple units of treatment systems and to isolate affected sections of the system implies the use of large numbers of smaller systems rather than single large units, thus reducing scale economies.

The possibilities of heating water in recycled systems may additionally involve the use of an enclosed and insulated structure together with a source of heat, and so increase capital cost, although operating costs of heated water is naturally significantly less than in open systems.

Operating costs are difficult to extrapolate from those in traditional water and waste treatment, dealing with different water qualities and design requirements, though costs for physicochemical processes are more easily established, as their effect is measurable in discrete units, relatively independent of treatment volume or concentration.

More importantly, the movement of water implies a considerable operating cost, dependent on the head of water to be recovered in returning the water to the holding tank inflow and whether water is moved by air or water. In this context, the use of air as an intermediate stage in inten-

TABLE 2. — Advantages of water reuse.

Close environmental control
Disease control
Freedom from site constraints
High security
Controlled feed rate
Management of production
High temperature operation

sifying production can be compared with movement of water on an energy basis, when it is normally found to be more economical.

Comparative Advantages

Suggested advantages of recirculated water systems are shown in Table 2. The relative economic advantages will depend considerably on the contribution individual factors will have in the overall economics of fish production. Thus, McFarlane and Varley (1976) note that disease contributes approximately 10% of losses in fish culture, and thus there may be economic benefits if disease is eliminated. On a risk basis, however, the possibility of contamination by disease organisms, once within the system, must be balanced against this possible advantage.

The same authors noted husbandry and management to cause more than 40% of losses in their survey and so the possibility of close environmental control has apparent advantages. Again, the attendant risks in complex systems are greater and recycled systems may lack inherent stability and so increase risk in poorly managed conditions. It must be noted that both disease and husbandry losses are considerably reduced by good management in any system, and so reduced risk in these cannot fully be claimed by either system. It is interesting to consider that Varley (1977) and Lewis (1979) (Fig. 6), in studying economics of trout culture, both note a considerable variation in profitability, often only attributable to management differences. At an extreme case, an averaged production from a poorly managed system could be perhaps 50% of that from a well-managed system of similar configuration and operating inputs.

Matters of security and freedom from site constraints are likely to be of more measurable economic significance, where unaccounted losses in fish culture operations may reach 10% in open site areas (Varley, personal communication) and freedom from deliberate or accidental contamination is correspondingly greater in totally enclosed systems. Losses in more compact, highly intensive open-flow systems are, however, likely to be similar to those in enclosed systems. Site constraint value may be expressed first in relative transport costs between suitable open-flow sites distant from centres of population, and recycled systems near market centres. This may be offset, however, by the relatively high costs of land in urban areas. At extremes, however, these factors may be indicated by attributing zero transport cost to recycled systems in suitable locations.

A further area of interest lies in multi-stage fish culture (e.g., of anadromous species), where otherwise marginal sites may be rendered suitable for ongrowing by the use of recycled water for early stages in the same locality. Examples of this may be found in salmon production in Norway (Edwards 1978) and the northeast American Atlantic Coast.

Benefits of controlled feeding rate and closer management and production scheduling are shared to some extent with intensive open-flow systems, and are

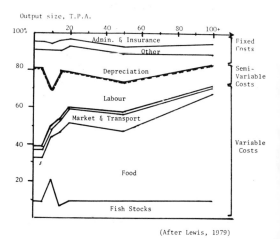

(After Lewis, 1979)

FIGURE 6. — Energy costs, pumping and aeration.

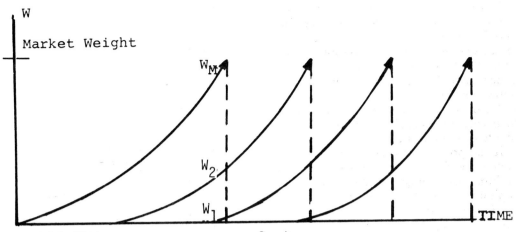

W_M = Market Weight
T_M = Time to attain W_M (months)
C   = Crops/yr
F   = No. of crops held for C crops/yr
      $(C \cdot T_M)/12$

Capacity
= EW = Sum of crop weights held =
  $(W_1 + W_2 + \ldots \ldots W_M)$
= $W_{AV} \cdot F$, (W_{AV} = Average weight of crops held)

Production
= CW_M ∴ Production : Capacity ratio
= $CW_M/$ EW = $CW_M/W_{AV} \cdot F)$ = $12 \, W_M/(W_{AV} \cdot T_M)$

FIGURE 7. — Multiple grouping, production/
   capacity ratios.

usually greater in comparison with cage or open-pond systems. These benefits can be relatively easily assessed on the basis of food cost contribution and production relative to holding capacity, but are generally much more significant when temperature control is available, in that with a known (and possibly optimised) temperature regime, a greater control over feeding rate and growth results is possible.

Improvements in food utilisation can be considered in terms of food conversion ratio, while growth results can be expressed in production:capacity ratio (P:C), based on growth rate:temperature relationships. Where these are known, theoretical maximum P:C ratios can be calculated using the analysis in Figure 7 (Muir 1977). Thus, for rainbow trout, it can be seen that a P:C ratio of 1.5-1.7 (Shepherd 1972), based on typical water flows, may be raised to a theoretical maximum of about 8 at a constant 16 °C. In practice, however, the need for grading, provision of separate holding areas, and imperfections in size distribution and market supply may reduce this to about 4.

As with production rate, temperature control may also confer significant benefits in producing early growth stages, where appropriate production timing may allow fullest advantage from seasonal growth in open systems and where risks in early feeding may be considerably reduced. The latter may not be highly significant, losses typically being below 30% of usually a relatively low-value stage, but growth programming may reduce ongrowing time by up to 50% if a single season can be used to reach market size, and so the entire capital cost of ongrowing can be proportionately distributed to the cost of a recycled system for early rearing.

Economic Assessments

Analyses in economic and energy terms have been applied to a number of intensive fish culture systems (Edwardson 1976; Varley 1977; Lewis 1979). A bibliography of cost information, with reference to water reuse, has also been prepared (Parker and Broussard 1977). Because most information on intensive systems has been developed on salmonids, it is appropriate to use cost distributions for their production as typifying relationships to be expected in recycled systems and modify accordingly for other species.

Distributing between capital and operating cost (variable, semi-variable, and fixed) shows the relative contribution of factors in intensive salmonid culture (Fig. 8). Costs of tank-based units are generally higher ($1574.54/t produced or approximately $2360/t capacity at 1979 U.K. prices [Lewis, 1979]), though costs in all cases are notably variant between individual systems and sizes.

It is possible to compare recycled systems with open-flow systems by examining the range of likely additional capital and operating costs and evaluating these against expected benefits. Additional costs can to some extent be related to degree of recycling (Fig. 5). Thus, where capital costs are set at a range above those for conventional systems, a graph as in Figure 8 can be developed, relating capital costs per tonne with additional overall cost and the expected increase in production/capacity ratio. The use of this ratio may also be extended to reflect any reduced losses to disease or poor security, in addition to gains obtainable through heating and insulation.

The cost of insulated buildings may be based on expected area per tonne capacity (approximately 70 m²). Costed at appropriate rates ($60-$200/m² for simple timber buildings at current U.K. rates) and added to capital cost according to expected production/capacity ratio at the higher temperature, this represents an extra 100% on open system capital costs.

Assessment of the effect on operating costs will depend on the method of depreciation employed and the distribution of depreciation periods. For simplicity it can be assumed that facilities in recycled systems will depreciate in a similar manner to those in an open system, and so depreciation costs can be adjusted directly as capital costs vary.

Cost of labour, and food, can be adjusted according to the expected changes; labour requirement is likely to be approximately similar, as benefits of increased production may be offset by increased maintenance of treatment facilities. Economies of scale in labour requirements may be used as a guide to the maximum

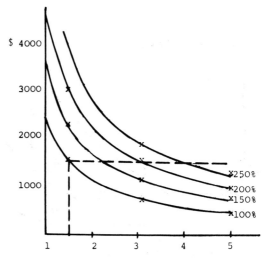

FIGURE 8. — Capital costs and production capacity (costs $/t).

benefit likely; Varley (1978) rates labour cost to reduce to 43% of the 20 tonnes per annum (tpa) level at 100 tpa, and this figure may be taken as a maximum reduction.

Cost of fish stocks, for ongrowing operations, may be reduced if recycled systems offer less disease risk. Mortalities over a complete production cycle are usually taken as 10-20%, or for ongrowing 5-10%. Assuming for example that losses in a well-controlled system are 3-5%, costs of fish stocks are reduced by perhaps 5%.

If a case can be made for risk reduction, the effect is most likely to be felt on level of return required for investment, which may be reduced from the 30-40% demanded of higher risk investments to basic commercial rates of 15-25%. A reduction in overall costs may be obtained, but this must be balanced against the increased overall capital cost tied up in facilities that may not be easily realisable.

Food costs can be adjusted on a conversion rate basis; quoted figures normally reflect an average conversion of 1.4-1.6; an improved conversion due to efficient feeding may be taken at 1.2-1.4, giving a maximum reduction to 80-90% of normal costs. This is also a convenient form of expressing reduced disease losses, where attainable.

Marketing and transport costs typically

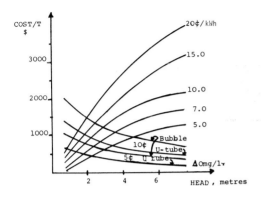

FIGURE 9. — Energy costs, pumping aeration.

At present energy prices, pumping costs will be between $100 and $1000 per tonne capacity, depending on design head. Energy costs for pumping and aeration are given in Figure 9, shown relative to energy prices. Similarly, heating costs can be assessed on the basis of temperature difference and degree of recirculation. Temperature difference may often in turn affect the production/capacity ratio, in which case the graph in Figure 10 can be modified as shown.

A recirculated system may thus be evaluated against an equivalent open system. It can be seen immediately that the major contributions to cost are variable or semi-variable costs, which will be closely related to production, and so benefits in terms of production may be minimal. Taking together the factors outlined it is convenient to compare their effect on a production/capacity basis, where food and labour costs decrease, while capital depreciation

represent some 5% of total operating costs (Varley 1977). In the extreme case of recycling facilities being perfectly placed for distribution, these can be reduced to zero. Against this, however, may be placed increased land and water costs.

The greatest difference between systems may occur in pumping and heating costs.

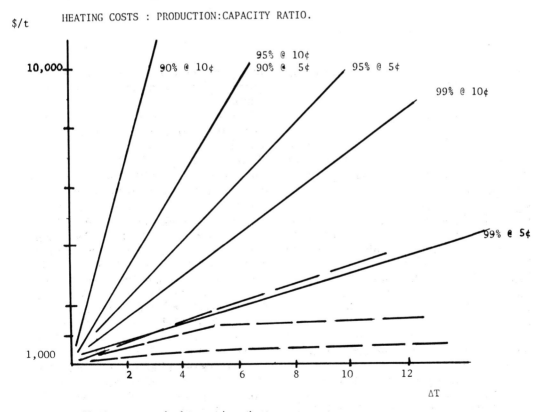

FIGURE 10. — Heating costs: production:capacity ratio.

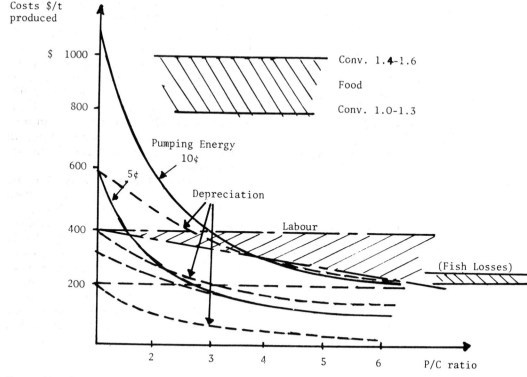

FIGURE 11. — Recycling and costs of production.

costs, heating, and pumping tend to increase (Fig. 11).

Discussion

The effect of increased cost of energy is particularly apparent and it can be seen that at both energy levels costs of pumping (and heating, where required) can be significantly greater than the possible savings in food and labour and of fish stocks, and that market/distribution costs are relatively insignificant. This is supported by Edwardson's (1976) analysis, in which fuel contributes 86.9% of total energy inputs in a recycled system, and totals in GJ/t are 15 times as high as those in flowing-water intensive systems.

The main method by which recycled systems are likely to compete are in increasing the P:C ratio, where the reduced capital and energy costs per tonne produced start to compete with conventional costs. It is interesting to note, however, the comparative costs of an available natural or heated effluent supply to an open system, giving comparable P:C opportunities.

The total costs for a particular set of circumstances are illustrated in Figure 12. This is shown at two energy costs, and allows for the spread of labour, food and fish stock costs. On the basis shown (capital costs double those of open-flow systems), lowest recycled costs only approach highest open flow cost at P:C ratios of 2.0 and 3.2 for 5¢ and 10¢/kWh respectively, while the cross-over point for average costs is not attained at either energy cost. The operating cost of heating must be added to this where necessary; at 99% recycle and 5°C temperature rise, this would typically reach $400 and $800 per tonne at 5¢ and 10¢/kWh respectively. Thus, it can be seen that in most circumstances the additional cost of a recycling system is likely to exceed considerably those of open-flow systems, even where supposed advantages, which are by no means well proven, are included. For the

124

production of rainbow trout, on which much of this analysis is based, it is clear that the marginal returns currently available are unlikely to justify its use.

Pumping costs have considerable effect on the overall operating cost of recycled systems. Means to reduce head loss are likely to be important in reducing cost. It may be difficult because of pipe, valve, fitting, and filter media resistance to reduce head loss much below 1 meter, in which case pumping represents some $200 and $400/tonne capacity at 5ᶜ and 10ᶜ/kWh. At this level, lowest recycle costs meet highest open-flow costs at P:C ratios of 1.5 and 2.0, respectively. This suggests that if the full advantages in food and labour costs are attained, the operation may just compete at 5ᶜ/kWh.

For a non-heated system, operating at a conventional P:C ratio of approximately 1.5, the case in Figure 12 suggests an increased average operating cost of 44% and 69% at 5ᶜ and 10ᶜ/kWh, respectively, and at 20% and 40% if a 1-m pumping head is attainable. This may be used to identify those types of culture (i.e., of higher value species) that may be considered. Such species at present might include Atlantic salmon, *Salmo salar*; eel, *Anguilla anguilla*; sole, *Solea solea*; and turbot, *Scophthalmus maximus* — though competition with open-flow technologies where margins are considerably higher may prevail against the relatively high capital investment involved. Where some form of water treatment is already required on an open-flow system, the overall costs are likely to approach the higher level shown in Figure 12, and the average or low recycle costs may be compared directly. In this even the low recycle costs at 1-m pumping head may just start

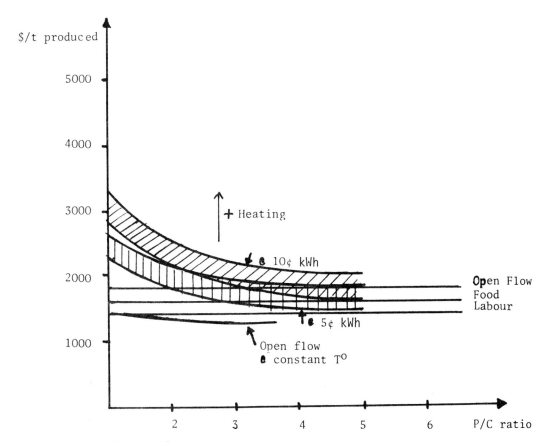

FIGURE 12. — Recycling and total costs.

to compete, though this neglects possible savings in food and labour in open-flow systems.

In the case where use of recycling makes available a cheaper form of ongrowing (e.g., water re-use for salmon smolt production for ongrowing in sea cages), its use may be considered on the contribution to total cost of the final product. In the example given, the weight of young fish is some 2-3% of the crop weight, and so addition to overall cost is likely to be negligible. Against the additional cost of recycling, however, must be laid the alternative of transporting the young fish from elsewhere.

In the instance where disease risks in fry production are unacceptable in open-flow systems, the value of recycled systems may be assessed on the margin in overall costs available for fry supply and on the alternative cost of providing sterilisation facilities sufficient for treatment of an open-flow system. If UV treatment is to be employed, pre-filtration may be necessary for satisfactory operation, and so capital costs are likely to approach those for closed systems. Thus, the alternatives may be compared largely on pumping costs.

Where use of heated water in the early stages reduces the overall time period for ongrowing, in highly seasonal production, the cost of heating and recycling may be balanced by the shorter use of ongrowing facilities (i.e., increased P:C ratio in ongrowing). The effects of increased P:C ratio are similar to those shown in Figure 12 for constant temperature open-flow systems, in which a doubling of P:C ratio reduces total operating cost by some 10%. The size to which early stages must be grown will depend on the species and location, though it is probable that the weight fraction of the total crop will be higher than in the case of salmon smolt production. Assuming a P:C ratio of 1 in the production of early stages, and heating costs of $400 and $800/t at 5¢ and 10¢/kWh, respectively, the maximum weight fraction of early stock in recycled water in the case above is 12.3% and 7.3% at the two energy costs, and 16.0% and 8.9% in a 1-meter head system.

Conclusions

Although much of this analysis is based on the relatively well-defined production costs of rainbow trout, the relationships shown are unlikely to differ significantly with other species in that cost requirements are likely to produce similar effects. At the present stage of development most differences occur in higher capital, food, and labour costs, and the effects of recycling will be similar to those outlined earlier.

It is clear that the importance of energy costs, 20-40% of operating costs in the cases shown, stresses the need to reduce pumping and heating costs to a minimum and the design of low head-loss systems using waste heat is indicated. The effects of capital cost (3-10% in the cases shown) and hence improvements in treatment system efficiency are less significant; given the preference for operating a number of small systems for risk reduction, the addition of 50% to conventional capital costs is unlikely to be significantly improved upon.

Given the diversity of present fish culture operations there is certainly an area in which recycled water systems can contribute to fish production. In the present climate of energy cost increases, it is unlikely, however, that recycled systems will gain acceptance for complete production cycles. In view of the uncertainty of future prices it is difficult to identify clear advantages in recycled water systems sufficient to outweigh the higher capital costs involved. Although the present high value of certain species makes their use currently feasible, the availability of lower-cost alternatives places the market position in some jeopardy.

References

COLLINS, M.T., J.B., GRATZEK, D.L. DAWE, AND T.G. NEMETZ. 1975. Effects of parasiticides on nitrification. J. Fish. Res. Board Can. 32:2033-2037.

————. 1976. Effects of antibacterial agents on nitrification in an aquatic recirculating system. J. Fish. Res. Board Can. 33:215-218.

EDWARDS, D.J. 1978. Salmon and trout farming in Norway. Fishing News (Books), London. 195 pp.

EDWARDSON, W. 1976. Fish farming. Report Number 6. Systems Analysis Research Unit, Univ. Strathclyde, Glasgow, Scotland.

ELLIOTT, J.W. 1969. Oxygen requirements of chinook salmon. Prog. Fish-Cult. 31(2):67-73.

FORSTER, J.R., J.P. HARMON, AND G.R. SMART. 1977. Water economy: Its effect on trout production. Fish Farming Int. 4(1):11-13.

HASKELL, D.C. 1955. Weight of fish per cubic foot of water in hatchery troughs and ponds. Prog. Fish-Cult. 17(3):117-118.

LEVINE G., AND T.L. MEADE. 1976. The effects of disease treatment on nitrification in closed system aquaculture. Proc. 7th Annual Meeting of the World Mariculture Society.

LEWIS, M.R. 1979. Fish farming in Great Britain: An economic survey with special reference to rainbow trout. University Reading Dep. Agric. Conserv. Manage. Misc. Study No. 67.

LIAO, P.G. 1970. Pollution potential of salmonid fish hatcheries. Water Sewage Works 117(8):291-297.

LIAO, P.G., AND R.D. MAYO.1972. Salmonid hatchery water reuse systems. Aquaculture 1:317-335.

_____. 1974. Intensified fish culture combining water reconditioning with pollution abatement. Aquaculture 3:61-85.

MAYER, F.L., AND R.H. KRAMER. 1973. Effects of hatchery water re-use on rainbow trout metabolism. Prog. Fish-Cult. 35(1)9-11.

McFARLAND, R.S., AND R.L. VARLEY. 1976. Risks, mortality and insurance on European trout farms. FAO Technical Conference on Aquaculture, Kyoto, Japan. AQ/CONF/76/E-8.

MUIR, J.F. 1978. Aspects of water treatment and re-use in intensive fish culture. Ph.D. thesis. Strathclyde Univ., Glasgow, Scotland.

_____. The use of threshold value for combined environmental effects on fish. ms.

PARKER, N.C., AND M.C. BROUSSARD. 1977. Selected bibliography of water re-use systems for aquaculture. Texas Agric. Exper. Sta., College Station, Texas. 33 pp.

PIPER, R.G. 1970. Know the proper carrying capacity of your farm. American Fishes and U.S. Trout News (May-June):4-7.

ROBERTS, R.J. 1979. Fish pathology. Bailliere Tindall, London. 328 pp.

SHEPHERD, C.J. 1973. Studies on the biological and economic factors involved in fish culture with special reference to Scotland. Ph.D. thesis. Univ. Stirling, Stirling, Scotland.

SPEECE, R.E. 1973. Trout metabolism characteristics and rational design of nitrification facilities for water reuse in hatcheries. Trans. Am. Fish. Soc. 103(2):323-334.

VARLEY, R.L. 1977. Economics of fish farming in the U.K. Fish Farming Int. 4(1):17-19.

WESTERS, H. 1970. Carrying capacity of salmonid hatcheries. Prog. Fish-Cult. 32(1):43-46.

WHEATON, F.R. 1977. Aquacultural engineering. Wiley-Interscience, New York. 708 pp.

127

Bio-Engineering Symposium for Fish Culture (FCS Publ. 1):128-130

SESSION IV INSTRUMENTATION AND AUTOMATION IN FISH CULTURE

Measurement and Control of Hatchery Functions

K.B. Jefferts

Northwest Marine Technology, Inc.
Shaw Island, Washington 98286

ABSTRACT

Hatchery operation is examined from a process control viewpoint. A number of variables affect hatchery output. Generation of a specific response by a hatchery requires control of those variables, and control implies measurement. It is not clear that adequate measuring techniques exist for all variables.

Introduction

One can view a fish hatchery as a factory, having as input raw material fertilized eggs, water, and feed. The output from this factory is fish, hopefully in groups of specific size, on specific dates, and in a specific condition of maturity or size.

If this view were in fact entirely correct, hatcheries might well be managed by process control engineers. On the other hand, it appears to contain enough truth to be worth pursuing.

At this point I need to qualify my remarks slightly. They are extracted from the context of Pacific Salmon Hatchery Management, but I think you will find the basic points broadly applicable.

Bergman, Johnson, and Rasch (1975) have pointed out the economic importance of some systematic approach to management of Pacific salmon hatcheries. Broadly speaking, the replacement cost of West Coast salmon hatcheries, in aggregate, is a significant fraction of a billion dollars, and the annual operating expense is perhaps 10% of the replacement cost. From our point of view, we have a valuable and expensive set of salmon factories to consider.

The basic point made by Bergman, Johnson, and Rasch, and since, adequately demonstrated, is that systematic analysis and planning of hatchery operations can provide impressive improvements in hatchery efficiency. Implicit in the scheme is the assumption that a hatchery is capable, within limits, of delivering specific outputs, groups consisting of specific numbers of fish, of specific size, on specific dates. The basic means for accomplishing this result is essentially ration control.

Nevertheless, let me observe that, even with this kind of elegant management, hatcheries often provide disappointing performance. One can view that fact as simply being due to the perversity of nature but that may not be entirely true.

Present Hatchery Management

In simple terms, I assert that hatchery management at present is largely an "open loop" process. By that I mean that management proceeds from an estimated input of fertilized eggs to a presumed output, with little process control information between. A review of primary and process variables, with some comment on measurement methods and crude confidence limits, will illustrate the point (Table 1).

If one is to take the table at face value, it is apparent that the outstanding difficulty lies with number determinations. The immediate question, however, is "How did I assign the probable accuracy?" and perhaps "What do I mean by 'probable accuracy'?"

I assigned estimated values for the accuracy of these determinations, in some

TABLE 1. — Hatchery process variables.

Variable		Measurement Method	Estimated Accuracy — Production Hatchery	Conclusion
FERTILIZED EGGS		Weight / Volume + Sample	±5%	Satisfactory
FEED	Quant.	Weight	±5%	Satisfactory
	Qual.	Assay / Hope	?	Usually Satisfactory
WATER	Quant.	Various	±5%	Satisfactory
	Qual.	Various / Hope	?	·Usually Satisfactory
N_i (t) (Survival Rates)		Wet Weight + Sample, Mark / Recapture	±50%	Unsatisfactory
W_i (t) (Growth Rates)		Sample	±20%	Probably Satisfactory
N_i (t$_f$) Release Numbers		Wet Weight + Sample Mark / Recapture	±50%	Unsatisfactory
W_i (t$_f$) Release Size		Sample	±20%	Probably Satisfactory

small part on the basis of personal observation, but primarily through conversation with many salmon hatchery technologists. I emphasize that the estimate is pertinent for production hatcheries, and not for experimental circumstances, where careful determinations can be made. Furthermore, I mean by "accuracy" something much stronger than standard deviation; more like the range within which 90% of determinations will lie.

In addition, the physical difficulty and fish mortality associated with the usual procedure imply that these estimates are infrequent as well as inaccurate.

Assuming that my pessimistic point of view is at least somewhat true, the implications for the hatchery manager, and later the fishery manager, are serious. If numbers are in error, so immediately are planned rations, water requirements, and rearing densities. Even if biomass estimates are more nearly true than number estimates — e.g., more fish than estimated in a group but smaller average size — the non-linearity of the growth equations (Stauffer 1973) and mother nature, which they represent, will leave a non-trivial effect.

The final uncertainty in the number released destroys the precision of otherwise effective techniques for evaluating contributions to fisheries — e.g., coded wire tag techniques.

Simply said, it appears that lack of timely, accurate number data, and, to a lesser degree, size data, is the most serious process information problem facing a hatchery manager. Furthermore, it appears that the problem has consequences that are at least serious if not sometimes worse.

Conclusions

The problem is, as usual, easier to define than is the solution. However, in the recent past, two different devices have appeared that should become useful tools in a broad attack on this problem. They are, respectively, a device produced by Bob Foster of the Washington Department of Fisheries for the purpose of extracting reliable random samples from a pumped stream of fish, and a multi-channel counter, designed to count fish in flowing water, produced by Northwest Marine Technology, Inc. The fact that I have been involved with the generation of both of them will, I hope, be attributed to my genuine concern over the problem.

I do not intend to describe these devices in detail here, but note only that, taken singly or in combination, they appear to offer hope for improving the precision of number and size determinations. On the other hand, I believe that the general problem is common to many kinds of hatcheries in widely differing circumstances, and is going to require more ingenuity and effort before it is adequately controlled.

References

BERGMAN, P.K., F.C. JOHNSON, AND A.J. RASCH. A systematic approach to salmon hatchery management. Washington Department of Fisheries.

STAUFFER, G.D. 1973. A growth model for salmonids reared in a hatchery environment. Ph.D. dissertation. Univ. Washington.

Bio-Engineering Symposium for Fish Culture (FCS Publ. 1):131-137

An Air-Operated Fish Culture System with Water-Reuse and Subsurface Silos

Nick C. Parker

Southeastern Fish Cultural Laboratory
U.S. Fish and Wildlife Service
Marion, Alabama 36756

ABSTRACT

A water-reuse system with two 42,000-liter subsurface silos, primary and secondary clarifiers (plate separators), and a rotating biological contactor (RBC) with 1394 m² of disc surface has been installed to provide seasonal environmental control for striped bass *(Morone saxatilis)* and priority warmwater fishes. Fish tanks extend 5.8 m into the ground for maximum thermal stabilization. Air-lift pumps remove water from the bottom of the silos and pass it to the filtration system. Gravity flow returns filtered water to the surface of the silo. All components are arranged to be bypassed for maintenance or operational evaluation. The RBC can be operated as a 1-, 2-, 3-, or 4-stage unit to establish performance characteristics for each stage. Flow rates can be varied to establish optimum hydraulic loading of the RBC with aquaculture effluents. Further evaluations are underway to assess performance of the system components and their application to aquaculture.

Preliminary tests with 10 warmwater species indicate that most perform well in the silos when operated as flow-through units. Eight of ten species were readily harvested from the silos with 0.5-2.0 mg/l sodium cyanide; all fish quickly recovered when placed in fresh water. Other harvest techniques include use of lift screens, bag nets, and rod and reel. Striped bass were overwintered in the silos and successfully spawned in the spring.

Introduction

Environmental control is either non-existent or prohibitively expensive in conventional fish culture systems, i.e. ponds and raceways. Innovative fish culture systems with environmental control have received recent research attention (Broussard et al. 1973; Parker and Simco 1973; Calbert and Huh 1976) but these systems have not been successfully adopted by commercial producers. The upflow silo developed by Buss et al. (1970) has been the basis for several aquaculture units with either single-pass water flow (Steinbach 1977) or water reuse capability (Slone et al. 1974; Meade 1974). Silos are attractive because of their small space requirements and the ease with which environmental control can be achieved. Subsurface silos result in lower operational cost for environmental control because of the thermal stabilization provided by the earth (MacDonald 1974).

A recirculating fish culture system consisting of subsurface silos, two primary clarifiers, a rotating biological contactor (RBC), and a secondary clarifier is partially installed at our laboratory. When complete, this system with low energy requirements for pumping, filtration, and thermal control will be operated throughout the year to efficiently produce fish in a controlled environment. Environmental and physiological requirements for optimum growth, production, maturation, and spawning will be evaluated for several species. Responses to chronic stress (i.e. stocking density, low DO, NH_3) associated with intensive culture will be monitored and related to species performance.

System Configuration

Two subsurface 42,000-liter fiberglass fish tanks, silos, 3 m in diameter × 6 m deep, have been installed at our laboratory. Each tank has a concave bottom with a 15-cm-diameter drain in the center. The drain is connected with two 90° elbows to a 15-cm-diameter air-lift pipe located on the outside of the tank (Fig. 1). A 3-m-diameter screen located 30 cm above the

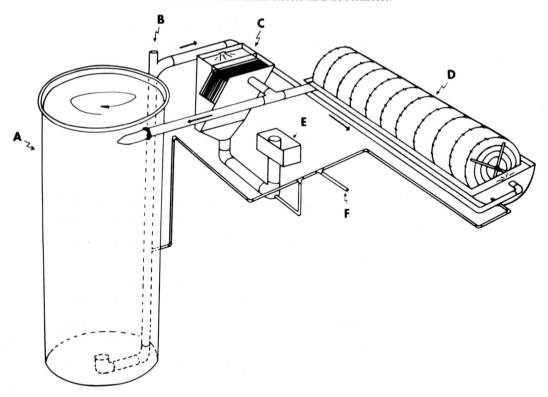

FIGURE 1.—Main components of the water reuse system connected to a subsurface silo. A, silo fish tank; B, air-lift pump; C, plate separator; D, rotating biological contactor; E, sludge hopper; and F, air input. Arrows indicate direction of water flow.

concave bottom keeps fish from entering the drain. Fresh water is added to the top of the tank and recycled water returns through a 15-cm port located in the tank wall.

The silos have been installed with their water levels extending 5.8 m below ground surface to provide temperature stabilization by using the earth as a thermal buffer. Subsurface water would float the empty silos out of the ground; therefore, they can not be drained. For emergency purposes, maintenance, etc. the bottoms of both silos were encircled with perforated sewer pipe placed in a gravel bed and connected to a 6-m-deep sump located between the two tanks. The sump would allow ground water, normally about 1 meter below the top of the silos, to be pumped down if the water level in the silos must be lowered.

Each tank is connected to a primary clarifier (a plate separator) for removal of suspended solids and to the RBC for nitrification of dissolved metabolites (Fig. 2). The secondary clarifier (a third plate separator) removes suspended solids from the RBC effluent before it returns by gravity flow to the tanks. The clarifiers are triangular-shaped boxes (Fig. 3) containing 62 m^2 of internal plate surface. Plates were constructed of 1.20 × 1.23-m corrugated fiberglass sheets spaced 2.7 cm apart. The corrugated sheets were assembled into a unit in which the cross-sectional area for water flow was 0.67 m^2. The plates are designed for operation at either a 45° or 60° angle. Water flow and sludge movement through the plates is from top to bottom. Solids are air lifted from the bottom of the settling chamber below the plates and discharged into a sludge hopper for disposal.

The operating head for water flow to the plate separators, and then to the RBC, is provided by air-lift pumps designed as integral units of the tanks. With an air flow of

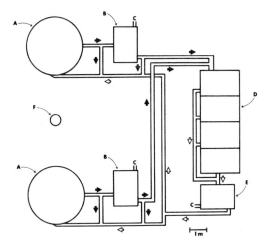

FIGURE 2.—Top view of the water reuse system connected to subsurface silos. A, silo fish tanks; B, primary clarifier (plate separator); C, sludge removal port; D, 4-stage rotating biological contactor (RBC); E, secondary clarifier (plate separator); and F, sump connected to perforated pipes surrounding each silo. Solid arrows indicate direction of flow after the RBC.

Recycled water from the secondary clarifier enters each silo at the water level through a port located in the side 30 cm from the top. Water flow within the tank is from the top to the bottom, out the drain, and back up the air-lift riser beside the tank. Flow patterns in acrylic models of this system have been compared with patterns produced by other tank configurations employing conventional water pumps and an upflow design. Circular flow patterns similar to that observed in models have been verified in the 42,000-l tanks by observing dye distribution from the surface.

Concept

In upflow silos water is discharged against the flat bottom, upwells to the surface and overflows from the top of the tank into a peripheral gutter (Buss et al. 1970). Upflow silos have excellent "plug flow" characteristics but since surface waters are

0.15 m³/min introduced 3 m below water into a 15-cm-diameter PVC pipe, the lifts and flow rates were 7.5 cm, 784 l/min; 23 cm, 485 l/min; and 38 cm, 223 l/min. These flow rates were developed with a 0.33-hp rotary-vane air pump. A 2.7-hp regenative-vane blower will be used when the system is fully operational. A 15-cm-diameter air-lift should produce 1900-3400 l/min with 70% submergence when air delivery is through a 5-cm pipe (Spotte 1970).

The commercial RBC has 1394 m² of disc surface. Discs are 2 m in diameter and arranged in four stages so that operation can be as a 1-, 2-, 3-, or 4-stage unit. Disc rotation at 2 rpm is provided by chain-drive from a 1.5-hp electric motor that simultaneously powers four rotating buckets. The rotating buckets produce a hydraulic lift of 76 cm and insure a uniform influent flow to the first stage of the RBC. The 76-cm lift provides the head pressure necessary for water to flow from the RBC through the secondary clarifier and into the silos. The RBC, a self-contained unit, is equipped with a fiberglass hood to protect microflora from harsh environmental conditions.

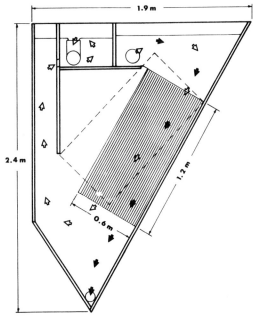

FIGURE 3.—Side view of the clarifier (width = 1.25 m) with a plate separator unit operable at either 45° (dashed lines) or 60° (as shown). Water (open arrows) and particulates (solid arrows) flow down through the 1.2-m plates. Settleable solids are collected and discharged from the bottom while clarified effluent overflows through an adjustable outlet at the surface.

133

discharged all dissolved oxygen that enters the tank must be introduced with the inflowing water. The downflow silo moves water from the surface and down through the tank prior to discharge. Dissolved oxygen picked up at the air-water interface is retained in the tank.

Flow patterns in acrylic models of the downflow tanks did not have the plug-flow characteristic of the upflow silos. "Deadwater" areas were eliminated by directing the inflow to the center of the tank, and circulation patterns were further improved by installing four additional air lifts to move more water from the bottom to the upper center of the tank. In a flow-through system the plug-flow water movement is the most efficient method for removing waste metabolites and introducing fresh water. In a recirculating system the plug-flow design is important as a method for maintaining optimum oxygen profiles. Waste metabolites are not produced uniformly throughout the day but are synchronous with the feeding cycle. The biofilter in this system will process about 7 tank volumes of water per day, but the water cannot be delivered as a plug flow because of the need for aeration at a rate of 27 tank volumes per day. Oxygen is expected to be limiting long before metabolites become a problem. An oxygen generator or liquid oxygen could be used in the system if production loads were extremely high. A U-tube (Speece 1969) could easily be added to the system either to saturate the water with atmospheric air or to supersaturate it efficiently with gaseous oxygen.

Waste material and dead fish are swept to the surface in upflow tanks only when water velocity is sufficiently high. In downflow tanks water velocity is greatest in the riser tube of the air-lift pump. Wasted material drops to the bottom of the tank and is swept by the circular current to the center, where it is picked up by the air-lift pump intake. A disadvantage of the design is retention of dead fish on the screen above the concave bottom. Dead fish will usually float to the surface in one day at the temperature at which the system is to be operated.

The bottom screen can be raised to remove either live fish, or dead fish, from the downflow silo. In the upflow silos a similarly placed screen would probably disrupt the flow pattern. Fish can be harvested from silos with purse seines (Buss et al. 1970) or other trap and netting devices. Removing fish through valved drains located in the bottom of the tank has not proven effective. Shear forces, produced at the drain by the head pressure of the silo water column, damage fish as they pass through the drain (Keen Buss, personal communication). A harvesting technique applicable to both silo designs is the use of chemical agents to bring fish to the surface. A disadvantage of chemical harvesting is that all fish must be removed from the tank at one time; thus, partial harvesting, or cropping, is not possible.

Air-lift pumps were chosen to provide the low lift necessary to operate the system because they simultaneously move and aerate water. Flow rates can easily be adjusted by regulating air volume. At a pumping rate of 2100 l/min the turnover time for the 42,000 l of water in the tank would be once every 20 minutes, or three times per hour. Of this 2100 l/min, 1350 l/min will move through the tank to maintain aeration and 750 l/min will pass through the primary clarifier for solids removal. Of the 750 l/min, 550 l/min will return directly to the tank and 200 l/min will pass through the RBC and secondary clarifier before returning to the tank. At this pumping rate one tank volume will pass through the primary clarifier 26 times per day and through the RBC 7 times per day. Fresh-water inflow has been tentatively set at 3 l/min to produce a 10% exchange per day.

Plate separators were designed into the system for water clarification before and after the RBC. Plate separators, similar in design theory to tube clarifiers, are commonly operated between 45° and 60° from the horizontal; 60° tubes are usually self-cleaning but settling capacity is less than with 45° tubes (Culp and Conley 1970). Therefore, plates in this unit were designed for operation at either 45° or 60° angle. Water flow is typically from the bottom to the top of the plate pack, while solids move in the opposite direction (Souther and Forsell 1971). The upflow system has encountered problems in at least one fish culture application (Bob Piper, personal communication) and the decision was made to operate our plate separators as downflow units. The cross-sectional area of the plate unit is 0.67 m². With the designed flow of 750 l/min in the primary clarifier, water velocity between the plates will be 112 cm/min and particles with settling rates greater than about 3 cm/min will be removed. The secondary clarifier, placed after the RBC, has a design flow of 200 l/min and should remove particles with settling rates greater than 0.8 cm/min.

Chessness et al. (1975) reported settling velocities in rectangular clarifiers equal to or greater than 3.048 cm/min for 40% of the filterable particles from catfish raceways. The microflora attached to the plate surfaces are expected to trap an additional proportion of the suspended solids and result in solid removal greater than 40%.

Preliminary determinations of waste particle settling velocities were made at our laboratory for settleable solids from channel catfish, *Ictalurus punctatus*, and striped bass, *Morone saxatilis,* culture tanks. Solids were collected from the fish tanks when water temperature was 14.5 °C and immediately placed in Imhoff cones. The time for 90% of the solids to settle was recorded and settling rates were calculated. Greater than 90% of the solids from catfish tanks settled at an average rate of 13 cm/min based on four trials; the smallest visible particles settled at 1.7-2.1 cm/min. Waste from striped bass tanks settled at an average rate of 16.7 cm/min in six trials and the smallest visible particles settled at 3.4 cm/min. Based on these observations it is expected that the plate separators will remove at least 90% of the settleable solids.

Biological filtration for removal of dissolved metabolites is provided by the RBC unit. RBC's were first proposed for wastewater treatment in the early 1900's in Germany and are now being adapted to several classes of industrial organic wastes (Antonie 1976). Only in this decade have they been investigated for aquaculture application (Parker and Simco 1973; Neal 1975; Lewis and Buynak 1976; Mock et al. 1977). Optimum hydraulic loading rates have not been established for RBC's receiving aquaculture effluent. The biochemical oxygen demand (BOD) of aquaculture effluents is at least an order of magnitude less than that found in municipal sewage. Lewis and Buynak (1976) used a hydraulic loading of 0.36 l/min·m^{-2} (12.8 gpd/ft²) of disc surface to establish a fish load of 2 kg/m² of disc surface area in a two-stage RBC. With influent from a sewage treatment plant, hydraulic loadings of 0.042, 0.056, and 0.070 l/min·m^{-2} (1.5, 2.0, and 2.5 gpd/ft²) of media surface produced NH$_3$-N removal efficiencies of approximately 94%, 80%, and 53%, respectively (Hao et al. 1975). BOD and NH$_3$-N in waste-water treatment plants usually range from 80 to 250 and 15 to 40 mg/l, respectively, while hydraulic loads range from 0.014 to 0.113 l/min·m^{-2} (0.5 - 4.0 gpd/ft²) of media. In recirculating systems for catfish, BOD has averaged 23 and 40 mg/l when ammonia averaged 2 and 8.3 mg/l (Simco 1976). With BOD and ammonia levels within these ranges the optimum hydraulic loadings from fish culture units are expected to be considerably greater than hydraulic loadings for wastewater treatment plants.

The four-stage RBC in our system is designed so that hydraulic loading and number of stages in operation can be varied to establish optimum nitrification efficiency with aquaculture effluents. Components of the system are arranged so that any component can be bypassed for either testing or maintenance. The silo tanks can both be operated with the RBC, or either tank can be operated as a flow-through unit isolated from the RBC.

The fiberglass tank wall averages 5.6 mm thick and thermal conductivity through the wall to the soil has been calculated using standard equations (Weast 1967). Assuming a water temperature of 16 °C and a soil temperature of 15 °C, the thermal loss through the tank wall would be 1.09×10^6 cal/h (4,334 BTU/h). With a soil density of 1.86 and specific heat of 0.33 the calculated energy required to raise soil temperatures 1 °C within an area of 0.6 m around the tank is 26×10^6 cal (103,000 BTU). Based on these calculations, ground temperature around the tanks and water temperature within the tanks were expected to remain relatively stable.

The mean annual temperature, based on daily measurements over a 3-year period at our laboratory, is 16 °C. Ground temperature on 11 January 1978 was -5.0 °C at the surface, and 6.5, 10.0, 13.0, and 16.3 °C at depths of 0.3, 0.6, 1.3 and 2.7 m, respectively. During winter operation with no well water inflow the water temperature in the silos was as low as 8 °C. The lowest water temperature recorded in the sump located between the two silos was 12.5 °C. Thermal variation with depth was 0.5 °C in both silos when circulation and aeration was provided by a 114 l/min air-lift pump. The reduced temperature in the silo was attributed to heat lost to the atmosphere as cold air was pumped through the air-lift. Waste heat produced by the electric motor and the blower will be captured by placing a shroud over the blower unit and connecting the air blower intake to the shroud. Silo tanks will be covered with transparent air-supported structures to capture solar energy for additional thermal control. Water temperature on 25 June 1979 was 23.5 and 21.0 °C, respectively, in a silo receiving 38 l/min of 21 °C well water and in a silo receiving no well water inflow.

Applications

Striped bass brood fish were overwintered in one silo and successfully spawned in the spring. The silo was operated as a closed system receiving no freshwater inflow and without a filter system. During this period striped bass were fed a pelleted diet supplemented with goldfish forage. Harvesting techniques included removal of individual fish by hook and line and removal of several fish by netting after the screen in the bottom of the tank was raised.

Yellow perch, striped bass, striped bass · white bass hybrids, largemouth bass, smallmouth bass, bluegill sunfish, goldfish, channel catfish, blue catfish, and tilapia have all been harvested from the silo by using sodium cyanide. The catfish, goldfish, and tilapia required cyanide concentrations above 2 mg/l for removal; while yellow perch, striped bass, and hybrid bass could be removed with 0.5 mg/l. Fish were netted as they swam to the surface and immediately placed in fresh water. Sodium cyanide quickly flushed from the gills and all fish survived with no apparent side effects. Striped bass · white bass hybrids are currently being reared in one silo equipped with an automatic feeder and operated as a flow-through system.

Commercial applications of the silo system could probably first be justified for cash crops of high volume. Applications might include fingerling production, brood stock maintenance, and production of marketable size food fish during the off-season. Similar systems could be used by commercial firms and research organizations to produce exotic species in a controlled environment. Endangered species could similarly be produced in this system. Some fish will not perform well in the deep silo configuration. Those that require shallow riffle areas or other special habitats could be cultured in raceways or other tank configurations arranged to receive water from the silo system. Systems such as this will become increasingly important to the aquaculture community as our production demands grow and water and land become more limited.

References

ANTONIE, R.L. 1976. Fixed biological surfaces — Wastewater treatment: The rotating biological contactor. CRC Press, Inc., Cleveland, Ohio. 200 pp.

BROUSSARD, M., N.C. PARKER, AND B.A. SIMCO. 1973. Culture of channel catfish in a high-flow recirculating system. Proc. Southeast. Assoc. Game Fish. Comm. 27:745-750.

BUSS, K., D.R. GRAFF, AND E.R. MILLER. 1970. Trout culture in vertical units. Prog. Fish-Cult. 32(4):187-191.

CALBERT, H.E., AND H.T. HUH. 1976. Raising yellow perch, *Perca flavescens*, under controlled environmental conditions for the upper Midwest market. Proc. World Maricult. Soc. 7:137-144.

CHESSNESS, J.L., W.H. POOLE, AND T.K. HILL. 1975. Settling basin design for raceways fish production systems. Trans. Am. Soc. Agric. Eng. 18(1):159-162.

CULP, G., AND W. CONLEY. 1970. High-rate sedimentation with the tube-clarifier concept. Pages 149-166 *in* Water quality improvement by physical and chemical processes. Vol. 3.

HAO, O., AND G.F. HENDRICKS. 1975. Rotating biological reactors remove nutrients. Water and Sewage Works. Oct.-Nov. Repr.

LEWIS, W.M., AND G.L. BUYNAK. 1976. Evaluation of a revolving plate type biofilter for use in recirculated fish production and holding units. Trans. Am. Fish Soc. 105(6):704-708.

MACDONALD, C.K. 1974. Production cost analysis for intensive salmon culture employing water reuse systems. Univ. R.I. Sea Grant Rep. 21 pp.

MEADE, T.L. 1974. The technology of closed system culture of salmonids. NOAA Sea Grant. Univ. R.I. Mar. Tech. Rep. No. 30. 30 pp.

MOCK, C.R., L.R. ROSS, AND B.R. SALSER. 1977. Design and preliminary evaluation of a closed system for shrimp culture. Proc. World Maricult. Soc. 8:335-369.

PARKER, N.C., AND B.A. SIMCO. 1973. Evaluation of recirculating systems for the culture of channel catfish. Proc. Southeast. Assoc. Game Fish. Comm. 27:474-487.

NEAL, R.A. 1975. Shrimp culture research in controlled environments. Pages 60-66 *in* Proc. Fish Farming Conf. Annu. Conv. Catfish Farmers of Texas, Texas A&M Univ.

SIMCO, B.A. 1976. The reuse of water in commercial raising of catfish: Phase 2. Univ. Tenn. Water Res. Res. Ctr. Res. Rep. No. 52. 66 pp.

SLONE, R.J., P.R. TURNER, AND D.B. JESTER. 1974. A new closed vertical raceway fish culture system containing clinoptilolite as an ammonia stripper. Proc. West. Assoc. Game Fish. Comm. ms. 14 pp.

SOUTHER, G.P., AND B. FORSELL. 1971. The lamella separator: A novel high rate sedimentation device for water treatment plants. Pages 180-182 *in* Proc. 32nd Inter. Water Conf. Eng. Soc. West. Penn.

SPEECE, R.E. 1969. U-tube oxygenation for economical saturation of fish hatchery water. Trans. Am. Fish. Soc. 98(4):789-795.

SPOTTE, S. 1970. Fish and invertebrate culture: Water management in closed systems. John Wiley and Sons, Inc., New York. 145 pp.

STEINBACH, D.W. 1977. An intensive culture system for channel catfish fingerlings. Proc. Southeast. Assoc. Game Fish. Comm. 31:484-492.

WEAST, R.C., ed. 1967. Handbook of chemistry and physics. Forty-eighth edition. Chemical Rubber Co., Cleveland, Ohio. F-65.

Bio-Engineering Symposium for Fish Culture (FCS Publ. 1):138-148

Design of Aeration Systems for Aquaculture

JOHN E. COLT AND GEORGE TCHOBANOGLOUS

Department of Civil Engineering
University of California
Davis, California 95616

ABSTRACT

The maintenance of adequate dissolved oxygen levels in aquatic culture systems is a significant environmental problem. Low dissolved oxygen levels can reduce growth, increase disease problems, and may cause massive mortality. The design of aeration systems for aquaculture based on published mass transfer rates is not valid because of differences in operating conditions. It is the purpose of this paper to present and document the design of aeration systems for aquaculture.

In an aquaculture system, the design of an aeration system depends on a knowledge of the dissolved oxygen requirements of the animals, the oxygen consumption of the animals, and the kinetics of oxygen transfer under the culture conditions. Other important design considerations may include the hydraulic mixing characteristics of the aeration system, noise production, and prevention of supersaturation levels of nitrogen gas (N_2) and carbon dioxide (CO_2). The economics and reliability of various aeration systems under different operating conditions are discussed. In terms of cost and reliability, pure oxygen and U-tube systems may be the most desirable.

Introduction

The maintenance of adequate dissolved oxygen levels in aquaculture is a serious environmental problem. In ponds, low levels of dissolved oxygen are more common during the summer, a period of high growth. In the past, if critical dissolved oxygen levels were expected, the producer could stop feeding or add fresh water to the ponds. Artificial aeration may offer a more effective means of maintaining adequate dissolved oxygen in ponds. In raceways, aeration can be used to increase the carrying capacity during the latter part of the production cycle or may be operated over the whole production cycle.

The design of aeration systems for aquaculture can not be based on published mass transfer rates for waste treatment applications because of significant differences in the objectives and requirements. The purpose of this paper is to examine the design of aeration systems for aquaculture based on operating characteristics, cost, and reliability.

Oxygen in Aquatic Systems

The design of aeration systems for aquaculture must be based on a knowledge of the lethal and sublethal effects of low dissolved oxygen levels, and the rate of oxygen consumption within the system. Dissolved oxygen criteria are formulated in terms of concentration (mg/L), but aerators must be designed on a mass flux basis (kg O_2 transferred per hour).

Lethal Levels

The lethal dissolved oxygen levels of fish range from 2 to 3 mg/L for the salmonids (Hermann et al. 1962) and to 0.5 to 1.0 mg/L for warmwater fish (Askerov 1975; Mahdi 1973; Saad et al. 1973; Moss and Scott 1961). The lethal dissolved oxygen level will decrease with increasing temperature due to increasing metabolic oxygen requirements and decreasing oxygen solubility. In ponds, the critical period for dissolved oxygen depletion is the nighttime, when the phytoplankton are consuming oxygen rather than producing oxygen. Therefore, the 3- or 6-hour lethal level may be more significant than LC_{50}'s based on longer periods. Fish repeatedly exposed to low dissolved oxygen levels may be more susceptible to disease and parasites (Lee 1973).

Sublethal Effects

The most important sublethal effect of low dissolved oxygen in terms of culture will be the effect on growth or production. The effect of low dissolved oxygen on the growth of fish depends strongly on the feeding level. When fish are fed to satiation several times a day, any decrease in the dissolved oxygen concentration below the air saturation levels will decrease the food consumption and growth of coho salmon, *Oncorhynchus kisutch*, and largemouth bass, *Micropterus salmonides* (Fisher 1963; Herrmann et al. 1962; Steward 1962). Under normal production levels of 1 to 3 percent of body weight per day, the growth of rainbow trout and channel catfish is not reduced until the dissolved oxygen drops below 5.0 mg/L (Andrews et al. 1973; Larmoyeux and Piper 1973).

Prediction of Oxygen Demand

The prediction of oxygen demand for flow-through and pond systems is considered in the following discussion.

Flow-through system

The prediction of oxygen demand in a flow-through system is relatively simple, as the fish are the major consumers of oxygen and reaeration is not significant. The sources and sinks of oxygen are presented in Figure 1. The oxygen consumption of various aquatic species under different feeding regimes is presented in Table 1. The oxygen consumption of fish per unit mass will increase with increasing temperature and decreasing size. The mean oxygen consumption of fed fish may be 20 to 30 percent higher than that of unfed fish. The oxygen consumption of the various species ranges from 6 to 40 kg O_2/1000 kg fish·d. The oxygen consumption of sockeye salmon fed 3 percent varied from 170 mg O_2/kg·h just before dawn to 370 mg O_2/kg·h just before the start of feeding (Brett and Zla 1975). In raceway systems, it may be necessary to design the aerators to deliver the maximum oxygen rate rather than an average value.

Static pond system

The static pond system is commonly used for the culture of warmwater species such as channel catfish, tilapia, and the freshwater prawn, *Macrobrachium rosenbergii*. The sources and sinks of oxygen are shown in Figure 2. The major oxygen

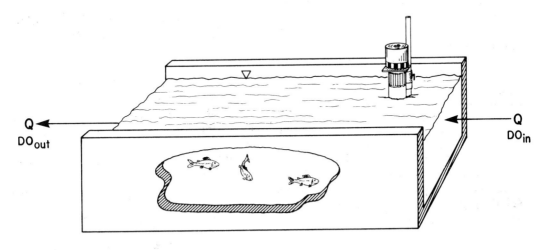

SOURCES OF OXYGEN

- •INFLOW WATER
- •AERATION

CONSUMERS OF OXYGEN

- • OUTFLOW WATER
- • FISH

FIGURE 1.—Source and sinks of oxygen in a flow-through system.

TABLE 1.—Oxygen consumption of aquatic animals.

Species	Size	Temperature, °C	Feeding level	Oxygen consumption, kg O_2/1000 kg fish·d	Source
Rainbow trout *Salmo gairdneri*	100 g	15	unknown	7.2	Liao 1971
Rainbow trout *Salmo gairdneri*	100 g	15	production levels	7.2	Muller-Feuga et al. 1978
Sockeye salmon *Oncorhynchus nerka*	28.6g	15	unfed	5.6	Brett and Zala 1975
Sockeye salmon *Oncorhyncus nerka*	28.6g	15	3%	6.6	Brett and Zala 1975
Channel catfish *Ictalurus punctatus*	100 g	30	unfed	13.4	Andrews and Matsuda 1975
Channel catfish *Ictalurus punctatus*	100 g	30	satiation	19.5	Andrews and Matsuda 1975
Malaysian prawn *Macrobrachium rosenbergii*	0.5g	24	unfed	36.0	Nelson et al. 1977
Malaysian prawn *Macrobrachium rosenbergii*	0.5g	24	satiation Purina marine ration #2	43.0	Nelson et al. 1977

consumers may be the phytoplankton or bacteria (Boyd et al. 1978; Romaire et al. 1978; Schroeder 1975).

In ponds, dissolved oxygen depletion is most likely to occur in the early morning hours because of phytoplankton respiration. Boyd et al. (1978) found that the nighttime dissolved oxygen decrease may be as high as 0.5 mg/L·h in the southeastern United States. This oxygen consumption rate may not be valid for other areas. At a stocking rate of 2,500 kg/ha of 400-g channel catfish, the oxygen consumption of the fish would be less than 0.11 mg/L·h. The computation of system oxygen demand in ponds may require the use of computer simulation analysis (Boyd et al. 1978; Romaire et al. 1978). The use of a coupled hydrodynamic-water chemistry model such as WQRRS (Hydrologic Engineering Center 1978) may increase the predictive capability of these models.

Oxygen Transfer Under Culture Conditions

The mass transfer of gases and prediction of mass transfer rates under field conditions are considered in this section.

Mass Transfer of Gases

The rate of oxygen transfer is function of the oxygen deficiency of the solution. In differential form this can be written as

$$\frac{dC}{dt} = K(C^* - C) \qquad (1)$$

where

C = concentration of O_2 in liquid
C^* = saturation concentration of O_2 in liquid
t = time
K = proportional constant.

The proportional constant depends on the chemical characteristics of the water and the geometry and mixing characteristics of the system. The saturation concentration of oxygen can be determined using Henry's law:

$$C = \beta X P_t \qquad (2)$$

C^* = saturation concentration of oxygen
β = Bunsen's coefficient at a given temperature and salinity
X = mole fraction of oxygen
P_t = total pressure

The solubility of a gas can be increased by increasing the mole fraction of the gas or

140

by increasing the total pressure. The solubility of nitrogen and oxygen can be computed as a function of temperature, salinity, gas composition, and pressure from the equations presented by Weiss (1970). The saturation concentrations of dissolved oxygen as a function of temperature and salinity are presented in tabular form in Standard Methods (1976) for moist air at 1 atmospheric pressure.

Prediction of Oxygen Transfer Rates Under Field Conditions

Except for some special cases (Schroeder 1977) it is impossible to compute mass transfer rates from theoretical considerations. In most cases, the mass transfer rate of a given aeration device is measured in the laboratory under standard conditions. Then, the transfer rates are computed for field conditions. An equation commonly used is listed below:

$$N = N_0 \left[\frac{\beta C^* - C_d}{9.17} (1.024)^{T-20} \alpha \right] \quad (3)$$

N = kg O_2/kW·h transferred under field conditions
N_0 = kg O_2/kW·h transferred under standard conditions
β = salinity-surface tension correction
C^* = saturation oxygen level in field liquid
C_d = design dissolved oxygen level
T = temperature, °C
α = oxygen transfer correction for waste

The standard conditions are 0.0 mg/L dissolved oxygen, 20 °C water temperature, and tap water. In most biological waste treatment systems the design oxygen levels are 0.5 to 1.0 mg/L, so the mass transfer rates under these conditions are not too much lower than those measured in the laboratory. The mass transfer of aeration systems for aquaculture will be lower than those in waste treatment because the dissolved oxygen level must be maintained at 5 to 9 mg/L and therefore the driving force is lower. For diffused aeration

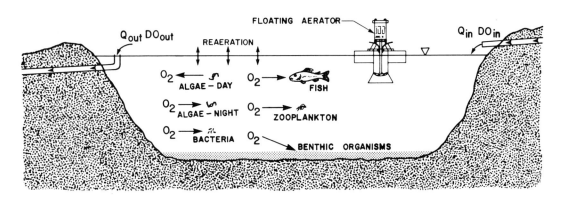

SOURCES OF OXYGEN
- REAERATION
- ALGAE (DURING THE DAY)
- INFLOW WATER
- AERATION

CONSUMERS OF OXYGEN
- TRANSFER ACROSS AIR – WATER BOUNDARY
- BACTERIAL DECOMPOSITION OF UNEATEN FOOD
- BENTHIC ORGANISMS
- OUTFLOW WATER
- ALGAE (DURING THE NIGHT)
- ZOOPLANKTON
- BACTERIA
- FISH

FIGURE 2.—Sources and sinks of oxygen in a pond system.

141

TABLE 2.—Typical oxygen transfer rates of various devices used in culture systems.

Aeration system	Transfer rate kg O_2/kW hr	
	Standard[a]	6 mg/L O_2[b]
Diffused-air systems		
fine-bubble	1.2 - 2.0	0.25 - 0.42
medium-bubble	1.0 - 1.6	0.21 - 0.34
coarse-bubble	0.6 - 1.2	0.13 - 0.25
Pure-oxygen systems		
fine-bubble		1.2-1.8
mechan. surf. aeration		1.0-1.2
turbine-sparger		1.2-1.5
Low-speed surface	1.2 - 2.4	0.25 - 0.50
Low-speed surf. w/draft tube	1.2 - 2.4	0.25 - 0.50
High speed floating aerator	1.2 - 2.4	0.25 - 0.50
Floating rotor aerator	1.2 - 2.4	0.25 - 0.50
U-tube aerator		
zero head available	4.5	0.95
1-foot head available	45.6	9.58
Gravity aerator	1.2 - 1.8	0.25 - 0.38
Venturi aerator	1.2 - 2.4	0.25 - 0.50
Static tube systems	1.2 - 1.6	0.25 - 0.34

[a]20 °C, tap water, zero dissolved oxygen, $\alpha = \beta = 1.0$

[b]20 °C, $\alpha = 0.85$, $\beta = 0.9$, dissolved oxygen = 6 mg/L

Note: kg/kW·h × 1.6440 = lb O_2/hp·h

systems, the mass transfer rate increases as the depth increases. Some municipal aeration basins are 3 to 5 m deep, while most ponds are less than 1 to 2 meters deep. Also, in aquaculture applications it will not be possible to duplicate the mixing and turbulence used in wastewater applications. Therefore, the mass transfer rates will be lower.

The mass transfer rate has a strong dependency on the oxygen deficient. The transfer rate can be increased by increasing C* by use of pure oxygen or by hydrostatic pressure. The temperature and waste-specific effects have a secondary effect on mass transfer. Typical transfer rates in terms of kg O_2/kW·h are presented for various aeration devices in Table 2. The transfer rates for field conditions were computed from Equation 3 using 20 °C, $\alpha = 0.85$, $\beta = 0.9$ and $C_d = 6$ mg/L. Under these conditions, Equation 3 reduces to

$$N = 0.21 N_0 \qquad (4)$$

Therefore, the mass transfer rates computed from Equation (4) for aquaculture conditions may be no greater than 21 percent of the rate measured under standard conditions. From Table 2, the transfer efficiencies range generally from 0.13 to 0.50 kg O_2/kW·h. Two systems are more efficient at transferring oxygen at high dissolved oxygen levels: the pure oxygen system and the U-tube aerator.

The U-tube aeration (Speece et al. 1970; Speece 1969a; Speece et al. 1969) consists of a deep hole (15-20 m) with a baffle that extends from above the water surface to the near bottom of the hole. The water flow is directed down one side of the baffle and back up the other side. If adequate head (20-30 cm) is not available, the water would be pumped through the system with a low-head pump. Air bubbles are injected on the down-flow side. The pure oxygen systems may consist of a fine bubble diffused air system or a coarse bubble and high speed floating aerator working together. For most aquaculture applications, pure oxygen would be purchased from a commercial supplier as a liquid.

Considerations in the Design of Aeration Systems for Aquaculture

The design of aeration systems for aquaculture will depend primarily on matching the system demands with the mass transfer characteristics of the aerators. Other important considerations are carbon dioxide removal, prevention of nitrogen gas supersaturation problems, noise production, and the mixing characteristics of the system.

Oxygen Transfer

Under the conditions present in aquaculture (Table 2) most aerators have mass transfer efficiencies that range from 0.13 to 0.50 kg O_2/kW·h (kg O_2/kW·h × 1.644 = lb O_2/hp·h). The U-tube and the pure oxygen systems are more efficient at transferring oxygen under high dissolved oxygen conditions. The transfer efficiency of the pure oxygen systems ranges from 1.0

to 1.5 kg O_2/kW·h. The cost of the liquid oxygen may range from \$0.08 to \$0.30 (FOB) depending on the quantity purchased (Kulperger 1978; Pacific Oxygen Sales, personal communication). The transfer efficiency of the U-tube aeration is approximately 0.95 kg O_2/kW·h. If 20 to 30 cm of head is available, the efficiency may be as high as 9.58 kg O_2/kW·h.

In raceway systems, the aeration system should be sized to supply the maximum oxygen demand after feeding (Brett and Zala 1975). The use of multiunit systems may be economic. One unit could provide the base aeration, and the second unit would be turned on only after feeding. If the feeding time could be staggered, peak power consumption could be reduced. To supply the maximum oxygen demand of 1000 kg of rainbow trout or channel catfish (Table 1), using high speed floating aerators (0.25 kg O_2/kW·h) would require 1.17 and 3.25 kW of power (1.6 and 4.3 hp), respectively.

The design of aeration systems for ponds may need to be based on some type of simulation analysis using a coupled hydrodynamic-water chemistry model. The oxygen transfer efficiency of aerators used in ponds will vary widely because of the changes in the dissolved oxygen levels. Therefore, it will be necessary to compute the oxygen transfer efficiency at each time step in the simulation as a function of oxygen and temperature.

Effects of System Configuration on the Use of Aeration

The operational use of aeration systems will strongly depend on the configuration of the culture system. Because of the large area and volume in ponds, the time response of these systems to aeration will be much larger than in raceway systems.

Raceway

The typical raceway used by the California Department of Fish and Game (Leitritz and Lewis 1976) is 30.48 m × 3.05 m × 0.76 m (100 ft × 10 ft × 2.5 ft). The standard flow is 0.085 m³/s (3 ft³/s) for less than 5 raceways in series. The mean hydraulic detention time (volume/flow rate) is equal to 0.23 h. Because there should be a complete water change every 0.23 hours, the response of the system is relatively rapid.

Pond system

A pond system is characterized by a large total volume and surface area. The detention time of these systems may vary from 60 to 120 days or more. In the following example, it will be assumed that the contents of the pond are mixed completely (no thermal or chemical stratification). This analysis is intended to give the reader a basic understanding of the time response of the pond system. The assumption of a well-mixed pond, though not strictly correct, will have little effect on these computations.

A 10-ha (24.71-acre) pond with a depth of 1 m (3.28 ft) is used in all the examples. The volume of this pond is 100,000 m³ and at 8 mg/L contains 800 kg of oxygen. Two sources of oxygen are available: a source of water and two floating aerators. A pump can supply 5.00 m³/min (1320 gal/min) to this pond. At 8 mg/L dissolved oxygen, this water would add 2.40 kg/h. The dissolved oxygen level of most well water is low (<0.5 mg/L). Therefore, well water would have to be aerated before use.

Two aerators that can add a total of 4 kg/h may also be used. The oxygen uptake of the pond during the nighttime is assumed to be 5 kg/h·ha or 50 kg/h for the total system (Romaire et al. 1978). It can be assumed that the dissolved oxygen level is 8 mg/L at 2000 hours. The critical dissolved oxygen level will be just before dawn at 0600 the next day. The dissolved oxygen level at time $t_0 + \Delta t$ can be computed from the following equation:

$$\text{Mass of Oxygen}/t_0 + \Delta t = \\ \text{Mass of Oxygen}/t_0 - \\ \text{Uptake of Oxygen} \times \Delta t \\ + \text{Mass Transfer Rate of Aerators}/ \\ t_0 \, \Delta t \qquad (5)$$

Therefore, the oxygen remaining in the system at 0600 without aeration will be

800 kg - 50 kg/h × 10 h = 300 kg or 3 mg/L

If the pump is started at 2000, the dissolved oxygen would be

800 kg - 50 kg/h × 10.0 h + 2.40 kg/h x 10h

- 1.20 kg/h x 10 h =
312 kg or 3.12 mg/L

The dissolved oxygen level of the effluent water was assumed to be 4 mg/L or 1.2 kg/h. If the aerators were started at 2000 hours the dissolved oxygen would be

800 kg - 50 kg/h × 10.0 h + 4 kg/h × 10 h =

340 kg or 3.40 mg/L

In a real pond, aeration may create a localized area of high dissolved oxygen that may prevent oxygen depletion problems. Depending on the wind and the configuration of the ponds, surface reaeration may add a significant amount of oxygen. The critical dissolved oxygen level would probably occur during a low wind period, and therefore this source of oxygen should be neglected during analysis.

The overall conclusion of this analysis is that aeration in ponds may have only a limited effect on the dissolved oxygen level. Under any conditions, it is necessary to start the aerators far before the dissolved oxygen levels drops to a critical level (Boyd et al. 1978; Romaire et al. 1978).

Carbon Dioxide Removal

In raceway and fish transportation systems, carbon dioxide may accumulate to toxic levels if pure oxygen systems are used without surface aeration (U.S. Fish and Wildlife Service 1978). Trout show distress when the free carbon dioxide level reaches about 25 mg/L (Leitritz and Lewis 1976), although fish can acclimate to higher carbon dioxide levels (U.S. Fish and Wildlife Service 1978). At air saturation levels, the volume of carbon dioxide produced by goldfish and rainbow trout is very nearly equal to the volume of oxygen consumed (Kutty 1968). The carbon dioxide production is then equal on a weight basis to

CO_2 produced =
$1.375 * O_2$ consumed   (6)

The equilibrium value of free carbon dioxide in water is approximately 0.5 mg/L (Stumm and Morgan 1970). Therefore, if the free carbon dioxide level is 15 mg/L, the driving gradient is 14.5 mg/L (Equation 1). Since the gradient for oxygen may range from 3 to 4 mg/L, carbon dioxide buildup will not be a problem if surface agitation or diffused aeration is provided (Speece 1973).

Dissolved Gas Supersaturation

The use of the U-tube aerator with air, the downflow bubble contact aeration system with air, and the venturi aerator have the ability to produce some level of nitrogen gas supersaturation. Because of the analytical difficulties involved with the measurement of nitrogen gas levels (Fickeisen and Schneider 1976), nitrogen gas concentrations are not measured in aeration tests. Pure oxygen systems (Ruane et al. 1977; Imhoff and Albrecht 1977; Speece 1969) can also produce supersaturated levels of oxygen. The aeration systems that have the greatest ability to produce gas supersaturation also have the highest mass transfer efficiency. These systems increase the saturation level by pressurization (or increase the partial pressure) and therefore may also result in supersaturation of nitrogen or oxygen with respect to the surface saturation level.

Gas Bubble Disease (GBD) appears to be a result of the total gas pressure rather than due only to the supersaturation of nitrogen gas (Weitkamp and Katz 1977; Fickeisen and Schneider 1976). Total gas pressure of 115 to 125 percent may prove lethal to fish. Dissolved oxygen levels of in excess of 300 percent saturation are required to produce GBD (Weitkamp and Katz 1977). Therefore, the supersaturation levels produced by pure oxygen systems may not have an adverse effect on fish.

There has been no documentation on the production of GBD from aeration systems. Since the U-tube, downflow bubble contact system, and the venturi aerators may produce supersaturated levels of nitrogen and oxygen, this parameter must be considered in their design and performance evaluation.

Noise Production

Diffused air systems, venturi systems, and mechanical aerators may produce high levels of noise. Under laboratory conditions aeration noise can have a significant effect on the hatching of *Cyprinodon variegatus* and the growth rate of *C. variegatus* and *Fundulus similis* (Banner and Hyatt 1973). It is not unreasonable to suspect that in recycled or raceway systems aeration noise may have an adverse effect on the fish. In pond systems, the fish may be able to avoid the noise. Additional research and prototype evaluation will be required to determine the effect of noise on culture organisms.

Mixing

The mixing characteristics of the aerator may be an important design consideration. In pond systems, it may be desirable to prevent thermal and chemical stratification during the summer months. Mixing during the day may evenly distribute the oxygen produced by the algae and decrease the loss of oxygen across the air-water interface (Busch and Flood 1979). Diffused aeration systems have been commonly used in the destratification and mixing of lakes (Zieminski and Whittemore 1970). The optimum air flow and bubble size for mixing (Zieminski and Whittemore 1970) may not produce high mass transfer rates (Mauinic and Bewtra 1976). The use of diffused aeration with the static mixing tube (Campbell and Rocheleau 1976) may produce good mass transfer and mixing. Surface aerators with draft tubes or other systems that pull water from the bottom may result in effective mixing.

In the culture of fish larvae, strong aeration and the resulting turbulence reduces larval mortality (Barahona-Fernandes 1978) and may aid in the normal inflation of the swim bladder (Doroshev and Cornacchia 1979).

Safety

The safety of these systems must be an important design parameter. All of these systems will require some energy input, most likely in the form of electricity. The hazard of electrical shock will be greatest in the floating aerators, paddlewheel aerators, and other types that may require waterproof housing and waterproof electrical cable. These aerators should be removed from the water for maintenance and repair.

The low pressure air used in the diffused aeration systems should present no serious safety problems. The use of pure oxygen systems will require some additional care to avoid fire and explosion.

The paddlewheel aerator (Rappaport et al. 1976; Busch et al. 1974) may present some additional safety problems if the paddles are unshielded. Excessive spray may result in ice formation during the winter, which could be a hazard to those walking in the area.

None of these aerators present any significant safety risks if common sense and normal safety procedures are followed. Serious injury may result, however, if these procedures are not followed.

Reliability

The reliability of these systems will be a critical factor in recycled or raceway systems. Under these conditions, it may be necessary to have a backup system that can provide aeration for several days. This could consist of a pure oxygen system or spare aerators and (or) a gas turbine-generator system.

The local availability of spare parts and service will be an important consideration in selection of aerators. Pure oxygen systems may be more reliable than mechanical systems, but depend on a reliable supply of oxygen.

In some applications, the aerators will be used only during part of the day. Most of these systems will be powered by electrical motors and therefore can be turned on and off at will. The intermittent use of diffused aeration systems, especially the fine bubble type, may result in clogging (Mitchell and Kirby 1976).

The reliability of these systems will depend strongly on their care and maintenance. The cost of a preventative maintenance program will be insignificant to the overall production costs.

Costs

The economics of the different aeration systems will depend on the relative capital and operating costs in relationship to the aeration demand. In raceway or recycled systems where aeration must be provided at all times, operating costs may be the critical factor in total costs. In application when aeration is provided intermittently, capital costs may be more important.

Capital costs

To transfer 1 kg O_2/h, approximately 2 to 4 kW of power is required (2.7 to 5.4 hp) (Table 2). This amount of oxygen would support 3400 kg of rainbow trout or 1200 kg of channel catfish. Total capital costs including freight, electrical installation, and site preparation for this sized system may range from $1500 to $2000 (ENRCC = 3000) for the common mechanical aerators such as the Fresh-flow (BTE type). Mitchell and Kirby (1976) found that capital costs required to transfer 1.15 kg O_2/h ranged from $600 to $5,100 for four systems. The paddlewheel aerator was the cheapest and diffused aeration was the most expensive. The cost of feeder lines may be a major expense in use of diffused aeration systems in large ponds. The capital cost of 12.2-m (40-ft) U-tube aerators in 1970 ranged from $700 in loose soil to $3,100 in rock (Speece and Orosco 1970). If the water had to be pumped, pump costs would add another $1,800.

These costs should be considered very approximate; local conditions may change the cost significantly. Capital costs rather than total systems cost (capital + operating costs) may be the critical design consideration if the operator is unable to pay for the higher cost capital items.

Operating costs

Energy costs will probably represent the largest single item in the operating budget. Assuming electricity to cost $0.05/kW·h (July 1979), the cost to transfer 1 kg O_2/h would range from $880 to $1750/yr. In-

creases in the cost of power will make this cost even more important in the future. The cost of pure oxygen ranges from $0.08 to $0.30/kg (July 1979) depending on the quantity purchased. To supply 1 kg O_2/h would cost $700 to $2600/yr. If 30 to 40 cm of head is available, the annual operating costs of a U-tube aerator may be as low as $46/yr. If the water must be pumped through the U-tube, electricity would cost $460/yr. Annual system maintenance may cost $5 to $20/kW, but the major work might be done on the off-season.

Total system costs

The total costs of an aeration system will depend strongly on its function. Based on a 10 year operating life, Mitchell and Kirby (1976) found the present worth of the total cash outlay (10 percent interest rate, electrical power costs = $0.03 kW·h) for four systems ranged from $9,400 to $30,600. The paddlewheel and diffused air systems were the cheapest, followed by floating aerator and a venturi system. Detailed cost comparisons between these systems is beyond the scope of this paper, although a few generalizations may be presented below.

For use in raceway or recycled systems the U-tube or the downflow bubble contact system (with pure oxygen) may prove the most efficient. A surface aerator with draft tube could be used to provide supplemental oxygen transfer during peak demand.

In ponds, diffused air systems with coarse bubbles or surface aerators appear to be most economical. Pure oxygen could be used to supply additional oxygen during critical periods. This oxygen could be injected in the diffused air system or below a surface aerator.

Conclusions

The design of aeration systems for aquaculture will depend on a knowledge of the oxygen requirements of the culture species, the oxygen demand of the system, and the mass transfer efficiency of the aeration system. The mass transfer efficiency of aerators under aquaculture conditions will be much lower than published data because of a reduction in the driving force, depth of aerator submergence, and turbulence. The transfer efficiency of aerators for aquaculture should be measured and rated under their normal operating conditions (dissolved oxygen = 5 to 6 mg/L). The U-tube aerator and the pure oxygen systems are more efficient at transferring oxygen at a high dissolved oxygen level.

References

ANDREWS, J.W., AND Y. MATSUDA. 1975. Influence of various culture conditions on the oxygen consumption of channel catfish. Trans. Am. Fish. Soc. 104(2): 322-327.

ANDREWS, J.W., T. MURAI, AND G. GIBBONS. 1973. Influence of dissolved oxygen on the growth of channel catfish. Trans. Am. Fish. Soc. 102(4):835-838.

ASKEROV, T.A. 1975. Survival rate and oxygen consumption of juvenile wild carp maintained under different conditions. Hydrobiol. J. 11(3):67-68.

BANNER, A., AND M. HYATT. 1973. Effects of noise on eggs and larvae of two estuarine fishes. Trans. Am. Fish. Soc. 102(1):134-136.

BARAHONA-FERNANDES, M.H. 1978. Effects of aeration on the survival and growth of sea bass (Dicentrarchus labrax l.) larvae: A preliminary study. Aquaculture 14(1):67-74.

BOYD, C.E., R.P. ROMAIRE, AND E. JOHNSTON. 1978. Predicting early morning dissolved oxygen concentrations in channel catfish ponds. Trans. Am. Fish. Soc. 107(3):484-492.

BRETT, J.R., AND C.A. ZALA. 1975. Daily pattern of nitrogen excretion and oxygen consumption of sockeye salmon (Oncorhynchus nerka) under controlled conditions. J. Fish. Res. Board Can. 32(12):2479-2486.

BUSCH, C.D., AND C.A. FLOOD, JR. 1979. Water movement for water quality in catfish production. Pap. No. 79-5041, presented at the Joint Meeting of the American Society of Agricultural Engineers and the Canadian Society of Agricultural Engineering, June 24-27, 1979, Winnipeg. 8 pp.

BUSCH, C.D., C.A. FLOOD, JR., J.L. KOON, AND R. ALLISON. 1977. Modeling catfish pond nighttime dissolved oxygen levels. Trans. Am. Soc. Agric. Eng. 20(2):394-396.

BUSCH, C.D., J.L. KOON, AND K.R. ALLISON. 1974. Aeration, water quality, and catfish production. Trans. Am. Soc. Agric. Eng. 17(3):433-435.

CAMPBELL, H.J., JR., AND R.F. ROCHELEAU. 1976. Waste treatment at a complex plastics manufacturing plant. J. Water Pollut. Control Fed. 48(2):256-273.

CHESNESS, J.L., L.J. FUSSELL, AND T.K. HILL. 1973. Mechanical efficiency of a nozzle aerator. Trans. Am. Soc. Agric. Eng. 16(1):67-68, 71.

CHESNESS, J.L., J.L. STEPHENS, AND T.K. HILL. 1972. Gravity flow aerators for raceway fish culture systems. Coll. Agric. Exp. Sta., Univ. Georgia, Athens. Res. Rep. 137. 21 pp.

CHESNESS, J.L., AND J.L. STEPHENS. 1971. A model study of gravity flow aerators for catfish raceway systems. Trans. Am. Soc. Agric. Eng. 14(6):1167-1169.

DOROSHEV, S.I., AND J.W. CORNACCHIA. 1979. Initial swim bladder inflation in the larvae of Tilapia mossambica (Peters) and Morone saxatilis (Walbaum). Aquaculture. 16(1):57-66.

FICKEISEN, D.H., AND M.J. SCHNEIDER. 1976. Gas bubble disease. Proceedings of a workshop, Richland, Washington, October 8-9, 1974. Energy Research and Development Admin., Oak Ridge, Tennessee. CONF-741033. 123 pp.

FISHER, R.J. 1963. Influence of oxygen concentration and its diurnal fluctuations on the growth of juvenile coho salmon. M.S. thesis. Oregon St. Univ., Corvallis. 48 pp.

HERRMANN, R.B., C.E. WARREN, AND P. DOUDOROFF. 1962. Influence of oxygen concentration on the growth of juvenile coho salmon. Trans. Am. Fish. Soc. 91(2):155-167.

HYDROLOGIC ENGINEERING CENTER. 1978. Water quality for reservoir systems. U.S. Army Corps of Engineers, Davis, California. 279 pp.

IMHOFF, K.R., AND D. ALBRECHT. 1977. Pure oxygen aeration in the Ruhr River. J. Water Pollut. Control Fed. 49(9):1959-1967.

KULPERGER, R.J. 1978. Oxygen supply considerations. Pages 209-218 in J.R. McWhirter, ed. The use of high-purity oxygen in the activated sludge process, Vol. II, CRC Press, West Palm Beach, Florida. 274 pp.

KUTTY, M.N. 1968. Respiratory quotients in goldfish and rainbow trout. J. Fish. Res. Board Can. 25(8):1689-1728.

LARMOYEUX, J.D., AND R.G. PIPER. 1973. Effects of water reuse on rainbow trout in hatcheries. Prog. Fish-Cult. 35(1):2-8.

LEE, J.S. 1973. Commercial catfish farming. Interstate, Danville, Illinois. 263 pp.

LEITRITZ, E., AND R.C. LEWIS. 1976. Trout and salmon culture. Calif. Dep. Fish Game, Fish Bull. 164. 197 pp.

LIAO, P.B. 1971. Water requirements of salmonids. Prog. Fish-Cult. 33(4):210-215.

LOYACANO, H.A. 1974. Effects of aeration in earthen ponds on water quality and production of white catfish. Aquaculture 3(3):261-271.

MAHDI, M.A. 1973. Studies on factors affecting survival of Nile fish in the Sudan. III. The effect of oxygen. Mar. Biol. 18(2):96-89.

MAVINIC, D.S., AND J.K. BEWTRA. 1976. Efficiency of diffused aeration systems in wastewater treatment. J. Water Pollut. Control Fed. 48(10):2273-2283.

METCALF AND EDDY, INC. 1979. Wastewater engineering: Treatment, disposal, reuse. Second edition. McGraw-Hill, New York. 920 pp.

MITCHELL, R.E., AND A.M. KIRBY, JR. 1976. Performance characteristics of pond aeration devices. Pages 561-581 *in* J.W. Avault, ed. Proc. 7th Annu. Meeting World Maricult. Soc., Louisiana St. Univ., Baton Rouge. 729 pp.

MOSS, D.D., AND D.C. SCOTT. 1961. Dissolved-oxygen requirements of three species of fish. Trans. Am. Fish. Soc. 90(4):377-393.

MULLER-FEUGA, A., J. PETIT, AND J.J. SABAUT. 1978. The influence of temperature and wet weight on the oxygen demand of rainbow trout (*Salmo gairdneri* R.) in freshwater. Aquaculture 14(4):355-363.

NELSON, S.G., A.W. KNIGHT, AND H.W. LI. 1977. The metabolic cost of food utilization and ammonia production by juvenile *Macrobrachium rosenbergii* (crustacea: palamonidae). Comp. Biochem. Physiol. 57A(1):67-72.

RAPPAPORT, A., S. SARIG, AND M. MAREK. 1976. Results of tests of various aeration systems on the oxygen regime in the Genosar experimental ponds and growth of fish there in 1975. Bamidgeh. 28(3):35-49.

ROMAIRE, R.P., C.E. BOYD, AND W.J. COLLIS. 1978. Predicting nighttime dissolved oxygen decline in pond used for tilapia culture. Trans. Am. Fish. Soc. 107(6):804-808.

RUANE, R.J:, T.Y.J. CHU, AND V.E. VANDERGRIFF. 1977. Characterization and treatment of waste discharged from high-density catfish cultures. Water Res. 11(9):789-800.

SAAD, M.A.H., A. EZZAT, AND M.B. SHABANA. 1973. Effect of pollution on the blood characteristics of *Tilapia zillii* G. Water, Soil, Air Pollut. 2(2):171-179.

SCHROEDER, E.D. 1977. Water and wastewater treatment. McGraw-Hill, New York. 370 pp.

SCHROEDER, G.L. 1975. Nighttime material balance for oxygen in fish ponds receiving organic wastes. Bamidgeh 27(3):65-74.

SCOTT, K.R. 1972. Comparison of the efficiency of various aeration devices for oxygenation of water in aquaria. J. Fish. Res. Board Can. 29(11):1641-1643.

SPEECE, R.E. 1969a. U-tube oxygenation for economical saturation of fish hatchery water. Trans. Am. Fish. Soc. 89(4):789-795.

——————. 1969b. The use of pure oxygen in rivers and impoundment aeration. Pages 700-712 *in* Proc. 24th Industrial Waste Conference, Vol. 1. Purdue Univ., Lafayette, Indiana. Engineering Extension Ser. No. 135. 788 pp.

SPEECE, R.E., AND M.J. HUMENICK. 1973. Carbon dioxide stripping from oxygen activated sludge systems. J. Water Pollut. Control Fed. 45(3):412-423.

SPEECE, R.E., M. MADRID, AND K. NEEDHAM. 1971. Downflow bubble contact aeration. Am. Soc. Civil Eng. 97(SA4):433-441.

SPEECE, R.E., AND R. OROSCO. 1970. Design of U-tube aeration systems. Am. Soc. Civil Eng. 96(SA4):715-725.

SPEECE, R.E., J.L. ADAMS, AND C.B. WOOLDRIDGE. 1969. U-tube aeration operating characteristics. Am. Soc. Civil Eng. 95(SA3):563-574.

STANDARD METHODS for the examination of water and wastewater. 1976. American Public Health Association, American Water Works Association, and Water Pollution Control Federation, Washington, D.C. 1193 pp.

STEWARD, N.E. 1962. The influence of oxygen concentration on the growth of juvenile largemouth bass. M.S. thesis. Oregon St. Univ., Corvallis. 44 pp.

STUMM, W., AND J.J. MORGAN. 1970. Aquatic chemistry. Wiley-Interscience, New York. 583 pp.

U.S. FISH AND WILDLIFE SERVICE. 1978. Manual of fish culture. Sect. 6. Fish transportation. U.S. Government Printing Office, Washington, D.C. 88 pp.

WARREN, C.E., AND G.E. DAVIS. 1967. Laboratory studies on the feeding, bioenergetics and growth of fish. Pages 175-213 in S.D. Gerking, ed. The biological basis of freshwater fish production. Blackwell, Oxford. 495 pp.

WEISS, R.F. 1970. The solubility of nitrogen, oxygen, and argon in water and seawater. Deep-Sea Res. 17(4):721-735.

WEITKAMP, D.E., AND M. KATZ. 1977. Dissolved atmospheric gas supersaturation of water and the gas bubble disease of fish. Environmental Information Services, Inc., Mercer Island, Washington. Prepared for the Water Resources Scientific Information Center, Dep. Interior, Washington, D.C. 107 pp.

ZIEMINSKI, S.A., AND R.C. WHITTEMORE. 1970. Induced air mixing of large bodies of polluted water. U.S. Environ. Prot. Agency, 16080 DWP 11/70. U.S. Printing Office, Washington, D.C. 46 pp.

Bio-Engineering Symposium for Fish Culture (FCS Publ. 1):149-155

Keeping Hatchery Design Simple

IVAN C. QUINATA

Missouri Department of Conservation
P.O. Box 180
Jefferson City, Missouri 65102

ABSTRACT

Today's concern for energy conservation systems applies not only to the large industrial sectors but also to fish hatchery facilities. Energy conserving systems can also increase hatchery production efficiency. Energy conservation is defined here as the efficient use of available resources to accomplish the goal. Simplifying these systems to permit hatchery personnel, whose specialized training is fish production, to operate and maintain the hatchery facility is vital in the design stage of a project. This paper presents a case history of a simple hatchery design utilizing available water energy from a dam, a well-water supply, and water supply from the dam to increase production efficiency.

Feasibility Study

In 1972 the Department of Conservation contracted the services of Kramer, Chin & Mayo, Inc., (1973) to develop a conceptual report on the feasibility of increasing hatchery production. The report concluded that production could be increased to 181,600 kg (400,000 lb) of 25.4-cm (10-in) trout per year. There was one concept that was adopted from this study, and that was production scheduling.

Design for the renovation of this facility was done by the Engineering Section of the Department. The location of the hatchery prompted some innovative ideas in an effort to maintain a simple but functional facility. These include a jet pump for recirculation and aeration, mud valves for raceway cleaning, and fiberglass starter tanks for maintenance ease.

Introduction

Shepherd of the Hills Trout Hatchery is the major rainbow trout (*Salmo gairdneri*) and brown trout (*Salmo trutta*) hatchery operated by the Missouri Department of Conservation. Shepherd of the Hills Trout Hatchery was developed to mitigate the impact to fisheries resources resulting from the construction of Table Rock Dam (Fig. 1). On June 12, 1957, the U.S. Corps of Engineers granted the Missouri Conservation Commission a permit ''for development and management of a fisheries area and a public access area with public recreational facilities incidental thereto.'' The hatchery was built the same year. The yearly production of 108,960 kg (240,000 lb) of 25.4-cm (10-in) trout from these inefficient pools was insufficient to meet the requirement of Missouri's trout-stocking program. Additional trout were transferred from nearby hatcheries to meet the increasing fishing pressure.

The design objective of the Missouri Department of Conservation was to increase production of this facility at a minimal cost and with a minimum of sophisticated and mechanically dependent automated systems that have become the trend of today's modern hatcheries.

Lake Water Supply

The lake water supply is limited to 0.623 m³/s (22 ft³/s) by law through an agreement with the Corps of Engineers. This limited water supply is just one of the factors that dictate the production capabilities of the hatchery. Other limiting factors include low dissolved oxygen content and higher water temperatures between August and December. Water temperature varies from 5 °C to 13.89 °C.

The lake water supply to the building is treated for low dissolved oxygen content, which drops to 0.2 mg/l, high iron content, and hydrogen sulfide traces at the water

Figure 1. — Shepherd of the Hills Trout Hatchery. Table Rock Dam is in the background.

tower. The water tower also reduces system pressure for distribution inside the hatchery building.

Well Water Supply

There are two wells that can be used in addition to the lake water supply. One well, using an oil-lubricated turbine pump, has a capacity of 567.75 l/min (150 gpm). The second well was drilled in 1976 to supplement the older well, but the capacity of the new well far exceeds the estimated capacity. This well produces 4920 l/min (1300 gpm) when free flowing. It is an artesian well and is capable of supplying up to 1892 l/min (500 gpm) to the well water tower. The iron content of this well water supply is 0.2 mg/l.

Well water supply to the hatchery building is treated at the well water tower for dissolved oxygen, iron content, and hydrogen sulfide traces by utilizing the available artesian pressure.

Spawning Facility

The spawning facility includes raceway units for year-round broodstock holding, raceway units for broodstock separation during the spawning activities, and a spawning building to maintain a comfortable working environment for hatchery personnel during this activity. The spawning building is designed to permit tour groups to observe the spawning operation. Fish transfer is accomplished through two gated openings in the raceway wall.

There are two spawning periods at the hatchery, approximately six months apart. Each spawn will produce six groups of growth rates. Each group will produce 13,620 kg to 18,160 kg (30,000 to 40,000 lb) of 25.4-cm (10-in) rainbow trout per month for stocking. Each group's growth rate is dependent upon the water temperature.

Hatchery Building

The hatchery building contains the administrative offices, the laboratory, and the initial stages of the trout rearing process. Water temperature control will be accomplished in the hatchery building facility from egg incubation to 6.35 cm (2.5 in) in length. The desired water temperature is achieved by mixing chilled well water at 5.0 °C (41 °F), well water at 15.56 °C (60 °F), and lake water, which varies from 5.0 °C (41 °F) to 13.89 °C (57 °F). Mixing is performed manually at the particular unit in operation and will be monitored periodically to compensate for sudden lake water temperature changes.

Egg Jars, Incubation Trays, and Swim-Up Troughs

Egg jars, incubation trays, and swim-up troughs will be supplied with chilled well water, chilled lake water, and lake water. The water supply temperature will be controlled in accordance with production scheduling (Fig. 2) normally by mixing well water and chilled well water at the unit of service.

Egg jars are used to the eye-up stage. Each jar will hold 100,000 eggs and require 18.93 l/min (5 gpm). There are 10 egg jars. The incubation trays are used to the hatched stage. Each tray will hold 12,000 eggs and will require 1.89 l/min (0.5 gpm). There are 18 stacks, 8 trays per stack.

Eighteen aluminum troughs are used to the swim-up stage. Each trough will hold 50,000 sac fry and require 26.5 l/min (gpm).

Fiberglass Starter Tanks

Twenty-eight fiberglass starter tanks, for the last stage inside the building, are used to orientate the fish to raceway culture. Each tank will hold 17,500 fingerlings and require 75.7 l/min (20 gpm). Some of the advantages of the fiberglass tanks include ease of cleaning, resistance to bacterial growth on the walls, movability, attractiveness, smooth surfaces. The tanks were designed to be self-cleaning, in that settleable solids would accumulate in the sump at the end of the tank. The use of removable stand pipes permits easy cleaning of the sump.

The only disadvantage we have determined through our research is the initial cost of the tanks, which is approximately twice the cost of concrete tanks. However, this may be attributed to the specialized manufacturing process involved since the

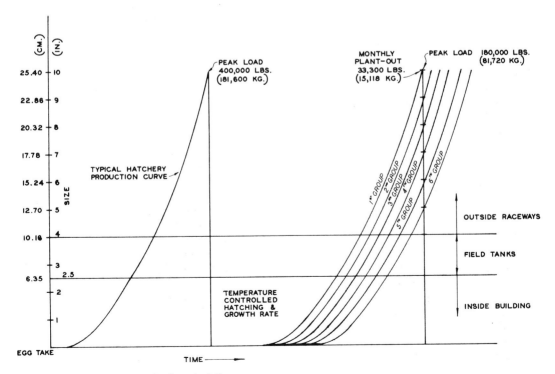

FIGURE 2. — Illustration of production scheduling.

tanks were constructed from our specifications. If these tanks would be mass produced, however, I have every reason to believe that the initial cost would be competitive to that of concrete tanks.

Chiller System No. 1

Chiller System No. 1 is designed to chill well water from 15.56 °C (60 °F) to 5.0 °C (41 °F) at a maximum flow rate of 151.4 l/min (40 gpm). The system will automatically compensate for lower flow rates to maintain chilled water at 5.0 °C (41 °F). The system will supply chilled well water to the egg jars, incubation trays, and aluminum troughs.

Chiller System No. 2

Chiller System No. 2 is designed to chill lake water from 13.89 °C (57 °F) to 5.0 °C (41 °F), at a maximum flow rate of 908.4 l/min (240 gpm). The system will automatically compensate for lower lake water temperatures and lower flow rates to maintain chilled lake water at 5.0 °C (41 °F). It will supply 5.0 °C (41 °F) chilled lake water to the fiberglass start tanks and can also be used instead of Chiller System No. 1, provided the lake water quality is acceptable.

The important feature of the chilled water system upon failure is its minimal effect on fish mortality. Should the chiller system fail, the chilled water supply can be transferred to lake water, thereby decreasing the shock factor of the water temperature change. This will minimize fish mortality, but will affect the production schedule timing.

Field Tanks

Field tanks are the first stage of the outdoor rearing units. They are an extension of the fish development to raceway culture. Trout will be transferred from the starter tanks inside the building to these concrete holding tanks, until the average fish length reaches 10.16 cm (4 in). There will be limited water temperature control for the 7.62-cm (3-in) to 10.16-cm (4-in) groups in the field tanks, but such control will be dependent upon the lake water temperature. Mixing is done with lake water and well water. As long as the lake water temperature is less than the fish-group temperature, water temperature in the tanks can be controlled. However, production scheduling was based on these groups' being raised at natural lake water temperature.

Each tank will hold 54,000 7.62-cm (3-in) fish, or 38,000 8.89-cm (3.5-in) fish, or 26,500 10.16-cm (4-in) fish. Each tank will require 378.5 l/min (100 gpm). These tanks, which are remnants of the previous hatchery facility and were kept for their useful capabilities, do not have the self-cleaning features of the new facilities.

Raceways

Because of our successful experiences with rectangular raceway units at our other trout hatcheries, we concluded that the same type of units will be constructed at this facility. The raceway units are constructed of reinforced concrete walls and footings and unreinforced concrete floors.

The following data are typical of each raceway unit:

Size: 30.48 m x 3.66 m x 1.07 m deep
(100 ft x 12 ft x 3.5 ft deep)
Maximum volume: 84.88 m³ (2997 ft³)
Minimum volume: 41.43 m³ (1463 ft³)
Maximum fish loading: 60.65 kg/m³
(3.78 lb/ft³)

The raceway units are designed as a self-cleaning system. Settleable solids are collected at the outlet sump and are cleaned by opening a mud valve and lowering the water level. The lower water level allows a faster water exchange, which increases the fish activity at the bottom of the raceway and thus creates a brushing effect.

The raceway unit may be emptied to allow algae and bacterial growth on the sidewalls to dry up. Once the raceway unit is ready to receive the next fish group, the dried flakes of algae can simply be rinsed away.

The water flow in each raceway depends upon the fish loading rate (Table 1), while maintaining a 2.5 water exchange rate per hour. The water exchange rate is maintained by varying the dam boards at the outlet end.

TABLE 1. — Loading rates at raceway units.

Size (cm)	Wt/Unit (kg)	Lake H$_2$O (l/min)^2	Recir. H$_2$O (l/min)^2	P. (kg/m^2)	Mixed flow (l/min)	Loading rate (kg/cm^2)	Loading rate (kg/l·min^{-1})	# Raceways required	Total lake H$_2$O (l/min)
11.43	1345	850	850	0.70	1700[a]	32.36	.79	1	850
12.70	1849	850	850	0.70	1700[a]	44.53	1.09	1	850
13.97	2479	1000	1000	1.05	2000	49.98	1.24	1	1000
15.24	3235	1266	1266	1.58	2532	51.90	1.28	1	1266
16.51	4118	1550	1550	2.39	3100	53.82	1.33	1	1550
17.78	2563	1040	1040	1.09	2080	50.14	1.23	2	2080
19.05	3151	1228	1228	1.48	2456	51.58	1.28	2	2456
20.32	3824	1323	1323	1.62	2646	58.63	1.45	2	2646
21.59	4601	1568	1568	2.46	3136	60.07	1.47	2	3136
22.86	5462	1814	1814	3.16	3628	64.23	1.49	2	3628
24.13	4285	1380	1380	1.76	2760	52.06	1.55	3	4140
25.40	5154	1588	1588	2.46	3176	62.15	1.62	3	4764

[a]These units may be operated from the metered fresh water line without recirculation.

The total lake water flow is 28,366 l/min (16.71 ft³/s).
The above loading rates and number of raceways required may vary according to lake water temperature and stocking practices.
The total legal lake water supply available is 37,340 l/min.
Broodstock units require 674 l/min.
Loading rates based on 10 °C water temperature.

Raceway Recirculating System

The raceway recirculating system consists of two eductors (water jet exhausters), two valves, and two aeration devices. The eductor is basically a jet pump with no moving parts and receives its motive energy from the static head available from Table Rock Reservoir (Fig. 3).

One eductor recirculates water from the outlet end of the raceway to the inlet end.

The recirculated water is mixed with fresh lake water, which is the motive force, on a 1:1 ratio. The eductor is designed to vary the flow rate in the raceway, depending on the size and weight of fish. The valves are used to control the differential and motive pressures of the eductor unit, varying the water flow through the system.

The other eductor unit entrains air for mixing with the recirculated water to raise

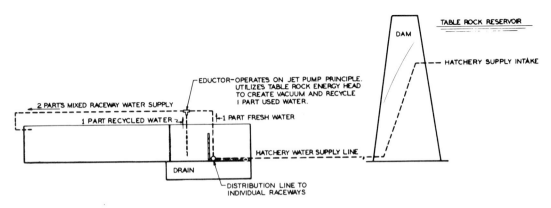

FIGURE 3. — Flow diagram. Typical raceway water supply. (*Shepherd of the Hills Hatchery, Missouri Department of Conservation Engineering Section.*)

the dissolved oxygen content of the mixed water. The efficiency of this unit is dependent on the system water pressure. System efficiency is further increased by the aeration devices at the inlet end of the raceway.

Alternate Water Line

An alternate water supply line is available at the inlet end of each raceway unit. This water supply line was provided to compensate for any efficiency loss of the eductor units. Since the raceway loading rate is based on one part fresh water and one part recirculated water, any loss of efficiency would reduce flow. The additional freshwater flow required for the fish loading rate may be obtained from this line. In addition, this line may be used for fish disease treatment when recirculation is undesirable. Another potential use for this line is in the theory of increased water exchange rate to flush the raceway during periods of high ammonia concentrations and low dissolved oxygen content after feeding.

Production Scheduling

The hatchery operates under production scheduling. This requires fish growth manipulation by controlling the water supply temperature. This concept is based on degree-days, which assumes that water temperature affects fish growth within limits (Fig. 2).

Eggs taken within each spawning period are divided into six fish groups, each group being hatched and raised at different water temperatures. For example, a fish group hatched and raised to 3.81 cm (1.5 in) in 7.22 °C (45 °F) water temperature will require 109 days. Another group hatched and raised in 9.44 °C (49 °F) water temperature will require only 69 days to grow to 3.81 cm (1.5 in).

Rearing inside the building, where water temperature is controlled, will take approximately 2 months for the first group and 5½ months for the last group to reach 6.35 cm (2.5 in) average length. The groups are then transferred to the field tanks where natural lake water temperature will be used. When the lake water temperature is lower than the

average annual water temperature, well water is used to temper water temperature as desired. It will take approximately 2½ months for each fish group to grow to 10.16 cm (4 in).

At an average length of 10.16 cm (4 in) the fish group is then transferred to the rearing raceways, where it spends approximately one year or until the fish average 25.4 cm (10 in) in length. Hatchery personnel then release the fish in Lake Taneycomo and other designated streams.

Hatchery Effluent

The normal raceway unit effluent will be discharged into two settling ponds totaling 25,488 m³ (900,000 ft³) in volume. This gives a 12-hour retention period for normal operations.

During the raceway cleaning operation the effluent is discharged into three settling ponds totaling 1699 m³ (60,000 ft³) in volume. This will give a 4-hour retention period for two raceways at a time. These ponds have to be dredged a minimum of once a year to remove accumulated solids.

The hatchery building facilities and field tank effluent will be discharged into one settling pond totaling 552 m³ (19,500 ft³) in volume. This will give a 2-hour retention period for normal operations.

All hatchery effluent is discharged into Taneycomo Lake. Flow rate is approximately 0.71 m³/s (25 ft³/s), including well water. At this time, the hatchery effluent is discharged without the use of the two major settling ponds, with a BOD level of 3.8 mg/l. Suspended solids are negligible.

Retention periods are designed to meet effluent discharge standards as set forth by Missouri's regulating agency (Department of Natural Resources) and by federal regulations.

Feed Storage

The hatchery facility is provided with 48.98-metric-ton (54-ton) bulk feed storage facility for dry pellet type feed. The storage facility is capable of handling three different sizes of feed. There are three individual bins, with one bin rated at 24.49 metric tons (27 tons) and two at 12.245

metric tons (13.5 tons) each. The storage facility was designed to handle 48.98 metric tons (54 tons) of feed transfer from truck to bins in 1.5 hours. For the hatchery building feed requirements a feed storage room is provided to store 45.4-kg (100-lb) bags of feed.

Costs

In keeping with the guidelines and philosophy of the Missouri Department of Conservation we have practically doubled the production capacity of this facility at a minimal cost of $0.18 per pound of fish produced over 30 years. The operating cost will be further increased by $0.22 per pound as a result of implementing production scheduling.

Conclusion

The limited lake water supply problem was overcome by designing a recirculating raceway system. During periods of low dissolved oxygen level, the lake water is aerated to maintain 80-90% saturation. Both systems utilize the available energy from the reservoir and do not require any external energy source.

The limited water recirculation does not greatly increase toxic parameters provided dissolved oxygen level is maintained at 80-90% saturation.

The loss or malfunction of the automated water chiller systems does not cause any fish mortality, but will affect the production schedule. Well water is used as the warmwater source. Use of the available energy from the artesian well will minimize use of the 40-hp pump, thus reducing external energy consumption.

Except for the water chiller systems all other hatchery features require little or no external energy source.

One of the most important factors not mentioned in this report is the hatchery manager. The facility was designed with sufficient flexibility to allow the hatchery manager to "do his thing," i.e. to produce fish. It is to be noted that the facility is but an instrument to the hatchery manager and not the other way around.

References

KRAMER, CHIN & MAYO, INC. 1973. Shepherd of the Hills conceptual plan report, prepared for the Missouri Dep. Conserv.

MISSOURI DEPARTMENT OF CONSERVATION. (Undated). Memorandum from Gary W. Camenisch to Charles E. Hicks. Preliminary water chemistry tests at Shepherd of the Hills Trout Hatchery.

U.S. ARMY ENGINEER DIVISION, NORTH PACIFIC CORPS OF ENGINEERS. 1973. Fisheries handbook of engineering requirements and biological criteria.

SESSION V   HATCHERY EFFLUENT TREATMENT

Introduction

WILLIAM A. GODBY

U.S. Fish and Wildlife Service
P.O. Box 25486, Denver Federal Center
Denver, Colorado 80225

Good morning. I am Bill Godby, moderator of this Session V Panel on Hatchery Effluent Treatment. I am the Regional Engineer for the U.S. Fish and Wildlife Service, Denver Regional Office. It is my pleasure to be invited to participate with this distinguished group of biologists and engineers.

The Hatchery Effluent Treatment Panel will discuss some of the methods used in treating fish hatchery effluents and the government regulations pertaining thereto. Emphasis will be placed in the areas of U.S. Federal regulations, British Columbia effluent treatment facilities, and effluent treatment facilities of the U.S. Fish and Wildlife Service.

Before I call on the panelists, I would like to set the stage by briefly commenting on the beginning of the need for hatchery effluent treatment and where I believe we are today.

In 1972, amendments to the U.S. Federal Water Pollution Control Act required the establishment of effluent standards for all point sources of pollutants permitted to discharge into waterways. For the first time in its long illustrious career of fish culture, its practices were challenged and considered an industry that produced point sources of pollutants to waterways.

The degree of treatment of fish hatchery wastes has been difficult to determine because of the lack of detailed guidelines and methods to describe accurately the pollutional characteristics of such wastes.

The studies conducted by Federal, State, and private entities on pollutional effects of fish hatchery wastes and the treatment thereof, to my knowledge, have never produced conclusive evidence and results that hatchery effluents are major sources of pollution to our waterways. In fact, I believe that the opposite has been shown.

Bio-Engineering Symposium for Fish Culture (FCS Publ. 1):157-161

Federal Regulation of Fish Hatchery Effluent Quality

JAMES HARRIS

U.S. Environmental Protection Agency
Region VIII, Montana Office
301 South Park Street
Helena, Montana 59601

ABSTRACT

Although fish hatcheries have been responsible for water quality degradation in some past instances, the overall impact created by these discharges has been less than significant. Control of hatchery effluents was authorized in 1972 with the passage of the Federal Water Pollution Control Act Amendments and the creation of the National Pollutant Discharge Elimination System (NPDES) permit program. Effluent guidelines that were to be promulgated pursuant to the 1972 amendments were never completed and nationwide guidance for writing discharge permits for hatcheries has not been prepared.

Regional offices of the Environmental Protection Agency have taken different approaches to limiting hatchery discharges ranging from using the incomplete draft guidance to not issuing permits at all.

The passage of the Clean Water Act of 1978 should not create any changes in the NPDES program relative to hatcheries and any additional guidance from EPA headquarters will probably not be forthcoming.

The good news is that for economic reasons hatchery effluent treatment beyond simple settling of cleaning flows should not be required.

Early Attempts at Effluent Control from Fish Hatcheries

There do not seem to be many documented early attempts by regulatory agencies to limit discharges from fish hatcheries. The Rifle Falls Trout Hatchery, operated by the Colorado Division of Wildlife, was closed in 1968 as a result of a civil action that was supported by the Colorado Supreme Court.

Complaints of alleged pollution had been lodged against the Fish and Wildlife Service's Jordan River National Fish Hatchery in Michigan since it began operation in 1964. In the fall of 1968, the Federal Water Pollution Control Administration conducted a study that concluded that the hatchery was contributing significant amounts of fish fecal materials and unconsumed fish food to the Jordan River. These materials were deposited on the river bottom and supported pollution-tolerant benthic organisms. These deposits were, at the time, in violation of the water quality standards of the State of Michigan.

Although most states had some type of water quality standards in the late 1960's, enforcement of those standards was seldom, if ever, evident. Federal law dealing with water quality was even weaker at the time with initial efforts based on an obscure section of the Rivers and Harbors Act of 1899. This enforcement tool was to become known as the "Refuse Act."

The U.S. Environmental Protection Agency's only mechanism for controlling wastewater discharges in the early 1970's was the Refuse Act. A permit program was begun for those who discharged to "navigable waters of the United States." Very few permits were actually issued under this program. Meanwhile Congress was in the process of creating a new means of Federal regulation to control point-source discharges.

The Federal Water Pollution Control Act Amendments of 1972

The Federal Water Pollution Control Act Amendments were adopted on October 18, 1972, and brought what might be termed as a "new era" to wastewater discharge control. The amendments created the National Pollutant Discharge

157

Elimination System (NPDES) program, which required that anyone discharging wastewater from a point source to a "water of the United States" apply for a discharge permit.

The permits were to contain limitations on the effluent quality which were to be met by a specified date in accordance with a compliance schedule. The amendments also required that effluent guidelines for certain "critical" industries be developed to provide the effluent limitations that would be incorporated into an NPDES permit. Fish hatcheries were one of the categories for which guidelines were to be developed. EPA had, in their promulgation of regulations pursuant to the 1972 amendments, placed hatcheries in the same category for guideline development as petroleum refining, pulp and paper processing, and others.

In addition to the effluent guideline development, EPA began an attempt to limit the number of applications that would have to be processed by requiring permit applications only from the large facilities within each category of discharges. The NPDES permit program regulations promulgated in 1973 revised the classification of fish hatcheries from the "critical industry" status to that of an agricultural facility. Their regulations required that a permit application be submitted for all hatcheries except those facilities discharging fewer than 30 days per year and those facilities producing less than 20,000 pounds of fish per year. Any facility, however, that was producing a water quality problem could be issued a permit.

The above mentioned regulations were challenged by the National Resources Defense Council and were revised for all discharger categories *except* hatcheries to include facilities of all sizes.

Although fish hatcheries were removed from the list of critical industries and classified as agricultural facilities, the development of effluent guidelines continued at EPA headquarters. The National Field Investigations Center (NFIC) at Denver, Colorado, (currently the National Enforcement Investigations Center) was contracted to develop the guidelines. A draft document was produced by NEIC in 1973 and circulated for comments from the fish farming industry and regulatory agencies.

The 1972 amendments outlined a strategy for achieving two levels of treatment and finally the goal of no discharge of pollutants by 1985. The first level of treatment for point-source discharges to waters of the United States was called Best Practicable Control Technology (BPT) and was to be accomplished by July 1, 1977. The next level of treatment was called Best Available Control Technology (BAT) and was to be achieved by 1983. The BPT and BAT limitations were to be proposed for each category of discharger by EPA's Effluent Guidelines Division in Washington, D.C.

The draft NFIC guidelines contained the BPT and BAT recommendations for fish hatcheries. The BPT effluent limitations were based on treating hatchery discharge during cleaning periods by simple sedimentation. Suspended solids and oxygen-demanding material reduction ranged from 15 to 20 percent for BPT.

BAT limitations were based on the addition of aeration equipment for the further reduction of volatile organics and solids. The guidance document containing the effluent limitations for BPT and BAT expressed the allowable effluent limitations in quantities that were based on the pounds of fish on hand at a given facility and made no allowances for variation in feeding rates. (Feeding rates vary with water temperature and fish sizes.)

The Effluent Guidelines Division received unfavorable comments on both the economic impact of the proposed limitations and the accuracy of the draft document. The possible installation of settling facilities to meet the BPT limits greatly disturbed the fish farming industry and further economic impact assessment was requested.

More importantly, the validity of the draft document itself was challenged. Data from operating hatcheries indicated that the proposed limits were, in some cases, an order of magnitude higher than existing discharge levels. If the draft guidelines were finalized in the then existing form, they would have allowed a degradation in watcr quality in some areas of the United States. Because of these concerns and others, the guidelines for fish hatcheries have not been published in final form as of this writing.

Hatchery Effluent Control Programs

Cold Water Facilities

The Environmental Protection Agency began to issue NPDES discharge permits shortly after the promulgation of regulations pursuant to the Federal Water Pollu-` tion Control Act Amendments of 1972. Although permits for fish hatchery discharges were not a high priority item, some regional offices began to develop a strategy for issuance.

An NPDES permit contains limitations that a facility's discharge must meet, a compliance schedule containing dates by which control procedures must be initiated, and a monitoring schedule for tracking compliance. All of these items were to be addressed in the effluent guidance document and were to be incorporated in the discharge permit when the document was finalized. As discussed earlier, however, no final effluent guidelines for fish hatcheries were published.

The Administrator of each EPA region could choose to issue NPDES permits without an effluent guideline's being published. The choice of each regional office or state that had assumed responsibility through delegation for administering the NPDES program was to issue permits to hatcheries with limitations from the draft guidance, issue permits with limitations developed by the region, or not issue permits. Obviously there was pressure from commercial producers and some government agencies to accept the last approach.

EPA Regions VIII, IX, and X include all of the Western states and a large number of coldwater hatcheries. Region VIII, which consists of the states of Colorado, Montana, North Dakota, South Dakota, Utah, and Wyoming, issued

discharge permits to all of the state, federal, and private hatcheries that were required to apply. The discharge permits contained effluent limitations formulated in the regional office providing for approximately 20 percent removal of suspended solids and organics.

Since the issuance of the first round of hatchery permits, four of the six states in the region have assumed responsibility for administering the NPDES permit program and have reissued discharge permits containing limitations based on the rationale used in the original permit. Some variation of the individual members may have occurred because of changes in operation at an individual hatchery. The EPA Region VIII office retains the responsibility for issuing permits to most of the federal facilities.

With the exception of the State of Utah, the state fisheries agencies have cooperated fully to assure that discharge limitations are realistic and that the permit requirements are complied with. Early monitoring efforts in Colorado and Wyoming indicated which hatcheries needed additional settling capacity and which facilities needed only modification of operational procedures. Recent data indicate that Colorado Division of Wildlife and the Wyoming Game and Fish Department hatcheries are consistently in compliance with their permit requirements. The Montana Fish and Game Department facilities are essentially in the same status. However, the Utah Division of Wildlife Resources has contended that EPA cannot issue discharge permits without the promulgation of effluent guidelines. This theory has been tested in court and proven incorrect.

EPA has also received excellent cooperation from the Fish and Wildlife Service in dealing with effluent control at federal hatcheries located within Region VIII. The Fish and Wildlife Service has participated in the development of several innovative treatment technologies.

Region IX of the U.S. Environmental Protection Agency is made up of the states of Arizona, California, Hawaii, and Nevada. In 1974 this region issued discharge permits to federally operated fish hatcheries using the draft guidance from EPA headquarters.

The State of California issued discharge permits to non-federal facilities using the same guidance through their regional boards. EPA and the State then became involved in litigation over authority to issue permits to federal facilities. Even though the court ruling was in favor of EPA, the program to issue permits to federal installations was delegated by Region IX to the State of California.

EPA's Region X consists of the states of Alaska, Idaho, Oregon, and Washington. The regional office has issued discharge permits to all federal hatcheries using the draft guidance mentioned earlier. The State of Washington was delegated the NPDES permit program and prepared fish hatchery permits for state facilities. Several of these permits were vetoed by the EPA regional office on the basis that the State was not requiring adequate treatment. This objection was based on the fact that the hatcheries in question did not have to provide a separate treatment unit to meet the permit limits. This is not an unusual situation in cases where hatchery permits are based on the Effluent Guidelines Division's guidance document. An agreement was reached between EPA's Region X and the State of Washington's Departments of Ecology, Game and Fisheries to provide monitoring at a specified sampling point. Best management plans were required from each facility.

Discharge permits issued in the State of Oregon were also based on the EPA guidance, and settling systems have been provided.

The EPA Idaho Operations Office has issued NPDES discharge permits for the fish hatcheries in that state. Approximately 50 coldwater commercial hatcheries and 15 federal and state facilities hold permits based on the draft development document. A state planning agency (Section 208 of the Clean Water Act) has completed a project to determine what limitations on fish hatchery effluents are economically feasible and environmentally acceptable. Although the final report is not available at the time of this writing, personal communication has indicated that the proposed limits will be more stringent than those in existing permits. This result seems to substantiate Region VIII's contention that the proposed guidance was not properly prepared.

Warm Water Facilities

The majority of warmwater fish farming and propagation is done in the South and southeastern United States. Fish farms in this part of the country are a major agricultural endeavor, and their representative organizations keep close watch on regulatory agencies and their programs.

EPA's Region VI office, located in Dallas, Texas, is responsible for issuing all discharge permits within the region, which consists of Arkansas, Louisiana, New Mexico, Oklahoma, and Texas. The regional office has chosen to rank all hatcheries except federal facilities as low priority for permit issuance. Since the states are not issuing NPDES permits in Region VI, only the federal facilities have received discharge permits.

Region IV of the EPA consists of Alabama, Florida, Georgia, Kentucky, North Carolina, South Carolina, and Tennessee. Region IV has issued approximately 150 discharge permits to tropical fish hatcheries which were based on the draft effluent guidance. This region has also issued permits based on the same document to federal hatcheries and catfish farms.

The commercial warm water lobby seems to be less involved with the development and issuance of discharge permits than the cold water faction. Hopefully this attitude will change in the future.

Future Regulation Under the Clean Water Act of 1977

The Federal Water Pollution Control Act was amended again in December of 1978 and renamed the Clean Water Act of 1978. The amended act required the promulgation of revised regulations for application for NPDES permits. Among the sections that were revised was the aquatic animal production category. The initial attempt by EPA to revise these regulations increased the permit application exemption level from 20,000 pounds production per year to 100,000 pounds production per year. The newly proposed figure would have eliminated over 90 percent of the coldwater hatcheries in the United States from regulation. The final revision published in mid-1979 retained the 20,000-pound limit for coldwater production facilities and raised the limit to 100,000 pounds for warmwater hatcheries. A minimum food requirement was also placed on the coldwater hatcheries to regulate short-term holding operations.

The revised NPDES criteria for facilities that must apply for a discharge permit will have little or no effect on coldwater hatcheries but may reduce the number of warmwater operations that need permits.

The current status of the effluent guidelines for aquatic animal production facilities leaves most EPA regions without adequate guidance to issue and enforce NPDES permits. There is some pressure either to complete the guideline process or to abandon it altogether in the near future. An adequate guideline would require a complete revision of past work on the guidance document. Currently, toxic substance guidelines are the top priority item in EPA headquarters and a revision of the hatchery document is not feasible.

Response to the original draft guidance document indicated that the proposed Best Available Treatment Technology alternative would create an economic hardship. If guidelines are promulgated, it is likely that there will be no BAT requirements. BPT levels, which are already in effect as of July 1, 1977, should be the only requirement. Those hatcheries in compliance with existing discharge permits based on BPT will not be required to install a higher level of treatment.

Future strategy for EPA regions and states with the NPDES responsibility should include the issuance of permits to all aquatic animal production facilities as required by the Clean Water Act of 1978. The BPT level of treatment can be met by most of the existing hatcheries with little or no construction.

Fish management agencies, whether they are federal or state, and commercial producers must communicate with regulatory agencies to assure that the industry is accurately represented when developing discharge permits. Call them; don't wait until they call you.

Conclusions

Federal regulation of fish hatchery discharges has been inconsistent in past years because of a lack of a properly prepared guidance document. EPA's regional offices and the states administering the NPDES permit program are using different criteria for discharge permits.

Effluent guidelines will probably not be forthcoming in the near future for hatcheries, and treatment beyond the simple settling of cleaning flows will not be required for economic reasons.

Eventually, a discharge permit will be issued to all hatcheries as required by the Clean Water Act. It is important that the federal or state fish management agency or the commercial grower communicate with the regulatory agency in this process.

References

ALBUS, CHARLES. Personal communication. U.S. Environmental Protection Agency (EPA), Region IV. June 1, 1979.

ANDERSON, DONALD. Personal communication. EPA, Washington, D.C. May 30, 1979.

HARRIS, JAMES C. 1972. Pollution characteristics of channel catfish culture. M.S. thesis. Univ. of Tex., Austin. 394 pp.

JOHNSON, WILLIAM. Personal communication. Sacramento Office, California. June 1, 1979.

KRAMER, THOMAS. Personal communication. EPA, Region IX, June 1, 1979.

MASARIK, MARK. Personal communication. EPA, Region X, Idaho Operations Office. June 4, 1979.

MOSBAUGH, KENNETH. Personal communication. EPA, Washington State Operations Office. June 1, 1979.

PETERS, DAVE. Personal communication. EPA, Region VI. May 31, June 1, 1979.

ROZACKY, GENE. Personal communication. EPA, Region VI, June 1, 1979.

STAMNES, ROBERT. Personal communication. EPA, Region X. May 30, 1979.

U.S. CONGRESS. 1972. The Federal Water Pollution Control Act Amendments of 1972, PL 92-500, October 1972.

U.S. CONGRESS. 1977. The Clean Water Act, PL 95-217, December 1977.

Bio-Engineering Symposium for Fish Culture (FCS Publ. 1):162-166

Hatchery Effluent Water Quality in British Columbia

R.A.H. Sparrow

Fish and Wildlife Branch
Ministry of Environment
Victoria, British Columbia
V8W 2Y9

ABSTRACT

In 1975 the Government of British Columbia established objectives to control the quality of wastewater discharged from existing and new fish hatcheries. These Pollution Control Branch (P.C.B.) objectives require monthly monitoring of BOD_5, SS, total phosphorus, ammonia and nitrate nitrogen as kg/100 kg fish per day.

Results of examination of hatchery water quality in British Columbia and elsewhere were used to compare the P.C.B. objectives to water quality from hatcheries using single-pass or water recirculation with reconditioning. Quality of wastewater from salmonid hatcheries employing single-use was similar to the P.C.B. objective specified for existing hatcheries. The more stringent objective required for new facilities approaches the level of water quality from hatcheries using water recirculation with reconditioning.

Short-term (10-min) retention of effluent from circular fish ponds was sufficient to remove in excess of 39% of the SS and allow efficient operation of a biofilter using 1 to 2 mm sand and a 1.59 $l/m^2 \cdot s^{-1}$ flow. Suspended solids and BOD_5 were reduced 80% and 69% respectively when biofilter backwashings were detained for eight hours in an aerated lagoon. However, it was expected that one-half of the N and P would remain in backwash subjected to this treatment prior to discharge.

Introduction

Within the last decade growth in fish propagation in the private and public sectors, coupled with concerns for a quality environment, have focused attention on the water quality of hatchery effluents. A survey of North American fisheries agencies (Liao 1970a) indicated one-third of all agencies had experienced pollution problems with hatcheries. Despite the widespread nature of the problem, there are few published papers that assist the bio-engineer in solving site-specific problems of pollution abatement.

Wastes from fish culture activities include various drugs and chemicals used for the control of fish diseases and parasites, the pathogens and parasites themselves, oxygen-demanding wastes, and various dissolved inorganic nutrients. The first two categories are usually considered to be transitory compared with the more continuous nature of the latter categories of waste.

British Columbia Pollution Control Objectives

In 1975, following a public inquiry and input from an advisory panel, the Provincial Pollution Control Branch (P.C.B.) issued guidelines for pollution control of hatchery effluents as part of the objectives for food-processing, agriculturally oriented, and other miscellaneous industries. These standards were intended to be minimum objectives reflecting effluent quality without reference to receiving waters. However, guidelines for the quality of receiving waters, although extremely general, are also incorporated in the objectives.

Effluent objectives are given in terms of pounds of contaminant per 100 pounds of fish per day (kg/100 kg fish per day). Two levels, A and C, are listed for hatchery effluent water quality with level C being the standard to which existing facilities must adhere (Table 1). Level A applies to all new facilities or as a level to which an ex-

TABLE 1. — Objectives for the discharge of effluent to marine and fresh waters from fish hatcheries.

Parameter[a]	Level	
	A	C
BOD$_5$	0.40	1.3
Suspended solids	0.40	1.5
Ammonia (N)	0.04	0.14
Nitrate (N)	0.12	0.12
Total phosphorus	0.02	0.035
pH range	6.5-8.5	

[a] As kg/100 kg fish per day.

isting facility may ultimately be upgraded. Parameters such as BOD$_5$ (5-day biochemical oxygen demand), suspended solids (SS), ammonia and nitrate nitrogen, total phosphate phosphorus, and pH must be monitored monthly and represent incremental increases above intake water values.

Prior to final selection of a site the potential discharger should provide the Pollution Control Branch with a report that shows an adequate measurement of physical, chemical, and biological parameters of the environment that may be subject to injurious alteration by the waste discharge. Also required is a comprehensive assessment of the quality and quantity of the proposed discharge as well as an assessment of the impact the hatchery will have on the receiving environment. This information is then used by P.C.B. to evaluate an application for a pollution control permit.

Wastewater from Single-Pass Facilities

In a single-pass or flow-through hatchery, wastes are more concentrated during raceway pond cleaning than in the normal operating mode (Mayo and Liao 1969; Bodien 1970; Brisbin 1970). Liao (1970a) stated the average BOD$_5$ was several times greater during pond cleaning, at times exceeding 35 mg/l when fish density surpassed 2.08 kg/l·min^{-1}. At a time of cleaning SS averaged 96 mg/l compared with a mean concentration of only 7 mg/l at other times. Evaluation of cleaning flows at the Summerland Hatchery in 1970 (Brisbin 1970) and more recent data from Abbots-

ford (Walden and Birkbeck 1974) show values for BOD$_5$ and SS in raceway cleaning flows that confirm or even exceed those given by Liao.

Cleaning flows accounted for 20-25% of the total hatchery output of BOD$_5$ and SS from Summerland (Brisbin 1970) and about 90% of these constituents could be resettled. A 12-minute retention has been reported (Jensen 1972) sufficient to remove 90% of settleable solids when the water velocity was 3.05 cm/s in a settling basin 0.46-m deep.

Since space and topography prevented development of settling lagoons at the Summerland Hatchery, raceway ponds are vacuumed daily to remove settled solids. After resettling, the sludge or slurry is trucked to a land-fill for disposal. For effluent from circular ponds Walden and Birkbeck (1974) have shown a detention time of 8 to 15 minutes is sufficient to reduce SS from 37 to 52%; BOD$_5$ to 16%; total Kjeldahl nitrogen to 27%; and total phosphorus to 14%; it is therefore likely that pond vacuuming will also reduce nutrient loads in hatchery effluents.

Waste characteristics for single-pass hatcheries in normal operations show the Pollution Control Objectives for level C would normally be satisfied. Data from several sources (Table 2) fall within the P.C.B. level C range of 1.3 BOD$_5$, 1.5 SS, 0.14 ammonia, 0.12 nitrate and 0.035 total phosphorus stated as kg/100 kg of fish per day.

Reconditioning Reuse Wastewater

Since Burrows and Combs (1968) described the first prototype recirculating system using biological filters, a number of hatcheries have been developed in North America (Mayo 1976) using this principal of water reuse (Pettigrew et al. 1978). Approximately 90% of the total flow in a hatchery may be recirculated and reconditioned in biological filters to convert toxic metabolites, with only the remainder, or 10%, of the flow discharged as wastewater. Therefore, in addition to increasing the water available for fish culture by reducing the inflow or make-up to 10% of

TABLE 2. — Waste characteristics from single-pass hatcheries. (As mean kg/100 kg fish per day with range in brackets.)

Source	Bodien 1970	Brisbin 1970	Liao 1970b	Walden and Birkbeck 1974
BOD$_5$	1.3[a]	1.0	1.34	
			(0.645-2.496)	(0.118-1.08)
Ammonia (N)	0.058		0.166	
		(0.05-0.13)	(0.031-0.404)	(0.014-0.043)
Nitrate (N)		0.20	0.401	
			(0.006-1.501)	(0.008-0.020)
Orthophosphate	0.015			
Total phosphorus	0.036		0.0333	
		(0.02-0.03)	(0.004-0.094)	(0.005-0.034)
Suspended solids				
		(1.0-1.5)		(0.118-0.941)

[a]Calculated from COD.

the total volume, less wastewater requires treatment from this type of hatchery.

Between 1972 and 1975, at the proposed site of the Abbotsford Hatchery (Walden and Birkbeck 1974, 1976), water reuse with reconditioning was extensively evaluated using pilot water recycling and a fish rearing system. Systems were operated at 90% recycle with provision for solids removal by settlement, biofiltration, reacration, and recycle. Biological oxidation of ammonia to nitrate took place in a flooded gravity sand filter (biofilter) prior to reaeration and reuse. The biofilter contained a 76-cm layer of Canada No. 1 chicken grit, a granite medium having 81% particles in the 1- to 2-mm range.

Waste from hatcheries recirculating and reconditioning water consists of the following:

1. Solids or sludge settled from fish pond effluent prior to biofiltration.
2. Intermittent backwashings from cleaning the biofilter.
3. A continuous discharge of water, usually as biofilter effluent, approximately equivalent to the amount of freshwater make-up.

Waste food and fecal material from circular fish ponds were continuously settled in rectangular tanks prior to entering the biofilter. At upflow and weir overflow rates of 1.89 l/m^2·s^{-1} and 3 m^3/cm·d^{-1} respectively, 37 to 59% of the SS were removed. Retention time in settling tanks was 8 to 9 minutes. At flows of 1.59 l/m^2·s^{-1} the biofilter functioned effectively when 40-50% of the solids were settled from fish pond effluent.

Filter backwashings in a recycle hatchery are high in BOD$_5$, SS, phosphate, ammonia, and nitrate (Table 3). Raceway cleaning wastes (Brisbin 1970; Liao 1970b; Walden and Birkbeck 1974) range from 31 to 122 mg/l BOD$_5$; 2.5 to 13.9 mg/l phosphate; 0.4 to 3.5 mg/l nitrate plus nitrite; 0.18 to 1.5 mg/l ammonia; and 55 to 230 mg/l SS. Therefore, filter backwashings are similar (Table 3) in content to pond cleaning wastes.

The BOD$_5$ to SS ratio in backwash waste at the Abbotsford pilot plant (Table 3) approximated 1.0, the ratio considered suitable for short-term aeration of suspended solids. A reduction of 80% of the SS and 69% of the BOD$_5$ was achieved when backwashings were treated for eight hours in an aerated lagoon (Walden and Birkbeck 1974). But the BOD$_5$ to total Kjeldahl nitrogen (TKN) and phosphorus ratios (Table 3) of 11.7 and 39 respectively were below the 20 and 100 values considered by Walden and Birkbeck (1974) to be necessary for good microbial nutrition in waste treatment. Therefore about one-half of the N and P will probably remain in backwash effluent subjected to short-term retention.

Metabolites are less concentrated in biofilter effluent (Table 4) than in effluent

164

TABLE 3. — Characteristics of filter backwash water. (Concentrations as mg/l.)

Source	Abernathy Bodien 1970	Dworshak Bodien 1970	Dworshak Liao 1970b	Abbotsford Pilot Walden & Birkbeck 1974
BOD$_5$	31.0	36.0	48.0	84.0
COD	91.0	81.0	158.0	
Ammonia (N)	0.1	0.13	1.34	
Nitrate (N)	3.6	2.4	1.03	
Total phosphate	4.4[a]	1.6[a]	0.99	2.13
SS	10.0	34.0	145.0	102.0
TKN	3.8	4.7		9.59
BOD$_5$/SS	3.1	1.1	0.33	0.82
BOD$_5$/TKN	8.2	7.7		11.7
BOD$_5$/P			48.0	39.4

[a]Orthophosphate

from a single-use hatchery (Table 2). About 80-90% reduction in BOD$_5$ and SS was reported by Liao and Mayo (1974) for various hatcheries recirculating water compared to a single-use wastewater. At Abbotsford, BOD$_5$ and SS were reduced at least 50% while ammonia was less than 10% of the concentration found in single-use wastewater (Table 2). It seems likely, with the possible exception of phosphorus, that the level A standard of Pollution Control Objectives (Table 1) would be achieved in wastewater discharged after biofiltration.

Options for Reducing Pollution

Hatchery wastewater characteristics will be modified by many parameters, some of which are water flow, fish size and density, feeding rates, fish health, activity of the animals, as well as temperature, oxygen, and other parameters representing influent water quality. Perhaps of all the parameters mentioned, feeding rates will have the greatest impact upon hatchery effluent water quality. Formulas for relating feeding rates to metabolite concentrations in effluents have been provided by Willoughby et al. (1972) and Liao and Mayo (1974). Use of these equations allows the fish culturist to establish appropriate feeding levels to achieve oxygen, ammonia, and other effluent standards that are not stressful to fish.

In British Columbia, examples of environmental degradation from hatchery effluents have not been obvious except

TABLE 4. — Waste characteristics of biofilter discharge from reconditioning reuse hatcheries. (Concentrations as kg/100 kg fish per day.)

Source	Bozeman[a] (upflow filter)	Abernathy[a]	Dworshak[a]	Abbotsford (pilot plant)[d]
BOD$_5$	0.199-0.144[b]			0.6286
Ammonia (N)	0.0075-0.028	0.0067	0.0097[c]	0.0066
Nitrite (N)	0.009-0.0016	0.0016	0.0006	0.0028
Nitrate (N)	0.0088-0.016	0.021	0.0357	0.161
Orthophosphate	0.034-0.070			
SS	0.220			0.285

[a]From Liao and Mayo 1974.
[b]Calculated from COD.
[c]Possible interference in analysis.
[d]Data on June 10, 1975, when loading was 0.607 kg fish/l·min^{-1}, 4.0% feed rate, 5.8 g fish, temperature 12-16 °C.

perhaps in small streams within the immediate vicinity of the point of discharge.

In order to achieve the level A standard of the Pollution Control Objectives new facilities will have to incorporate water reuse with reconditioning or provide aeration with retention of effluents in a lagoon or similar system. Even then it may not be possible to achieve the N and P concentration given in the level A standard.

References

BODIEN, D.G. 1970. An evaluation of salmonid hatchery wastes. U.S. Dep. Int. Fed. Water Qual. Admin., Portland, Oregon, 51 pp.

BRISBIN, K.J. 1970. Report on pollutional aspects of trout hatcheries in British Columbia. Prepared for Fish and Wildlife Branch, Dep. Recreation Conserv. Underwood, McLellan and Associates. 33 pp.

BURROWS, R.E., AND B.D. COMBS. 1968. Controlled environments for salmon propagation. Prog. Fish-Cult. 30:123-126.

DEPARTMENT OF LANDS, FORESTS AND WATER RESOURCES, 1975. Report on pollution control objectives for fish-processing, agriculturally oriented, and other miscellaneous industries of British Columbia. 35 pp.

JENSEN, R. 1972. Taking care of wastes from the trout farm. Am. Fishes U.S. Trout News. 16(5):4.

LIAO, P.B. 1970a. Pollution potential of salmonid fish hatcheries. Water Sewage Works 117:291-297.

_____. 1970b. Salmonid hatchery wastewater treatment. Water Sewage Works 117:439-443.

LIAO, P.B., AND R.D. MAYO. 1974. Intensified fish culture combining water reconditioning with pollution abatement. Aquaculture 3:62-85.

MAYO, R.D. 1976. A technical and economic review of the use of reconditioned water in aquaculture. F.A.O. Conference on Aquaculture. Kyoto, Japan, May 26 to June 2, 1976. FIR:AQ/Conf/76/R. 30. 24 pp.

MAYO, R.D., AND P. LIAO. 1969. A study of the pollutional effects of salmonid hatcheries. Kramer, Chin & Mayo, Inc., Seattle, Washington. 51 pp.

PETTIGREW, T., E.B. HENDERSON, R.L. SAUNDERS, AND J.B. SOCHASKY. 1978. A review of water reconditioning re-use technology for fish culture with a selected bibliography. Fish. Mar. Serv. Tech. Rep. 801. iv + 19 pp.

WALDEN, C.C., AND A.E. BIRKBECK. 1974. Feasibility and design studies for water reclamation and wastewater treatment at a trout hatchery. Prepared for British Columbia Department of Public Works by B.C. Research, Vancouver. 234 pp.

_____. 1976. Enhancement of the water reclamation system for Abbotsford Trout Hatchery. Prepared for the British Columbia Department of Public Works by B.C. Research, Vancouver. 169 pp.

WILLOUGHBY, H., H.N. LARSEN, AND J.T. BOWEN. 1972. The pollutional effects of fish hatcheries. Am. Fishes U.S. Trout News. 17(3):6.

Bio-Engineering Symposium for Fish Culture (FCS Publ. 1):167-173

Hatchery Effluent Treatment
U.S. Fish and Wildlife Service

Terry W. McLaughlin

Denver Engineering Center
U.S. Fish and Wildlife Service
P.O. Box 25486, Denver Federal Center
Denver, Colorado 80225

ABSTRACT

The U.S. Fish and Wildlife Service has evaluated several methods of removing pollutants from the discharges of salmonid fish hatcheries. Simple settling of cleaning wastes appears to be a satisfactory method of removing most pollutants from the discharges.

In 1966 the Lamar National Fish Hatchery recognized that extensive trout culture can cause pollution problems. Accumulation of waste deposits downstream from the hatchery was considered the major problem and an experimental settling basin was constructed. This initial settling basin had limited success.

In 1970 a new system was devised utilizing two adjacent 2.74- by 30.48-m raceways as settling basins. The flow from a series of four raceways was split beween them to achieve a velocity of 4.57 cm/s and a retention time of 12 minutes. This system was operated for 14 days, during which time the trout in the four raceways were fed 737.30 kg of feed. At the end of the test all the collected sludge in the settling basins was removed, weighed, and analyzed. Approximately 16% of the total food fed during the test period was recovered.

The system's influent and effluent were sampled and analyzed during various hatchery operations such as feeding and cleaning of raceways and resulted in more than 90% reduction in settleable solids under all conditions.

Following the testing at Lamar, several sedimentation systems were designed and constructed at various facilities. On August 7, 1973, Acting Deputy Director Loveless assigned an evaluation team to evaluate the operational characteristics of the effluent treatment systems at the Jordan River (Michigan), Jones Hole (Utah), Bozeman (Montana), and Quinault (Washington) National Fish Hatcheries. A brief summary of the evaluation team's report follows.

Jordan River National Fish Hatchery

Jordan River's treatment system consists of two linear clarifiers, 9.14 m by 30.48 m with a water depth of 1.22 m. The water flow averaged 378.50 liters per second (lps) and was divided equally between the two bays, providing an overflow rate for the unit of approximately 0.68 lps/m². (Fig. 1).

Three sample stations were established: (1) at the water source, Six Tile Creek; (2) where the effluent enters the system; and (3) at the discharge into the Jordan River. Both bays could not be tested simultaneously so one or the other was selected for each trial. Automatic composite samplers were used at the clarifier while grab samples were collected at Six Tile Creek.

FIGURE 1. — Settling basins at the Jordan River National Fish Hatchery.

Samples were iced and shipped by air to the laboratory.

Monthly samples were collected from September 1974 through August 1975. For the first five months, two composite samples were collected on the sample day — one during the working day, 0800 hours to 1600 hours, and the other during off hours, 1600 hours to 0800 hours.

All samples were analyzed for total suspended solids, settleable suspended solids, biochemical oxygen demand (BOD), chemical oxygen demand (COD), and total Kjeldahl nitrogen (TKN). Part of the settleable solids analysis was done at the station to avoid the effects of agitation in shipping. The results during this period were below test limits and of limited value. Therefore, the next seven months, one-hour composite samples were taken during fish feeding and raceway cleaning.

The settling basins were cleaned by drawing off the top water to expose the sludge, then moving it with a garden tractor to a channel across the center of each basin leading to manholes at the sides. It was then removed with a truck-mounted liquid manure spreader for land disposal. During this operation, sludge was collected from each load to form duplicate composite samples and the total volume of sludge removed was recorded. The samples were analyzed for dry weights of solids and the total dry weight of sludge was calculated. Volatile solids were also measured. A brass grain sampler, or "trier," was used to collect representative samples (Figs. 2, 3).

Data collected during raceway cleaning were fairly consistent compared to those collected during feeding and the 24-h cycles. The clarifier system removed 79.2% settleable suspended solids, 65.8% total suspended solids, 47.6% BOD, 62.6% COD, and 38.2% TKN.

Data collected during fish feeding showed lower efficiencies compared to measurements during cleaning. The system removed 54.5% settleable solids, 30.8% total suspended solids, 32.2% BOD, 26.7% COD, and 14.3% TKN.

Twenty-four-hour composite data proved to be inconsistent; sometimes

FIGURE 2. — Cleaning settling basin at the Jordan River National Fish Hatchery.

higher values were found leaving than entering the treatment system. Very low concentrations, approaching lower test limits, and loss of material on the sides of the container may have caused these errors.

The accumulation of sludge as related to food fed ranged from 10.8% to 38.0%. Digestion of sludge during the holding period may have contributed to differences in percent recovery of food.

Jones Hole National Fish Hatchery

Jones Hole has three linear clarifiers, 34.75 m by 12.70 m and 1.83 m. They received a total flow of 881.78 lps and had an overflow rate of 0.66 lps/m². Each unit has a sludge-scraper system for sludge removal; i.e., long redwood boards attached by chain to move and deposit sludge in sumps at the upper end of the clarifiers. The system was designed to pump the

FIGURE 3. — Cleaning the settling basin at the Jordan National Fish Hatchery.

168

FIGURE 4. — Jones Hole National Fish Hatchery with settling tanks in the foreground.

sludge to drying chambers; however, the pumping system was inadequate to handle the sludge. There was no alternate system for sludge removal; therefore, sludge quantitation was not possible (Fig. 4).

Water samples were collected at two stations: (1) entering the facility and (2) leaving the facility. Samples were collected from each station before the raceway cleaning (as a background sample) and during raceway cleaning. Initially, grab samples were taken from the influent and effluent of the three bays. The three samples were combined to form composite influent and effluent samples. Effluent sample collection was delayed thirty minutes, so the same water was sampled entering and leaving the facility. One-hour composite samples were taken July through December by automatic samplers. All samples were analyzed for total suspended solids, settleable suspended solids, BOD, COD and TKN.

Sludge samples were collected for analysis using a "trier" to characterize the compacted sludge.

The most consistent data were collected during raceway cleaning. Some intake concentrations were very high, between 100 and 404 mg/l, resulting in efficiencies for suspended solids approaching 100 percent. These high values came from grab samples rather than composites. The linear clarifiers removed 91.2% settleable suspended solids, 90.0% total suspended solids, 76.8% BOD, 67.9% COD, and 61.3% TKN.

Analysis of samples collected during feeding produced very low concentrations of suspended solids, BOD, and COD; however, TKN held close to 2 mg/l. Since few usable data were collected on settleable solids, and other results were highly variable, averages were not calculated. Background data collected before raceway cleaning were so low and inconsistent that true averages could not be obtained.

Sludge samples were collected and analyzed for characterization of compacted sludge. The results varied with depth, siltation, and degree of decomposition.

FIGURE 5. — Lamella plate packs before installation at the Bozeman Fish Cultural Development Center.

Bozeman National Fish Hatchery

The treatment facility at Bozeman is a Lamella separator designed to handle 315.42 lps at an overflow rate of 0.34 lps/m². It has a total of 400 plates in two bays, four 50-plate packs per bay (Fig. 5).

The facility is equipped with a sludge scraper system to move the sludge to two sumps at the end of each bay. The system was designed to pump the sludge to a concentrator at the side of the unit and from there into a storage vault. Unfortunately, the sludge pumps were too small to handle the load. The sludge had to be removed by a vacuum liquid manure spreader, which operates on the power take-off system of a tractor.

Sampling stations were located at the intake and outfall of the separator and one-hour composite samples were taken by automatic sampler during raceway cleaning. Background grab samples were taken prior to cleaning.

Samples were analyzed for total suspended solids, settleable suspended solids, COD, and TKN. The waterflow for the sampling period was approximately 75.70 lps and 200 plates were in use at an overflow rate of 0.18 lps/m².

To quantitate the amount of sludge that was recovered by the separator, the unit was cleaned, then put into service for a 30-day period. The accumulation of sludge was then removed, volumes recorded, and samples taken for solids analysis to determine the dry weight solids recovered.

Most data were collected during raceway cleaning, October through November 1975. With the exception of the first value on October 9, the percent removal of total suspended solids for the first half of the study was above 80%. In late October and November, the efficiency dropped to a low of 65.6% removal. The percent removal for settleable and total suspended solids was 83.3% and 77.9%, respectively, excluding the first value on October 9. The removal efficiencies for COD and TKN were 66.0% and 48.2%, respectively.

A preliminary test of the Lamella separator was conducted at two overflow rates to show the effects of changing the settling area, using a half bay (100 plates) as compared to a full bay (200 plates). The flow was approximately 88.32 lps with overflow rates of 0.41 lps/m² for a half a bay and 0.21 lps/m² for a full bay. When a full bay was used, the head loss increased 6.35 cm in eight days. A half-bay had an initial head loss of 10.16 cm and an additional head loss of 9.53 cm in five days. Growth of bacteria on top of the plates and in return tubes caused the head loss. The half bay was less effective in the removal of total suspended solids when the head loss reached 17.73 cm. At 10.16-cm head loss, the efficiency of the half bay was comparable to the average for total suspended solids reported.

The results of two 30-day sludge collections show that the percent recovery of food fed was about 15%.

Quinault National Fish Hatchery

Quinault (Fig. 6) has a Lamella separator with approximately 1,800 plates designed to treat up to 1,387.83 lps at an overflow rate of 0.34 lps/m² of horizontal settling area. There are three bays, each holding 12 50-plate units. Below the plates, there is a sludge-scraper system for moving the sludge to two sumps at the lower end of each bay. The sludge is pumped from the sumps to a concentrator, where it is partially dewatered before it is pumped to drying beds (Fig. 7).

Sampling stations were established (1) at the hatchery intake (Cook Creek); (2) at the intake to the separator; and (3) at the

FIGURE 6. — Aerial view of the Quinault National Fish Hatchery.

hatchery discharge to Cook Creek. Initially, two composite samples were collected on a 24-h cycle, one during working hours, 0800 hours to 1700 hours, and the other from 1700 hours to 0800 hours. Grab samples were collected at 2-h intervals during the day and combined to form a composite at 4- to 8-h intervals at night. The Cook Creek water supply sample was a single grab sample. All samples were analyzed for total suspended solids, TKN, and ammonia nitrogen.

Concentrations were below the limitations of the suspended solids tests and since the rearing units at Quinault are Burrows ponds, which are self-cleaning, raceway cleaning could not be used to bring concentrations up to measurable levels. Because of low concentrations, night sampling was discontinued after September 10, 1975.

Sludge sampling and quantitation were attempted. A small bypass line was installed on the discharge pipe from the sludge pumps and samples were collected as the units were being cleaned. The samples were analyzed for percent dry weight of solids per unit volume.

The data indicate fairly constant but very low concentrations entering the Lamella separator. Because of the low level of pollutants, measurements of settleable solids, BOD, and COD were omitted. The mean percent removal of total suspended solids and TKN was 40.3% and 20.4%, respectively. Ammonia nitrogen measurements were also very low but were reduced an average of 47.1% (Figs. 8, 9).

FIGURE 7. — Lamella separators and sludge concentrator at the Quinault National Fish Hatchery.

FIGURE 8. — Lamella separators at the Quinault National Fish Hatchery.

FIGURE 9. — Lamella plate pack, Quinault National Fish Hatchery.

In spite of difficulty in getting representative samples of sludge, some data were collected. Frequent cleaning of the separator produced percentages of food recovered from 4.4% to 23.8% with an average of 10.6%.

Discussion

Construction of Lamella separators and concrete settling basins are expensive, so alternatives are desirable from an economic standpoint. With cost being a consideration, the use of vacuum cleaning devices is often mentioned. During the summer of 1977 a team evaluated the practicality of using such devices.

Conclusions

Vacuuming systems can be an interim method of allowing hatchery discharges to meet NPDES permit requirements. Even with lightweight hoses, heads, booms, quick couplers, wheeled tractors, etc., vacuum cleaning is a labor-intensive device that requires much labor input resulting in the need for additional manpower (Fig. 10).

Many sludge-settling lagoons are presently serving national fish hatcheries, with the hatcheries being able to meet their NPDES discharge requirements. The most common problem that has occurred with the use of the settling lagoons is low dissolved oxygen in the effluent. Where this problem has occurred, installation of mechanical aeration has increased the dissolved oxygen to satisfactory levels.

Studies conducted at various hatcheries indicate that the largest pollution load is discharged when raceways are being cleaned. This fact is being utilized in the design of the Iron River National Fish

FIGURE 10. — Mobile vacuum pumper used at the Bozeman Fish Cultural Development Center.

Hatchery. Piping for the raceways will have separate reuse lines and cleaning waste lines. The system will be valved such that the cleaning waste can be discharged to one settling chamber and the flow through to a second settling chamber. The cells can be alternated daily to balance the sludge buildup in the two cells.

By having the cleaning waste separated from the normal flow through, the detention time that will be available for waste settling will be approximately sixteen hours, which should effectively remove the settleable solids from the effluent.

References

CAULFIELD, J.D. 1976. Hatchery pollution abatement, an environmental and economic assessment. UMA Engineers, Inc., Portland, Oregon. 24 pp.

GODBY, W.A., J.D. LARMOYEUX, AND J. VALENTINE. 1976. Evaluation of fish hatchery effluent treatment systems. Final report. Dep. Int. U.S. Fish Wildl. Serv., Washington, D.C. 59 pp.

MCLAUGHLIN, T., AND W. GESELBRACHT. 1972. Pages 34-45 in Annual report on development center activities 1971. Fish Cultural Development Center, Bur. Sp. Fish. Wildl., Lamar, Pennsylvania.

_____ . 1977. Evaluation of vacuum cleaning devices. Final report. Denver Engineering Center, U.S. Fish Wildl. Serv., Denver, Colorado. 20 pp.

Bio-Engineering Symposium for Fish Culture (FCS Publ. 1):174-182

Guidelines for Economical Commercial Fish Hatchery Wastewater Treatment Systems[1]

VINCENT A. MUDRAK

Pennsylvania Fish Commission
Benner Spring Fish Research Station
Bellefonte, Pennsylvania 16823

ABSTRACT

Four modern salmonid hatcheries, employing raceway culture with bottom cleanout drains, were evaluated as to the effectiveness of their effluent wastewater treatment facilities. Those systems studied included concrete clarifiers, earthen settling basins, and earthen stabilization lagoons. Analysis concluded that the best treatment technology consisted of a combined system of (1) a rectangular concrete clarifier designed to accept the raceway cleaning flow, coupled with (2) a "polishing" stabilization lagoon that received both the raceway and clarifier discharges. Design criteria were formulated for developers planning to implement these concepts into prospective hatchery design. Studies of the settling facilities were also conducted to determine the anticipated volume of solids accumulation. It was demonstrated that concrete clarifiers recovered a volume of sludge equivalent to about 20% of the dry weight of fish food fed; comparatively, the earthen settling basins recovered only about 3% of this weight. Four management-related problem areas that affected operational efficiency were noted and addressed; these were operational strategy, diet formulation, hydraulic loading, and solids handling.

Introduction

The U.S. Environmental Protection Agency, through the Federal Water Pollution Control Act (PL 92-500) and its associated program, the National Pollutant Discharge Elimination System (NPDES), has established formalized effluent guidelines for the fish culture industry, SIC #098. Likewise, it has listed treatment technologies capable of meeting these criteria. The technologies currently recognized by the EPA as sufficient to meet the needs of modern raceway fish culture are (1) settling of the cleaning flow with sludge removal, (2) vacuum cleaning of the raceways, or (3) an equivalent alternative. (Economic factors, however, have slowed down the implementation of this program.) The Pennsylvania Fish Commission combining environmental concern with professional prowess realized the state of the industry and developed a plan to establish both clean and productive fish culture stations. It was in this light that the hatchery wastewater treatment systems were constructed. Subsequently, representative hatcheries employing different treatment technologies were selected for study in order to evaluate the effectiveness of each component structure of the wastewater treatment complex (Table 1.)

Effectiveness of Treatment Systems

Four Pennsylvania Fish Commission hatcheries — Big Spring, Oswayo, Pleasant Gap, and Tionesta — were evaluated as to the effectiveness of their effluent wastewater treatment systems. Water sampling over a three-year period was conducted on a monthly basis, with each sample series representing a one-week composite. Sampling locations consisted of representative sites, so that each component waste treatment structure could be evaluated separately. Scheduling was arranged so that the time interval between sampling and analysis remained minimal. All samples were transported to the Benner Spring Fish Research Station, at

[1]The information contained in this paper was abstracted from research project #3-242-R, entitled "Guidelines for Economic Commercial Hatchery Wastewater Treatment Systems," sponsored by the U.S. Dep. Comm. Nat. Ocean. Atmosph. Admin., and the Pa. Fish Comm.

174

TABLE 1. — Treatment system technology.

PA Fish Commission Fish Culture Station	Type of settling facility							Type of stabilization lagoon		
	Detn.[a]	Earth.[b]	Conc.[c]	Circ.[d]	Rect.[e]	Mech.[f]	Sldg.[g]	Detn.	Earth.	Depth[h]
Big Spring F.C.S. (Cumberland Co.)	6.0		*	*		*	*			
Oswayo F.C.S. (Potter Co.)	0.5		*		*		*	4.0	*	1.5
Pleasant Gap F.C.S. (Centre Co.)	1.5	*								
Tionesta F.C.S.	0.5		*		*		*	2.0	*	1.1

[a]Detention time in hours
[b]Earthen
[c]Concrete
[d]Circular
[e]Rectangular
[f]Mechanical sludge scraper
[g]Provisioned for periodic sludge removal
[h]Depth in meters

which time chemical, physical, and bacteriological analyses were performed. The following discussions keynote the results.

(1) The Big Spring, Oswayo, and Tionesta clarifiers functioned consistently well respective to oxygen demand and solids removal. None was significantly more efficient than another, with the average percent removal during raceway cleanout being the following: biochemical oxygen demand (BOD), 84%; chemical oxygen demand (COD) 85%; suspended solids, 88%; and settleable solids per thirty minutes, 99%. Considering the effectiveness of each clarifier on each respective hatchery system over a twenty-four-hour period, the average percent removal dropped to BOD, 80%; COD, 82%; suspended solids, 85%; and settleable solids per thirty minutes, 98% (Table 2).

(2) The small stabilization lagoons at the Tionesta and Oswayo stations offered additional treatment (Fig. 1). Positive percent reductions occurred with respect to oxygen demand and settleable solids. A negative percent reduction, however, was often realized upon comparing the suspended solids and phosphate data. This was empirically evidenced by the periodic loss of solids as unicellular algae.

TABLE 2. — Efficiency of component wastewater treatment structures.[a]

Clarifier-Unit Efficiency (%), race cleanout

PFC hatchery	BOD	COD	Sus. sol.	Set. sol.	T.PO$_4$P
Big Spring	83.5	80.4	88.0	99.0	76.7
Oswayo	79.2	84.6	87.3	97.8	65.6
Tionesta	88.2	88.9	88.9	99.1	68.2

Clarifier-Unit Efficiency (%), 24 hours

PFC hatchery	BOD	COD	Sus. sol.	Set. sol.	T.PO$_4$P
Big Spring	73.7	72.8	80.2	98.4	63.1
Oswayo	76.7	83.0	86.0	97.6	62.2
Tionesta	88.2	88.9	88.9	99.1	68.2

Clarifier-System Efficiency (%), 24 hours

PFC hatchery	BOD	COD	Sus. sol.	Set. sol.	T.PO$_4$P
Big Spring	49.7	44.9	50.6	94.5	47.2
Oswayo	36.1	32.2	43.8	90.7	30.7
Tionesta	24.2	20.8	28.3	88.4	14.2

[a]The efficiency of each clarifier respective to itself has been defined as the Clarifier-Unit Efficiency and determined by evaluating the clarifier influent versus the clarifier effluent. These data have been developed and expressed as efficiency during the raceway cleanout period of the day and also over a twenty-four-hour cycle. The efficiency of the clarifier respective to the entire system has been defined as the Clarifier System Efficiency, determined by comparing its effect upon the hatchery output.

FIGURE 1. — View of the concrete settling chamber and earthen stabilization lagoon at Tionesta Fish Hatchery.

Average percent reductions of wastes passing through these lagoons over a twenty-four-hour period are as follows: BOD, 15%; COD, 13%; suspended solids, -2%; settleable solids per thirty minutes, 34%; and total phosphate, -1%. These percent reductions respective to the entire hatchery system over a twenty-four-hour period are averaged thus: BOD, 10%; COD, 8%; suspended solids, -2%; settleable solids per thirty minutes, 3%; and total phosphate, 0% (Table 3).

(3) Bacteriological analysis of water, as it passed through the hatchery system, demonstrated the occurrence of two interesting phenomena. One, there was a consistent increase in total coliform, *Aeromonas* species, and total bacteria counts; and two, there was no significant increase of those bacteria that would usually be considered as dangerous indicators (i.e., the fecal coliform and/or fecal streptococci groups).

(4) The earthen settling basins at the Pleasant Gap Hatchery were difficult to manage, erratic in efficiency performance, and periodically odorous (Fig. 2). Accumulated solids often washed out of the basins during the cold ice-cover months and the hot sunny months. Rummaging waterfowl also stirred up the settled material, hence reducing the system's efficiency.

Each wastewater treatment system was evaluated according to (1) relative efficiency in waste removal; (2) aesthetic appearance of the associated hatchery discharge; and (3) functional compatability with the ongoing fish culture processes. The small rectangular concrete clarifiers were adequately efficient while providing for very simplified operations (Fig. 3). Considering the reduction of the studied parameters, the stabilization lagoons offered very limited treatment. Aesthetically, however, each significantly enhanced the quality of the effluent

TABLE 3. — Efficiency of component wastewater treatment structures.[a]

Lagoon-Unit Efficiency (%), 24 hours

PFC hatchery	BOD	COD	Sus. sol.	Set. sol.	T.PO$_4$P
Oswayo	14.6	21.3	0.8	37.8	-2.6
Tionesta	14.9	4.4	-4.9	29.3	1.1

Lagoon-System Efficiency (%), 24 hours

PFC hatchery	BOD	COD	Sus. sol.	Set. sol.	T.PO$_4$P
Oswayo	7.7	13.0	0.4	2.7	-1.4
Tionesta	11.3	3.5	-3.5	3.4	0.9

Total System Efficiency (%), 24 hours

PFC hatchery	BOD	COD	Sus. sol.	Set. sol.	T.PO$_4$P
Big Spring	49.7	44.9	50.6	94.5	47.2
Oswayo	43.9	45.2	44.2	93.4	29.3
Tionesta	35.5	24.2	24.8	91.8	15.2

[a]Some of the wastewater treatment systems have included extended treatment in the form of small stabilization lagoons. The efficiencies of these have been expressed as Lagoon-Unit Efficiency and Lagoon-System Efficiency. The efficiency of the entire wastewater treatment system, i.e. the effect of all treatment components evaluated collectively, has been termed the Total System Efficiency.

discharge (minimized odor and reduced *Sphaerotilus*). Also provided by the lagoon was a "buffer zone" should any operational mishap have occurred. Analysis of these factors concluded that the best treatment scheme would be a combined system whereby (1) a concrete rectangular clarifier designed for at least

FIGURE 2. — Earthen settling basin at the Pleasant Gap Fish Hatchery.

FIGURE 3. — Internal view of the rectangular clarifier.

30-minute retention of the cleaning flow would be followed by (2) an earthen stabilization lagoon designed to accept both the clarifier discharge and the raceway overflow with a 4-hour retention period.

Composition of Solid Waste

Sludge composition and rate of accumulation were studied at the settling chambers of these hatcheries. Sludge samples were collected every time the settling facility was cleaned; and subsequently, representative aliquots were analyzed for solids composition. Fish feeding records were also maintained for future correlations.

Results of this work yielded a positive correlation (r = +0.9025) between the pounds of fish pellets fed at the Big Spring Hatchery and the amount of sludge that accumulated in the clarifier (Figs. 4, 5). At this site, the dry solids ratio of sludge to fish food fed was 0.258 : 1.00, or a recovery of 25.8%. Likewise, on the average, every thousand pounds of fish food (pellets) produced 244 gallons of sludge at 11.9% solids. Similar observations have also been made at the other settling chambers respective to percent return of dry solids (Table 4).

The earthen settling basins at the Pleasant Gap facility were evaluated with respect to solids recovery. Core sampling was conducted at preselected sites, so that information could be gathered as to the composition of the sludge with respect to

FIGURE 4. — Circular clarifier at the Big Spring Hatchery.

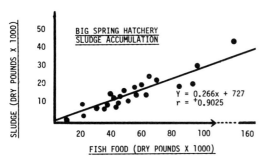

FIGURE 5. — Regressions analysis of accumulated sludge vs fish food fed.

depth and lateral distance from the basins' influent point. Sludge samples were also taken as the basins were mechanically cleaned. Both sample series were analyzed for solids composition, the results of which were compared for added verification.

Results of this earthen settling basin study have determined the dry solids recovery (sludge/fish food) to be 3.2%. An inverse relationship was also ob-

served, whereby settled sludge decreased in organic (volatile) content with respect to (1) the sludge depth; (2) the lateral distance away from the influent site; and (3) the length of time the sludge was permitted to remain prior to removal.

Design Criteria

The most important consideration of a proper wastewater treatment design should be the nature and composition of the waste, but consultants often overlook this point, using the premise that fish culture waste is characteristically typical to human domestic waste. As such, standard city-type treatment systems are often prescribed.

Pennsylvania studies have demonstrated that hatchery wastes, generated from fish culture processes utilizing raceways with bottom waste cleanout drains, are markedly different from domestic wastes. Hatchery waste solids are denser, settle faster, and tend to concentrate into a heavy viscous sludge. For this reason it is recommended that, where appropriate, the following design criteria be considered.

Clarifier Design Considerations

(1) Basic design: Make it simple, accessible, and drainable.
(2) Influent flume: Provide for uniform, laminar flow.
(3) Surface settling rate: Maximum of 40.8 m³/(d·m²) (1000 gpd/ft²) at peak hourly flow.
(4) Detention time: Minimum of 30 minutes at peak hourly flow.
(5) Weir loading: Maximum of 372

TABLE 4. — Solids composition of accumulated waste.

PFC hatchery	Settling facility	Sludge[a] % solids	Sludge % vol. sol.	Gal sludge per 1000 lb fish food	Dry sol. % recovery
Big Spring	concrete	11.9	51.7	244	25.8
Oswayo	concrete	5.7	74.4	301	16.8
Tionesta	concrete	4.6	61.2	431	18.3
Pleasant Gap	earthen	46.5[b]	17.3	N/A[c]	3.2

[a]Sludge from concrete chambers was pumped wet.

[b]Sludge from the earthen basins must dry out before removal by excavating machine.

[c]Not applicable.

m³/(d·m) (30,000 gpd/ft) at peak hourly flow.

(6) Sludge hopper: Maximum sludge depth of 1.5 meters; placement at influent end; size for six weeks' sludge accumulation.

(7) Sludge scrapers: Do not implement continuous mechanical scraping devices.

(8) Sludge pump: Provide accessibility should blockage occur (a hoist works well).

Stabilization Lagoon Design Considerations

(1) Placement: Should accept raceway overtopping flow and clarifier discharge flow.

(2) Detention time: Four hours at daily average flow.

(3) Average depth: Should range from 1.2 to 1.5 meters.

(4) Structure: Earthen, length less than three times the width; slope sides at 2 horizontal:1 vertical, and cover with crushed stone.

Management Considerations

Several years of study have revealed some problem areas regarding the treatment of fish culture wastes. These concerns fall into categories of (1) operational strategy; (2) problem diets; (3) hydraulic loading; (4) sludge removal; and (5) sludge disposal. Each of these areas is characterized below as a separate entity.

Operational Strategy

Over the last ten years fish culture has advanced to the point that it can be considered a science. Related elements of pathology, chemistry, pharmacology, nutrition, statistics, and genetics have faced the hatchery manager every day. Now a new element has been added — effluent standards. It is imperative that the hatchery manager understand the principles of wastewater treatment as it applies to his system. (That is the only way that optimum efficiency will be maintained.) It is recommended that all new construction or renovation projects incorporate with the engineering design module an operation and control manual for the hatchery

manager. Upon completion of the project, but before startup, the designer should visit the site and review the operations manual with the hatchery manager on a point-by-point basis.

Problem Diets

Modern technology has benefited the fish culture industry inasmuch as processed dry diet formulations of excellent quality are available from many reputable commercial vendors. However, even the most conscientious vendor may produce a batch of dry pelleted food that is inferior in nutritional or physical composition. Should the diet formulation contain too many fines (dust-like particles) or be lacking in a bonding agent (to hold the diet constituents together) feeding would probably result in excessive amounts of suspended solids in the hatchery's effluent waters. As such, monitoring of the dry fish food should be a routine function.

With the advent of bulk (silo) storage of fish food, excessive fines in the feed have become more prevalent. This results from fracturing during loading operations whereby dealers blow pellets into the overhead storage bin. Hatchery managers should be vigilant and incorporate a collection device for recovery of fines should the problem persist.

Hydraulic Loading

Excessive volumes of water entering the clarifier over an extended period of time can result in solids washout and hence reduced efficiency. This is especially true of a settling chamber that has accumulated a large column of sludge. It is imperative that designers consider and eliminate cross connection flow, groundwater infiltration, and stormwater runoff from entering the underdrain system leading to the clarifier. Hatchery managers should minimize shock loading by staggering raceway cleaning periods, and by using flow retaining shields to permit a concentrated flush of solids from the raceway cleanout area.

Leakage from the raceway drain plugs should be minimized. Two exceptions to this occur in extreme warm and cold weather periods. At these times it is advan-

179

tageous to maintain a constant low flow — just enough to keep ice from forming in the winter, and enough to inhibit excessive gas production in the sludge during the summer.

Sludge Removal

In order for clarifiers to function properly, accumulated solids must be periodically removed from the settling chamber. If sludge is permitted to stay in the chamber for too long a time, sludge digestion will occur. This results in a lower operational efficiency and a change in physical composition of the sludge. In fact, the sludge may become very viscous and extremely difficult to remove. Should there be an extended delay, solids will accumulate to the point where settling efficiency is lost and solids carryover will occur.

It is recommended that a sludge pumping schedule be implemented where the time interval between sludge pumpings be related to the time it takes to fill the clarifier's hopper. (This interval can be estimated for a well-operated rectangular clarifier, assuming that each 1000 pounds of dry food pellets will yield 1.14 m^3, or 300 gal, of waste sludge at 8% solids.) In Pennsylvania, the recommended design should permit hopper storage for six weeks' sludge accumulation during the maximum feeding period of the year.

Sludge Disposal

The wastewater treatment process is not complete until the accumulated solids are

FIGURE 7. — Vehicle used for sludge transport and disposal.

FIGURE 8. — Heavy sludge application to test strip.

removed from the system and wasted in an environmentally acceptable manner (Fig. 6). Direct land application of liquid sludge continues to be the recommended procedure, whereby solids are trucked to well-drained disposal sites and sprayed onto the cover vegetation (Figs. 7, 8). This process is essentially trouble free during clear weather conditions; however, rain and snow lead to problems.

FIGURE 6. — Sludge disposal area at the Big Spring Hatchery test site.

TABLE 5. — Nutrient composition of settled sludge.

| | Percent dry weight | | |
	Nitrogen	Phosphorus	Potassium
Big Spring sludge[a]			
(≤ 17 days old)	4.85	1.79	0.15
Pleasant Gap sludge[b]	2.17	2.99	0.46
(Top 7.5 cm)			
Pleasant Gap sludge[c]	1.41	1.49	0.71
(25 cm ≤ Depth < 30 cm)			

[a]Big Spring sludge is typical of fresh sludge that is pumped from a concrete settling chamber.

[b]Pleasant Gap sludge (top 7.5 cm) represents settled material that accumulated in the earthen settling basin over the last six months of operation before mechanical excavation. Sludge from this basin was permitted to dry out during the warmweather months.

[c]Pleasant Gap sludge (25 cm ≤ D < 30 cm) represents the settled material that accumulated in the bottom of the earthen settling basin (oldest sludge) during the first six months of operation.

Deep snows can make sludge transport or field disposal virtually impossible. Engineers should consider standby sludge storage capabilities in those regions where long winters prevail. This may be a small sand bed or nearby field strip to be utilized only during inclement weather conditions. Contracts with sewage treatment facilities may also be feasible if one is nearby.

Disposal of accumulated solids during rainy weather periods often leads to excessive odors. Though not any more offensive than livestock waste, local residents complain, at times bitterly. A successful fish culture operation, whether it is commercial or governmental, must receive public acceptance. As such, it makes sense for hatchery managers to be flexible, and to plan activities to utilize (and optimize) available clear weather conditions.

Just how much land is required for proper sludge disposal? This is a question for which we do not as yet have a firm answer. We know, however, that sludge heavy metal concentration is minimal and the nitrogen content is low (Table 5). These factors favor land disposal. Annual sludge application rates based upon nitrogen and heavy metal loading have also been calculated; nitrogen was the controlling factor (Tables 6, 7). For a mixed grass crop with a nitrogen uptake at 224 kg/ha (200 lb/acre) the annual sludge application would be limited to 0.86 kg/m^2 (3.85 tons/acre). This would equate to 123 m^3/ha (13,200 gal/acre) of sludge at 7% solids. Also, the estimated lifetime of the cropland based on a total permissible cadmium limit at 3.36 kg/ha (3 lb/acre) for a sludge content of 7.6 ppm cadmium would be 51 years. In practice, we have observed that Pennsylvania hatcheries require at least one acre of well-drained land per MGD flow (3,785 m^3/d) if odors are to be minimized. Factors such as weather conditions, volume and periodicity of dosage, and vegetation must be considered. In summary, careful study of the land disposal process is necessary if environmental intrusions are to be avoided.

Summary

A good design should reflect a sound engineering base coupled to established

TABLE 6. — Sludge metal limitations on land application rates.

Metal	Average concentration ppm	Permissible metal loading per year kg/ha	lb/acre	Allowable sludge loading[a] (dry solids/area/year) metric tons per hectare	tons/acre
Cd	7.6	1.1	1	148	60
Cu	49.0	22.4	20	457	204
Cr	91.0	22.4	20	272	110
Pb	92.0	22.4	20	269	109
Ni	60.0	4.5	4	81	33
Zn	342.0	33.8	40	143	58

[a]Annual sludge application rate based upon metal concentrations as analyzed in sludge samples from Big Spring and Pleasant Gap Fish Culture Stations.

TABLE 7. — Agricultural land area required for disposal of typical fresh fish hatchery sludge[a] based upon Pennsylvania Department of Environmental Resources criteria.

Crop	Yield/acre	Nitrogen uptake kg/ha	Nitrogen uptake lb/acre	Allowable available N limit kg/ha	Allowable available N limit lb/acre	Annual sludge application rate t/ha	Annual sludge application rate ton/acre
Corn	120 bu	168	150	125	112	6.42	2.87
	140 bu	207	185	156	139	7.97	3.56
Corn silage	32 tons	224	200	168	150	8.62	3.85
Grain sorghum	4 tons	280	250	209	187	10.72	4.79
Wheat	60 bu	140	125	105	94	5.40	2.41
	80 bu	208	186	157	140	8.04	3.59
Oats	100 bu	168	150	125	112	6.42	2.87
Barley	100 bu	168	150	125	112	6.42	2.87
Orchard grass	6 tons	336	300	252	225	12.92	5.77
Brome grass	5 tons	186	166	139	124	7.12	3.18
Tall fescue	3.5 tons	151	135	113	101	5.80	2.59
Bluegrass	3 tons	224	200	168	150	8.62	3.85

[a]Analysis of fresh sludge from the Big Spring Hatchery determined the total nitrogen content to be 4.85% and the ammonia nitrogen to be 1.35%. This relates to an available nitrogen value of 19.5 kilograms per metric ton (kg/t) or 39 lb/ton. One can also assume the concentration of hauled sludge to approximate 7% solids.

biological principles and a proven management plan — the concept of "biggest is best" often overcompensates. The design should be simple and manageable, and, above all, the wastewater characteristics must be considered.

Proper operation of the wastewater treatment system is just as important as the structural design. Therefore, in-house quality control testing and evaluation must be performed by the hatchery manager; it is imperative that he realize that these operations are an integral part of the modern fish culture process.

References

AMERICAN PUBLIC HEALTH ASSOCIATION, AMERICAN WATER WORKS ASSOCIATION, AND WATER POLLUTION CONTROL FEDERATION. 1975. Standard methods for the examination of water and wastewater. Fourteenth edition. American Public Health Association, Washington, D.C. 1193 pp.

COMMONWEALTH OF PENNSYLVANIA, DEPARTMENT OF ENVIRONMENTAL RESOURCES. 1977. Interim guidelines for sewage, septic tank, and holding tank waste use on agricultural lands. Chapter 75, subchapter C, section 75.32 of the Rules and Regulations of the Department of Environmental Resources.

MUDRAK, V.A. 1976. Guidelines for economical commercial hatchery wastewater treatment systems. Annual performance report 1975-1976. Commercial research project #3-242-R, U.S. Dep. Comm., NOAA, and Pennsylvania Fish Commission. 61 pp.

_____ . 1977. Guidelines for economical commercial hatchery wastewater treatment systems. Annual performance report, 1976-1977. Commercial research project #3-242-R, U.S. Dep. Comm., NOAA, and Pennsylvania Fish Commission. 77 pp.

_____ . 1977. Hatchery waste and wastewater — Clarification and stabilization. Trans. N.E. Soc. Conserv. Eng. 17 pp.

MUDRAK, V.A., AND K.R. STARK. 1978. Guidelines for economical commercial hatchery wastewater treatment systems. Annual performance report, 1977-1978. Commercial research project #3-242-R, U.S. Dep. Comm., NOAA, and Pennsylvania Fish Commission. 85 pp.

REGAN, R.W. 1978. An evaluation of wastewater and residual sludge handling and disposal at University Park. Internal report, prepared for the Office of Physical Plant, The Pennsylvania St. Univ., University Park 16802.

U.S. ENVIRONMENTAL PROTECTION AGENCY. 1974. Development document for proposed effluent limitations guidelines and new source performance standards for the fish hatcheries and farms point source category. Internal draft report, prepared by the EPA Office of Enforcement, National Field Investigations Center, Denver, Colorado. 237 pp.

Bio-Engineering Symposium for Fish Culture (FCS Publ. 1):183-189

Aquaculture as a Method for Meeting Hatchery Discharge Standards

C.M. BROWN AND C.E. NASH

Kramer, Chin & Mayo, Inc.
1917 First Avenue
Seattle, Washington 98101

ABSTRACT

The potentials and limitations in utilizing aquaculture as a technique to meet fish hatchery discharge standards are described. Existing systems for treating municipal wastewater and their applicability for treating hatchery effluent are discussed. Economic evaluations suggest in some cases there may be a considerable cost savings if aquaculture treatment systems are used in place of conventional methods. By-products from the treatment systems include forage for fish under cultivation in the hatchery as well as silage for livestock.

Introduction

As the aquaculture industry is being burdened with increasingly stringent regulations regarding wastewater discharge, many operations must consider for the first time the problems and costs of constructing and operating treatment facilities. Regulations that must be met will vary to some degree with the size and location of the facility but, in general, the parameters of primary concern are biochemical oxygen demand (BOD), settleable and suspended solids (SS), nitrogen (N), and phosphates (P). Meeting standards for these parameters will involve adapting techniques already developed for treating domestic and industrial wastes.

One of the wastewater treatment techniques currently being refined through experimental research and development projects in various locations is that of culturing other aquatic organisms, both plants and animals, to reduce nutrients, pathogens, solids, and BOD. Of all the treatment systems presently being proposed, this appears to be an attractive option, one that is obviously in keeping with the expertise of the fish farmer and that adapts well to the facilities of the industry.

It is the purpose of this paper to review the techniques of treating wastewater through aquaculture and to discuss the problems and potentials of adapting such techniques for treating effluents from modern fish hatcheries and fish farms.

Techniques

Fish Pond Systems

The practice of recovering nutrients from wastewater by using fish ponds is not new. For hundreds of years people have been adding household and farm animal wastes as fertilizers to fish ponds to stimulate production of acceptable food organisms for fish and waterfowl. Many aquatic animals have been raised as valuable farm products but have not been considered as part of a waste treatment mechanism per se. Many systems have been practiced for the integration of agricultural wastes with fish culture in the less developed countries where emphasis is necessarily placed more on food production than effluent treatment (Table 1). Recent research, however, has reexamined the optimization of aquaculture systems for a high degree of pollutant removal as well as for controlled production of cultured organisms.

The primary function of aquatic animals in a wastewater treatment system is to harvest algae, detritus, insects, and other particulate matter contained in the waste. Aquaculture systems have been integrated with wastewater stabilization

TABLE 1. — Annual production data of integrated aquaculture/agriculture waste utilization projects.

Annual production kg/ha	Fish under culture	Manure source	Country	References
4,900	Carp	Cowshed (fluid)	Israel	Helpher and Schroeder 1977
3,500	Polyculture	Ducks	Southeast Asia	Ling 1977
4,140	Carp, catfish, largemouth bass, buffalo fish	Swine	USA	Buck, Baur, and Rose 1976
1,646[a]	Tilapia	Cattle	USA	Collis and Smitherman 1978
2,729[a]	Silver carp, bighead carp	Sewage lagoons	USA	Henderson 1978
3,000	Polyculture	Domestic septic tank system	Java	Hickling 1962
3,700	Polyculture	Domestic wastewater storage reservoirs	Israel	Hickling 1962
1,000	Carp	Domestic sewage	Germany	Schaperclaus 1961
2,000	Tilapia	Pigs	Rhodesia	Van der Lingen 1960
4,000	Tilapia	Ducks	Rhodesia	Van der Lingen 1960
3,000	Tilapia	Compost and farmyard	Madagascar	Gruber 1960
1,000	Carp	Town sewage effluents	Poland	Wolny 1962

[a]Area equivalents.

ponds to benefit from the large quantities of aquatic food material produced as a by-product of microbial degradation and to assist in the stabilization of the pond. A typical cycle is illustrated in Figure 1. Depending upon the degree of nutrient removal required, effluent may be discharged directly from the fish ponds or may be further treated.

Although systems vary greatly in terms of configuration, species selected, and degree of removal effected, the general criteria required for maintenance of higher aquatic life remain relatively constant. Key elements to be considered in the design of any aquaculture wastewater system include water quality (particularly with respect to minimum dissolved oxygen and toxic components), temperature, food availability, substrate, and space.

Criteria for sizing facilities are not well established. In general, sewage oxidation ponds receive an average of 33.6 kg BOD per hectare per day. At this rate, the ponds will remain aerobic because of oxygen produced through primary productivity and transfer across the water/air interface. Without supplemental aeration or water exchange, insufficient oxygen surpluses exist to support a fish population. Therefore, trade-offs among BOD loading, aeration, water exchange, and fish populations must be made.

Figure 2 illustrates an approach to estimating the components of a balanced system. This theoretical model indicates some of the pathways and efficiencies typical of the nutrient transfer cycle within a pond. Obviously many assumptions have been made in this illustration, and many pathways generalized. Therefore, it should be considered only an example of an approach, rather than a guideline for sizing treatment facilities. It should be noted that fish in the treatment facility will also add to the BOD, solids, and nutrient

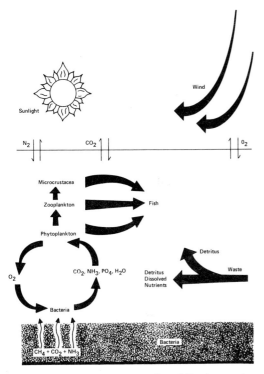

FIGURE 1. — Biological process of a stabilization pond.

concentration of the effluent. However, in a properly balanced system, many of the metabolites produced will reenter the food web and eventually be consumed by the organisms.

Some aquatic organisms utilized in wastewater treatment facilities include catfish, carp, goldfish, fathead minnows, golden shiners, mosquito fish, bluegill, green sunfish, largemouth bass, tilapia, muskellunge, freshwater mussels, marine bivalve mollusks, and shrimp. Some are filter feeders that directly reduce BOD and SS, while the carnivores keep the fish population in check. The integration of species (polyculture) takes advantage of selective feeding habits, thus optimizing overall treatment efficiency. Any species selected should be hardy and adaptable to periods of poor water conditions, especially fluctuations in dissolved oxygen and pH.

Achieving a properly balanced system is the key to a successful aquaculture treatment scheme. It is important to select species that not only will interact well, but

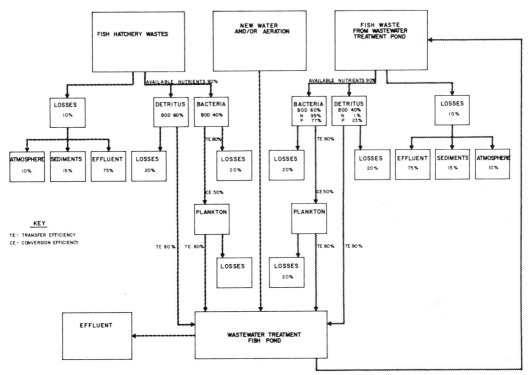

FIGURE 2. — Model of nutrient transfer in aquaculture wastewater treatment pond.

TABLE 2. — Suitable organisms for wastewater treatment in temperate climates. Minimum sustained D.O. per liter, 4 mg.

Organism	Type of food consumed	Food consumption rates (%) of body weight per day	Preferred temperatures	Compatible organisms for polyculture	Maximum biomass
Catfish[a]	microcrustacean, insect larvae	2-3%	15-32 °C	tilapia	1,993 kg/ha
Carp	plankton, detritus	2-5%	20-27 °C	tilapia, tadpoles	1,704 kg/ha
Tilapia	phytoplankton, insect larvae, microcrustacean	3%	16 °C min.; 22-30 °C is optimum	carp, catfish	504 kg/ha
Buffalo fish	phytoplankton, detritus, benthos	1% at 13 °C; 4% at 18 °C	13 °C min.; 18-24 °C is optimum	catfish	403 kg/ha
Freshwater mussels[b]	plankton, detritus	(4.9 l/h at 40% efficiency)	10-16 °C	sunfish, bass	50 kg/m³

[a]For catfish, 0.2 kg BOD and SS produced per kilogram of food consumed. For the other organisms named, the same amount is assumed.

[b]For freshwater mussels, velocity of 0.2 m/s maximum.

also will actively feed at the temperatures anticipated. Table 2 indicates some species that can be utilized in temperate climates. Since some of the effectiveness of the treatment pond is due to settling of detrital and plant material, behavior of the animals is extremely important to prevent resuspension of settled particles. If periodic harvesting of fish is required or is desirable, care must be taken to design traps that will minimize pond agitation and will allow continuous operation of the treatment facilities.

If environmental conditions are suitable for spawning, species must be selected that will not adversely impact receiving waters, as escapement of eggs and larvae may be possible. Monosex hybrids may be a recommended solution for many sites.

The effectiveness of wastewater treatment systems using fish varies with location and system design. The best performance recorded in a freshwater system using aerated domestic sewage was an overall reduction of 97% BOD, 4% SS, with final concentration of 6 ppm BOD and 12 ppm SS (Coleman et al. 1974). This same system obtained concentrations of less than 3 ppm total nitrogen and total phosphorus in the final effluent. On a routine basis, the system meets standards of 30 ppm BOD and 30 ppm SS (Hall 1978).

Marshes and Integrated Systems

Fish are usually the first consideration for integrated aquaculture/waste treatment systems, but some of the most effective systems will rely on marshes containing aquatic plants, alone or in combination with aquatic animals. Table 3 presents some of the marsh and marsh/pond systems that have been or are currently under examination.

Marsh treatment systems are more suited to temperate climates, but require harvest at regular intervals to protect against increase in BOD from decomposition of the plants themselves. Unfortunately, since the performance of such systems during winter and in freezing temperature conditions is low, the use of such systems in many locations is limited.

Because the purpose of most marsh treatment systems is the removal of nutrients, primarily nitrogen and phosphorus, optimization of performance is directed specifically at these elements.

TABLE 3. — Marsh-treatment average effluent quality.

Reference	Influent BOD mg/l	SS mg/l	Effluent BOD mg/l	SS mg/l	% Removal BOD	SS	Marsh type	Influent type	Retention time	Plants
Dinges 1976	90.0	35.0	3.5	7.0	96.0	80.0	Artificial M/P[b]	Stabilization pond effluent	5.3 days	Lemna / Hyacinth
De Jong 1976	257.0		11.0		96.0		Artificial	Raw	10 days	Bulrush Reeds
Spangler et al. 1975b	26.9	127.0	5.35	10.0	80.1	29.1	Natural	Secondary effluent		Sagittaria Typha Sparganium Elodia Potamogeton Ceratophyllum Lemna
	38.5	8.6	4.5	2.0[a]	88.0	77.0[a]	Artificial	Secondary effluent	5 hours	Bulrush
	28.0	54.0	25.0	18.0	10.0	67.0	Artificial	Secondary effluent	10 days[c]	Bulrush
	317.0	42.0	72.0	32.0	77.0	24.0	Artificial	Primary effluent	10 days[c]	Bulrush
Demgen 1977	30.0	18.0	12.0	10.0	60.0	45.0	Artificial M/P	Secondary effluent	4-5 days	(unspecified)
Small 1976	170.0	353.0	19.0	43.0	89.0	88.0	Artificial M/P	Secondary[d] effluent	10 days	Cattail

[a]Turbidity.
[b]M/P = Marsh/Pond.
[c]Probably somewhat less.
[d]No sedimentation provided before the marsh.

Reduction of solids and BOD is primarily due to low velocities in the marsh itself. Table 3 presents some of the performances of marsh and marsh/pond treatment systems on the removal of BOD and suspended solids in domestic wastewater.

A typical marsh system would probably have several species planted at approximately 106,250 plants per hectare and a retention time of about 10 days. Maximum suitable depth would be approximately 0.6 meter. Baffles may be required to insure adequate hydraulic characteristics and to permit harvesting without disruption of the entire marsh system.

Since accumulation of solids in the marsh reduces its effective volume, a system for pretreatment of solids may be advisable. A fish pond with selected species can serve this purpose. Such a combined system will undoubtedly provide the most reliable aquaculture treatment scheme for removing nutrients as well as BOD and SS, while providing valuable by-products such as forage fish and silage.

An example of an integrated system is one relying on marine algal, shellfish, and seaweed species sequentially to reduce nitrogen by 80% in domestic sewage before discharge into saltwater (Ryther 1975). A similar integrated freshwater system using a marsh followed by a fish pond containing carp, golden shiners, and freshwater mussels to treat domestic sewage removed 88% SS, 89% BOD, 62% total nitrogen, and 71% total phosphates (Small 1976). The average effluent SS concentration was 43 ppm with a maximum of 100 ppm, and for BOD the average was 19 ppm with a maximum of 46 ppm. The maximum application rate was 935,363 liters per day per hectare.

Integrating an aquaculture system directly into the marsh treatment process, although desirable in concept, offers extremely poor flexibility for harvesting plant or animal life independently. Also, it does not offer a means of monitoring the effectiveness of each element separately. In addition, should a major fish kill occur, the inability to bypass the aquaculture system could be particularly damaging in terms of effluent quality.

Limitations in Applying Wastewater Aquaculture Technology to Fish Hatcheries

Aquaculture wastewater treatment is in essence a form of nutrient recovery. Wastes that lend themselves most readily to nutrient recovery are those containing the highest concentrations of usable wastes in forms that are easy to extract. Based on experience with salmonids, normal discharges from most intensive fish hatcheries have extremely low concentrations of valuable nutrients and food items. It is only during periods of pond cleaning and pond harvesting that wastes are at suitable concentrations for treating in an aquaculture treatment scheme. The system is therefore most adaptable to facilities that have the capability of separating pond cleaning wastes from normal discharges.

Pond cleaning techniques that permit the greatest degree of concentration of wastes, such as vacuum systems and solids collection sumps, will allow for reduction in size of required facilities. As in any treatment system, scheduling of pond cleaning and harvesting is necessary to avoid overloading.

Cost Comparisons of Aquaculture vs Traditional Wastewater Treatment

Cost comparisons of wastewater treatment techniques are difficult to make, considering the variety of treatment systems, operating conditions, and scale of facilities. Since all wastewater aquaculture treatment schemes are pilot-scale, the comparisons with cost of full-scale traditional systems are tenuous at best.

Despite the obvious shortcomings, some idea of comparative costs is necessary to determine the value of pursuing such a treatment scheme. For this purpose an EPA study was undertaken to evaluate the cost of aquaculture versus traditional treatment methods to meet several ranges of effluent standards (Henderson and Wert 1976). The study indicated that, providing effluent limitations do not exceed 10 ppm BOD and 15 ppm SS, aquaculture treatment methods cost 28% to 62% of conventional treatment. Similarly the costs of removing nitrogen and phosphates to a level of 5 ppm may be as low as 6% of conventional tertiary treatment. The study cited the above effluent concentrations as being the limits of reliable treatment using aquaculture.

Other economic assessments of individual aquaculture wastewater schemes indicated similarly favorable cost/benefits, providing supplemental heating was not required (Smith and Huguenin 1975).

Conclusions

Based on experience with the treatment of domestic wastewater, reductions of approximately 90% BOD and SS may be possible with marsh and fish pond aquaculture treatment techniques. However, because of differences in the specific gravity and nitrogen concentrations of the solids, it is not possible to make a direct transfer of experiences gained through the treatment of domestic wastewater to treating fish hatchery wastes. Therefore, before utilizing aquaculture treatment methodology, additional research in designing systems compatible with hatchery operations, and on species that can utilize hatchery by-products must be undertaken to develop suitable treatment techniques and to insure system reliability.

Although much research and development in both biological and engineering aspects are required before endorsing aquaculture as a wastewater treatment method for hatcheries and fish farms, the potential is exciting. The added benefits of producing forage crops within low energy systems makes the method even more attractive for aquaculturists.

References

BUCK, D.H., R.J. BAUR, AND C.R. ROSE. 1976. Experiments in recycling swine manure in fish ponds. FIR:AQ/Conf/76/E.29. FAO, Rome, Italy. 12 pp.

COLEMAN, M., J. HENDERSON, H.G. CHICHESTER, AND R.L. CARPENTER. 1974. Aquaculture as a means to achieve effluent standards. Pages 199-214 in Wastewater use in the production of food and fiber. U.S. Environ. Protect. Agency. 660/2-74-041.

COLLIS, W.J., AND R.O. SMITHERMAN. 1978. Production of tilapia hybrids with cattle manure or a commercial diet. Pages 43-55 in Proc. symposium on culture of exotic fishes. Am. Fish. Soc. Auburn, Alabama.

DE JONG, J. 1976. The purification of wastewater with the aid of rush or reed ponds. Pages 133-139 in Biological control of water pollution. Univ. Penn. Press, Philadelphia.

DEMGER, F., AND B.J. BLUBAUGH. 1977. Mt. View sanitary district marsh enhancement pilot program. Mt. View Sanitary District, Martinez, California. Prog. Rep. No. 3. 36 pp.

DINGES, R. 1976. A proposed integrated biological wastewater treatment system. Pages 225-230 in Biological control of water pollution. Univ. Penn. Press, Philadelphia.

GRUBER, R. 1960. Considerations sur l'amelioration des rendements en pisciculture congolais. (In French.) Bull. Agric. Congo belge. 51(1):139-157.

HALL D. Personal communication. Oklahoma St. Dep. Health. December 1978.

HENDERSON, U.B., AND F.S. WERT. 1976. Economic assessment of wastewater aquaculture treatment systems. U.S. Environ. Protect. Agency, Ada, Oklahoma. EPA-600/2-76-293. 120 pp.

HENDERSON, S. 1978. An evaluation of the filter feeding fishes, silver and bighead carp, for water quality improvement. Pages 121-130 in R.O. Smitherman, W.L. Shelton, and H.J. Grover, eds. Proc. symposium on

culture of exotic fishes. Fish Culture Section, Am. Fish. Soc., Auburn, Alabama.

HEPHER, B., AND G.L. SCHROEDER. 1977. Wastewater utilization in Israeli aquaculture. Pages 529-599 in D'Itri, ed. Wastewater renovation and reuse. Marcel Dekker, Inc., New York.

HICKLING, C.F. 1962. Fish culture. Faber & Faber, London, 317 pp.

LING, S.W. 1977. Aquaculture in Southeast Asia. Contrib. No. 465. Coll. Fish., Univ. Washington, Seattle. 108 pp.

RYTHER, J.H. 1975. Preliminary results with a pilot-plant waste recycling marine aquaculture system. Woods Hole Institution, Mass. 16 pp.

SCHAPERCLAUS, W. 1961. Lehrbuch der teichwirtschaft. P. Parey, Berlin. 582 pp.

SMALL, M. 1976. Data report marsh/pond system (November). Brookhaven National Laboratory, Upton, New York.

SMALL, M., AND C. WURM. 1977. Data report meadow/marsh/pond system (April). Brookhaven National Laboratory, Upton, New York.

SMITH, L.J., AND J.E. HUGUENIN. 1975. The economics of wastewater aquaculture systems. Pages 285-293 in I.E.E.E. Conf. on Engineering in the ocean environment, Oceans '75, September 22-24, San Diego, California.

SPANGLER, F.L., W.E. SLOEY, AND C.W. FETTER. 1976a. Experimental use of emergent vegetation for the biological treatment of municipal wastewater in Wisconsin. Pages 161-171 in Biological control of water pollution. Univ. Penn. Press, Philadelphia.

_____. 1976b. Wastewater treatment by natural and artificial marshes. U.S. Environ. Protect. Agency. EPA-600/2-76-207. 184 pp.

VAN DER LINGEN, M.I. 1960. The use of animal manures and the integration of animal keeping and fish culture in tilapia culture. Publ. Cons. Sci. Afr. S. Sahara 63:212-216.

WOLNY, P. 1962. Przydatność oczyszczonych ścieków miejskich do hodowli ryb. [Value of purified municipal wastewater for fish culture.] (In Polish, English summary.) Rocz. Nauk Roln. Ser. B. 81(2):231-250.

Bio-Engineering Symposium for Fish Culture (FCS Publ. 1):190-191

SESSION VI   FISH PRODUCTION FACILITIES

Bio-Engineering and Facility Design

David W. McDaniel

National Fisheries Center, Leetown
U.S. Fish and Wildlife Service
Kearneysville, West Virginia 25430

The design of production facilities has been going on for thousands of years, but there are still problems associated with the process. It is very common, even with our high level of technology (or perhaps because of it), to find that the final product of the engineer and the biologist just doesn't work properly. This is especially true in the case of modern complex facilities. The ultimate purpose of these facilities is to provide a highly efficient system that reduces the need for costly manpower to produce large numbers of fish. The end result, however, is often just the opposite of what was intended.

It is understandable that the process of going from the simple to the complex facility will involve a period of experimentation that does not always produce desirable results. The nature of these facilities, large scale with liberal use of concrete and complicated machinery, does not lend itself to later modification to correct major deficiencies. It is usually necessary to manage around the problems rather than to correct them. For this reason the basic deficiencies remain and it is common to blame the involved participants for the problems. From the engineering standpoint the blame is associated with the criteria established by the biologist; and from the biologist's standpoint, it is associated with the failure of the engineers to design properly from adequate criteria. While there are no doubt inadequacies on both sides, the real need is to determine how the problem can be overcome once and for all. It is not necessarily correct to blame the lack of communication between the biologists and the engineers for the problem. There are many examples where the end results were less than acceptable, but where the biologists and the engineers were closely involved throughout the design and construction process. Everything went well until the water was turned on.

The term "bio-engineering" as it pertains to the construction of cultural facilities implies that there is such a process as bio-engineering. This assumption is very possibly the basis for the problems resulting in facilities that do not work properly. It is beneficial to discuss the way that biologists and engineers view production facilities in order to gain insight into how past problems can be prevented in the construction of future facilities.

From the biologist's standpoint the requirements of a production facility to raise a certain quantity of fish are understood. In most cases the environmental needs of the fish are known. The facility must first provide for these needs, and, secondarily, provide a situation that will allow the culturist to do the necessary manipulations to feed, clean, and move the fish in an efficient manner. Another consideration is the need to treat inflows and outflows to meet certain regulations or health criteria. Through experience most biologists develop certain biases as to what are the best pond design, water flow patterns, etc. The biologist views the facility as a system, not of valves and structures, but of conditions.

From the engineer's standpoint the criteria established by the biologist must be viewed from the restrictive considerations of site and materials. The design must incorporate considerations that have little concern for the environmental needs of the fish or the biologist. In many cases the engineer is expected not only to make the facility pleasant to the eye but to create a facility that is considerably more complex than what is needed to raise the desired quantity of fish. During the design process, certain trade-offs occur that more often than not do not favor the basic purpose of the facility, i.e, the production of fish. In the end the design is a compromise at best.

During the design and construction phase the biologist and the engineer can be in constant communication with one another without any "bio-engineering" taking place. In the eyes of both, the system will do what it is supposed to do, but when the water is turned on it just doesn't work.

While it is an oversimplification to say that there is an easy solution to the overall problem, there is a basic malfunction in the process that must be recognized and eliminated before progress can be made to resolve the overall problem. In the majority of situations, the biologist's knowledge of engineering and the engineer's knowledge of biology are such that there is

not sufficient overlap to create a bio-engineering relationship. Throughout the communication process, this area of misunderstanding prevents the integration of the two sciences. Through this crack fall the expectations of both sides that the design is correct and will function properly. In order to communicate, the biologists and the engineers must learn to speak the same language. How many times do we have to restage the building of the Tower of Babel before we see the significance of understanding what each other is saying?

While it would be a vast improvement if the biologist and the engineer would learn each other's language, the ultimate solution is to develop a new language. This new language will contain certain words that are found in both of the old languages, but it will integrate these words so that they will have new connotations. People learning this new language will at last begin to think in a new way and look at problems in a new way. The association of values and dissolved oxygen will begin to make sense, and bio-engineering will become a reality.

Bio-Engineering Symposium for Fish Culture (FCS Publ. 1):192-195

Channel Catfish Production in Geothermal Water

LEO RAY

Fish Breeders of Idaho
Route 3, Box 193
Buhl, Idaho 83316

ABSTRACT

Channel catfish, *Ictalurus punctatus,* have been raised by Fish Breeders in geothermal facilities for six years. The product is superior to other catfish on the market.

Concrete facilities and 7,000 gal/min (26.5 m³/min) of water allow densities of 5-10 lb/ft³ (200-400 kg/m³) of space and 10,000 lb/ft³·s⁻¹ (160,000 kg/m³·s⁻¹) of water. Oxygen and ammonia are the principal factors limiting production. Diseases are related to these.

Identification of the resource, facility design, financing, construction, production, personnel, processing, marketing, and distribution are the main problem areas preventing expansion of geothermal fish farming.

Introduction

Fish Breeders of Idaho, Incorporated, has been raising channel catfish in high density concrete raceways for six years. The water is supplied by artesian geothermal wells. Total flow is 7,000 gal/min at 90 °F (26.5 m³/min at 32 °C). Cold water from springs and streams is used to cool the hot water to between 80 and 85 °F (27 and 30 °C), the ideal production temperature.

The quality of channel catfish produced in the clean water is far superior to that of any other catfish on the market. A fish is like a sponge; it tastes like the water in which it is raised. The geothermal water produces a quality that has allowed Fish Breeders to introduce catfish into the gourmet markets and obtain high prices.

Facilities

Fish Breeders is located in the Snake River Canyon near Buhl, Idaho, at an elevation of 3,000 ft (914 m). Yearly temperatures are from -10 °F to 105 °F (-23 to 41 °C), and ambient water temperature seldom exceeds 75 °F (24 °C). The climate is too cold and the growing season too short to grow catfish commercially without hot water. Geothermal water changes a noncommercial area to a 365-day optimum growing season.

The fish farm is located on a hill. Approximately 80 ft of elevation is used in the farm. This is very important in aerating and reusing water.

The production facilities are concrete. Each section is 24 ft long, 10 ft wide, and 4 ft deep (7.3 x 3.0 x 1.2 m). The space utilized by the fish is 770 ft³ (21.8 m³). The sections are arranged four in a series with a 2-ft (0.6-m) drop between sections. The raceways are in pairs with a common center wall. The water passes through four sets of raceways, each raceway having four sections (16 sections total), from the top of the hill to the bottom. The upper end of the farm is used for catfish production and the lower end for tilapia production.

Water Supply

Five artesian geothermal wells supply 7,000 gal/min (26.5 m³/min) of 90 °F (32 °C) water. This water is mixed with cold water that varies from 32 °F to 74 °F (-1 to 23 °C) to obtain a temperature of 80-85 °F (27-29 °C). The geothermal water is used directly. No heat exhangers are utilized. The wells were all drilled by Fish Breeders and are approximately 700 ft (213 m) deep. The water flows through each raceway at 1,500-2,000 gal/min (5.7-7.6 m³/min).

Stocking Rate

There are two densities to consider in producing catfish: pounds per cubic foot of space and pounds per second-foot of water.

Both are interrelated but the degree of interrelationship is unknown. They will be considered separately in this report. The pounds per cubic foot of space is primarily limited by social factors. The pounds per second-foot of water is primarily limited by water quality.

Channel catfish are social animals and in their natural environment tend to congregate. Eggs are laid in a mass. Sac fry congregate after hatching. Fingerlings, and even adults, spend much of their time in schools. They do not establish individual territories as do shrimp and lobster. This is one major factor that makes them ideal for high density production. Channel catfish do establish a social pecking order. Stocking at high densities appears to interfere with this pecking order and reduces fighting.

Normal stocking densities are from 5 to 10 lb of fish/ft³ (80-350 kg/m³) of space. Lower densities are used for small fish. Densities up to 20 lb/ft³ of space (700 kg/m³) have been tested, but at this time are not recommended for commercial production.

Water Analysis

The pounds of fish that can be produced per second-foot of flowing water is limited by water quality. Water analysis of the water entering the raceway compared to water analysis of the same water leaving the raceway tells what the fish have put into the water and what the fish have taken out of the water. The factors of greatest importance are the oxygen removed and the carbon dioxide and ammonia added.

Oxygen removed is the first factor that limits production; however, oxygen is easily replaced by running water over waterfalls. At the same time oxygen is replaced, carbon dioxide is removed. Theoretically, oxygen can continuously be replaced and most of the carbon dioxide removed by a chain of waterfalls. A 2-ft (0.6-m) drop will replace approximately 50% of the oxygen removed. There are ways of increasing the break-up and aeration of the water to achieve saturation in a 2-ft drop; however, this is usually not done.

Ammonia is not easily removed by waterfalls and will continue to accumulate until it becomes the principal limiting factor in raceway production. The ammonia can be in an ionic state or a gaseous state. The gaseous state is very toxic to fish and can be partially aerated out of the water. The ionic state is less toxic, remains in the water, and changes to nitrates. The amount of ammonia that will be in the gaseous state is related to pH, temperature, and water chemistry. The amount of ammonia that fish can tolerate is dependent on these same factors plus the oxygen and carbon dioxide levels. The higher the oxygen level, the more ammonia fish can tolerate. The level of ammonia that fish can tolerate is between 0.5 ppm and 2.0 ppm.

The amount of oxygen removed, carbon dioxide produced, and ammonia produced is dependent on the amount of food fed in the raceway, not the amount of fish in the raceway. Five thousand lb (2,268 kg) of fish could be fed 1% body weight (50 lb—23 kg—of feed), and the ammonia, carbon dioxide, and oxygen levels would be basically the same as if 2,500 lb (1,134 kg) of fish were fed 2% body weight (50 lb—23 kg—of feed). The limiting factor is the amount of feed that can be fed per second-foot (28.3 l/s) of water, and the amount of fish that can be stocked is dependent on the percent of body weight that is fed.

Oxygen required to metabolize 50 lb (23 kg) of feed is approximately 2 ppm from one second-foot of water (28.3 l/s). The water will be aerated three times, once between each section. Fifty lb (23 kg) of feed can be fed in each section. This gives a total of 200 pounds (91 kg) of feed that can be fed in the entire raceway. Assuming 50% reoxygenation and saturation of 8 ppm, the oxygen at discharge would be 3 ppm. This is an absolute minimum level for production. A total of 8 ppm oxygen would be utilized in metabolizing the 200 lb (91 kg) of feed.

Approximately 0.2 ppm ammonia would be deposited in one second-foot of water from metabolism of 50 lb (23 kg) of feed. Feeding 200 lb (91 kg) of feed in one

193

second-foot (28.3 l/s) of water, the ammonia in the discharge would be 0.8 ppm.

Carrying Inventory

Maximum recommended inventory for commercial production of channel catfish on water at Fish Breeders is about 10,000-15,000 lb per second-foot (160,000-240,000 kg/m^3·s^{-1}) of water. Yearly production will usually be three to four times the carrying capacity.

In order for a commercial hatchery to obtain maximum profit, production must be pressed to the limit. This means carrying inventory will be increased to the point disease will develop because of overloading. Disease is usually a major concern in fish production. It should not be. Ninety-nine percent of all diseases are secondary expressions of poor water quality, poor feed quality, and poor management (the human element). If good water quality is maintained, good feed used, and laborers handle fish properly, disease should not be a problem. It is management's job to balance carrying capacity to water quality and stay below the point where disease is of significant concern.

Problem Areas Preventing Expansion

In starting a geothermal fish farm, one can expect several problems. The following factors are responsible for preventing faster expansion in geothermal fish farming.

Identifying the Resource

The extent of most geothermal sites considered for fish production has not been identified. Surface springs show a potential but do not tell the complete story. Wells must be drilled before the volume, temperature, and water chemistry are determinable. These must be known before a production potential can be analyzed. Considerable money must be spent before a decision can be made recommending a site for geothermal fish farming. The person with a geothermal resource must first identify the resource, and then see if it is suitable for fish production.

Facility Design

People interested in building fish farms usually examine government hatcheries for models to copy. This is a poor place to study a good design. Poorly designed commercial farms go broke and disappear. Poorly designed government fish farms continue to operate indefinitely, consuming tax dollars.

An engineer's dream is too often a fish culturist's nightmare. The KISS rule is the only rule to follow (Keep It Simple, Stupid). There has been little communication between fish culturists and engineers. This lack of communication has resulted in a situation where there are few, if any, engineers who understand the management of fish culture well enough to design a commercial fish farm. Commercially viable fish farms that were designed by an engineer are hard to find. The entire industry has been designed and built by the fish farmers themselves. This does not mean that there is no room in the design work for engineers. It means engineers are not in tune with the problems of fish culture and, therefore, behind the times.

Financing

Financing is difficult. Geothermal projects are considered high risk. Fish farming is considered high risk, and high density production even a higher risk. Capital expenditure is high; operating expenses are even higher. An executive of a large company looking at fish farming said they consider some businesses as cash generating and others as cash consuming. They considered fish farming as cash consuming. It is a good description of fish farming.

Contrary to common opinion, most fish farms that have gone broke did not do so because of underfinancing. Those farms starting underfinanced have had a high success ratio, while those adequately financed "spent it like they had it" and went broke. Too many mistakes were made too fast, and the investors ended up very disenchanted. Bankruptcy followed because of a lack of willingness, not ability, to refinance and do what was necessary to succeed.

194

Construction

If engineering follows the KISS rule, construction is simple. There is no substitute for good facilities. Concrete is usually the best. The facilities will probably be the first fish farm the contractor has ever built.

Production

Trying to achieve more production than the water quality will allow causes most production problems. If the rules laid down earlier in this paper are followed, production should not be a problem. Keep good water quality, keep good feed quality, and keep labor stress down.

Personnel

Experienced personnel are not available. Personnel must be trained. There are many degree students graduating, but most do not have any experience. They need experience if they are being hired to run a new farm.

Processing

A processing plant will not be available in most situations. A producer will need to build a processing plant. Very little machinery is available for processing. Most of the labor is hand labor.

Marketing

No money is made raising fish until the fish are sold. A common misconception is that there is an unlimited market for fish products such as channel catfish. The truth is that there is an unlimited potential market for channel catfish and other good fish. The potential is unlimited; the existing market is full. There is no room for additional production without additional marketing. Markets must be developed, and they cannot be developed without fish. The fish must be raised before the market can be developed.

There were approximately 150 million lb (68 million kg) of catfish sold in the United States in 1978. This production breaks down as follows: Brazilian imports, 30 million lb (13.6 million kg); wild catfish harvested, commercially in the United States, 30 million lb (13.6 million kg); channel catfish raised on farms, mainly in Mississippi and Arkansas, 90 million lb (41 million kg); and 500 thousand lb (227 thousand kg) raised in high density geothermal facilities. These catfish figures are listed in order according to quality and price, with the poorest quality listed first. There is a considerable difference in quality of product and price. The Brazilian fish sell as low as 60¢/lb ($1.32/kg), and the geothermally produced fish sell for $3.05/lb ($6.72/kg) wholesale. In catfish, as in any other product, one gets what he pays for.

Distribution

Most geothermal resources are not near large market areas for fish. Distribution can be a major problem. The easiest market to develop is the fresh market. This complicates distribution, for the fresh market delivery must be available on a dependable weekly basis. In Idaho, Fish Breeders is in a good distribution area because of the 25 million lb (11.3 million kg) raised within 15 miles (24 km) of the farm. Each production and market situation will have to be viewed individually to solve the distribution problem.

Summary

These are the main problems in establishing a geothermal fish farm. No phase can be left out. A project will need to master each phase to be successful.

If you want to raise fish to get rich, you will probably go broke. If you want to raise fish because you like the challenge of fish culture, you will probably get rich. The best advice is to start small and grow slow.

Bio-Engineering Symposium for Fish Culture (FCS Publ. 1):196-200
© 1981 by the Fish Culture Section of the American Fisheries Society

How to Develop Criteria
for Design and Operation of Fish Facilities

PHIL JEPPSON AND LEROY R. TAYLOR

CH2M HILL
P.O. Box 8748
Boise, Idaho 83707

ABSTRACT

A Federal program for restoring the summer *Oncorhynchus tshawytscha* (chinook salmon) fishery led to the decision to build a hatchery on the Payette River in McCall, Idaho. The facility will produce one million juvenile salmon annually to be released in the South Fork of the Salmon River for their seaward journey. From the initial call for proposal in May 1977 through completion of the final design in September 1978, the emphasis was on coordination of Federal and State agency people and the private consultants involved. The review required substantial agreement among the U.S. Army Corps of Engineers, which was building the facility; the Idaho Department of Fish and Game, which would operate it; other governmental agencies and private architects, biologists, and engineers. Hatchery operating personnel and administrators were involved in the review process. The coordination techniques addressed the frequent necessary communications and the problem areas as they were encountered.

Introduction

The need for a hatchery to enhance the depleted summer run of chinook salmon, *Oncorhynchus tshawytscha,* in the South Fork of the Salmon River was recognized as the first priority of the Federal Government's Lower Snake River Fish and Wildlife Compensation Plan. It was considered necessary to have the hatchery on line by October 1, 1979. The U.S. Army Corps of Engineers was responsible for the design of the hatchery, and the State of Idaho was responsible for operating the facility upon completion. The State became vitally concerned that the new hatchery be completed on schedule and be a satisfactory hatchery from both biological and operational aspects. Therefore, when the Corps failed to receive design money during its fiscal year 1977, the State applied for and received a grant from the Pacific Northwest Regional Commission (PNRC) to keep the project on schedule.

The project was assigned to the Idaho Department of Fish and Game (IDF&G) engineering staff to prepare design criteria and preliminary design for the new hatchery. It was apparent that this staff was too limited to accomplish the entire design, so CH2M HILL was hired in May 1977 to provide architectural and engineering services. It was felt that the State could provide the biological and anadromous hatchery expertise to accompany a multi-disciplined consulting firm in a very close cooperative team effort.

Methods

Hatcheries Tour — Step 1

A tour of a number of operating hatcheries in the Northwest was perceived as being a logical first step in development of design criteria.

The CH2M HILL project manager and biologist accompanied several IDF&G personnel on a tour of anadromous fish hatcheries throughout the Northwest. Hatchery operators were usually quite vocal when asked to comment on different design features of their hatcheries or when asked what they would like to see that would be different at their facilities.

A wide range of fish rearing techniques was observed and recorded at the different hatcheries. Vessel loadings and flow rates were recorded from incubation through rearing and adult holding. Particular attention was paid to the fish handling and "human engineering" aspects of the dif-

TABLE 1. — Hatcheries tour.

- First step in criteria development
- Tour objectives — observation of a wide range of fish rearing techniques
- Knowledge gained — good features and bad

ferent facilities. We found that some hatcheries were designed without significant input from the hatchery operators. Others represented the latest engineering technology in accord with the biological life cycle of the fish.

This tour was a most beneficial starting point toward establishing our bio-engineering criteria (Table 1). We also picked up other valuable information, such as the need to have the hatchery manager onsite during the construction of the new hatchery.

Bio-Engineering Criteria — Step 2

Step 2 began with a meeting of the key engineers and biologists responsible for developing detailed criteria (Table 2). Through this and later discussions, the private consulting team was able to develop a rapport with its counterparts in the IDF&G that proved invaluable in working through the criteria development and, later, the design. Notes of the conversations were kept, and a typed summary was distributed. This documentation procedure was followed throughout the project.

Preliminary criteria developed by IDF&G biologists, engineers, and hatchery operators were furnished as a starting point and detailed criteria developed as a joint effort. The criteria included derivation of theoretical time of hatch, growth rate based on water temperature, volume needed for raceways, flow rates, maximum allowable velocities, etc. These criteria

TABLE 2. — Development of bio-engineering criteria.

- Preliminary meetings of biologists, engineers
- Criteria development by fish biologist — growth rates, basin sizing, flow rates, etc.
- Publication of preliminary bio-engineering criteria
- Review by Idaho Department of Fish and Game (operator)

were checked against general criteria developed by a number of authorities in the fisheries field.

The detailed criteria, including charts, and other descriptive information were circulated in typed form for comment by the remainder of the consultant team and the IDF&G.

IDF&G personnel reviewed the preliminary draft of bio-engineering criteria, giving particular attention to the basis upon which the criteria were developed.

Meetings were held to discuss and settle problem areas in the criteria with input from fishery experts in other agencies. The criteria were then finalized, enabling work to commence on the preliminary design.

A difficult facet of this phase of the work, and one that is common to development of most projects, was the development of communications among engineers, biologists, operators, and other members of the team. In this project, communication was enhanced by documentation in writing of what was said at meetings, distribution of that documentation, and preparation of visual aids for a follow-up meeting.

Visual aids, rather than relatively complex drawings of the proposed facilities, depicted what the designers thought the biologists and operators wanted. The sketches, flow charts, and/or schematics were distributed before the review meeting so the attendees could have time to prepare their comments. Also, a list of topics for discussion was usually circulated before the meeting. This step saved time and effort because unsatisfactory solutions to various design problems could be identified earlier and discarded.

Design Report — Step 3

After general agreement was reached on bio-engineering criteria, development of criteria for the hatchery systems and support systems such as water supply, waste water, buildings, roads, and utilities was begun. Surveys and soils investigations were completed, and additional water quality and temperature investigations

TABLE 3. — Preparation of design report.

- Development of criteria for hatchery systems — water supply, wastewater, buildings, roads, etc.
- Weekly review sessions with IDF&G engineers
- Periodic review sessions with IDF&G hatchery operators — review methods
- Reviews and feedback from Federal agencies and other State agencies
- Final review
- Reconciliation of differences of opinion
- Publication of final predesign reports

dictated by the bio-engineering criteria were started.

Coordination and review on an almost daily basis were held between the consultants and the department engineers. IDF&G hatchery operators were frequently called upon for comments on particular aspects of the design. This ensured a design reflecting the needs and desires of the actual operators of the facility.

Copies of draft preliminary design were sent to all concerned State and Federal agencies and to recognized fishery authorities throughout the Northwest. Their comments were solicited along with their attendance at a review meeting. All agencies and authorities responded and by the time our review meeting ended, the problem areas were resolved and we were confident the preliminary design reflected the best fishery expertise available (Table 3).

Plans and Specifications — Step 4

Acceptance of the design report by the IDF&G and the Corps of Engineers and approval of funding for final design signaled the start of the next step. The Corps, following the prescribed Federal process, asked for statements of qualifications from interested design firms and selected CH2M HILL and our subconsultants to continue with the project.

Since preliminary layouts had been developed and approved, detailed designs by all disciplines could start at the same time on a relatively tight time schedule. The emphasis in this phase was on teamwork. Frequent internal meetings were held with representation from each discipline. Preliminary plans were fol-

lowed as closely as possible and changed only when detailing dictated such change.

One-day formal review sessions were scheduled by the Corps at the 25, 40, 60, and 90 percent complete points. Each major design discipline was represented at these sessions for the consultant and the Corps as reviewer. The IDF&G was also represented. At the meetings, discrepancies in pre-meeting submittals, differences in philosophy, and other topics were discussed. Where differences could not be resolved at the meetings, they were deferred and worked out subsequently between the designer and the Corps reviewer.

At the 90 percent review a few differences in design philosophy were still being found, probably because the differences were not apparent in the preliminary stages. The resulting changes affected each discipline in a domino effect, and because of the relatively tight time schedule some discrepancies were caused in the bidding plans. This occurred in spite of extensive coordination between the parties involved. It points out the importance of early coordination if changes in detailed drawings (and consultant's requests for more money to cover the changes) are to be avoided. Where such changes must be made, a major checking effort is required to ensure that each sheet and each discipline's affected design are subsequently changed.

Typical of the items identified at this stage were methods of drainage of tanks, details of dropgates, etc. It was obvious that the designers had not reviewed as many of these small but vital details with IDF&G personnel as would have been desirable.

The personal and regional likes and opinions of operators, biologists, and designers often had to be evaluated by the review team to ensure that the proposed methods were the best way to get the job done. Face-to-face meetings of the affected parties proved to be the most satisfactory means for resolution of conflicts.

A second problem area was uncovered, in terminology, at the 90 percent review stage. The consultant's designers had in some cases developed their designs using

TABLE 4. — Preparation of plans and specifications.

- Selection of architect-engineer
- Development of final designs
- Review with Corps/IDF&G at 25, 40, 60, and 90 percent
- Identification of differences in concept early on
- Publication of final plans and specifications
- Bidding

terminology from their own computer-based specifications system. Since the actual specifications were based on the Corps guide specifications, a considerable effort was required to convert drawing terminology to the Corps system. Earlier recognition of the problem would have helped.

Upon complete incorporation of the 90 percent review comments, contract documents were prepared. The project was bid and is under construction (Table 4).

Operation and Maintenance Manual — Step 5

On completion of the design, an operation and maintenance manual was prepared. Since the objective was to produce a manual the operators would use rather than keep on the shelf, we enlisted the aid of a number of senior hatchery operators and supervisors. A day-long session was spent with the operators, letting them tell us what they wanted in the manual.

Each of the major disciplines involved in the hatchery design then prepared its relevant section of the manual. The sections were edited and a first draft printed for review by the design team, Corps, and IDF&G officials, and, most important, the experienced hatchery operators.

After incorporation of comments received, a second draft was prepared. This draft is being printed and bound in

TABLE 5. — Preparation of operation and maintenance manual.

- Preliminary review of objectives
- Publication of preliminary draft
- Review by Corps. IDF&G, and IDF&G hatchery operators
- Publication of finished draft

looseleaf fashion. After a year of facility operation, the manual can be updated using operators' comments to become a more useful tool (Table 5).

Conclusions

Criteria for design and operation of your new hatchery should be developed in a methodical fashion. We suggest the following steps:
- Take a hatchery tour;
- Develop bio-engineering criteria;
- Prepare a design report;
- Develop plans and specifications; and
- Prepare operation and maintenance manual.

Prior to starting each step, you should identify a review process to be used for that step. Make sure you have a good variety of biologists, engineers, and hatchery operators involved in the process. Periodic reviews are the key to arriving at a consensus. Reviews must be as extensive as possible early in each step to avoid backtracking.

QUESTIONER: In Idaho and the West in general, there are a number of very talented fish culturists who are in the commercial sector. In carrying out a careful review process in preparing for a hatchery, was the private sector ever consulted in the process? You apparently went to the senior biologist, the man who has been around for many years and is very experienced and skilled. Was there an intentional decision not to include the private sector in the planning process?

LEROY TAYLOR: No, we did not. One of the reviewers was a senior hatchery operator from the Magic Valley who was familiar with most of the commercial operations there. This is a good idea.

CECIL FOX: When you are comparing and talking about responsibility and who establishes design criteria, you are talking about two sets of criteria, one of them being that commercial fish growers are primarily interested in cost per pound of fish per unit. They accept full responsibility for failure or loss for a single species or a product that they anticipated they could make money on. They relate the cost to success. They relate to that particular species they have selected and developed over the years and one that does well at their particular site. Many of the matters we discussed today were in terms of Federal and State hatcheries. We are talking about species that are not always selected by the station managers but by sportsmen, or as a result of mitigation. They demand some sort of predictability year after year. If the managers are not able to produce the fish that someone else requested, they are subject to severe criticism. I suggest that you go to the Magic Valley and look at the Federal, State, and commercial hatcheries, and you will find three types of hatcheries, all operated by different methods as a result of different criteria. This should always be remembered. The criteria that you are trying to achieve must be identified.

LEO RAY: When we establish a hatchery, we need to establish the criteria that we are trying to meet. If the criteria are wrong regardless of whether the engineers or the biologists made the determinations. We get into problems with Federal and State governments regarding buck passing and the buck just keeps getting bigger. I find it very difficult to justify government hatcheries. A reference was made yesterday where the capital expenditure was amortized at 50 cents a pound of fish. I can justify this for a research facility. For any type of production facility, this is out of sight. I was asked yesterday what I felt that the capital expenditure would be to afford a commercial hatchery. I answered it this way. If the interest on your investment is more than 10 cents per pound of production, you are going to have difficulty competing in private industry. If your interest is over 10 cents per pound, you had better be cutting labor somewhere to justify this expense.

DALLAS WEAVER: One of the matters that have received very little consideration so far is system reliability. One of the hatcheries we visited yesterday is using a pump system. What do you do when the power goes off? This factor, expecially from a commercial standpoint, could be disastrous. Leo Ray is lucky. His water supply is artesian and he does not have to depend on the power company. This problem in a system design seems to be absolutely critical, especially if you are talking about pump systems.

(UNIDENTIFIED): You can have reliability for pumping systems by installing standby generators.

[McCall facility uses gravity flow throughout for hatchery water.—LRT]

RAY: Many people come to our farm with suggestions for improvement. The most common improvement, or almost every improvement, involves the use of pumps. I have a unique system of artesian water supply and plenty of elevation. Pumps are not needed. However, whenever an engineer comes around and recommends improvements, he always recommends the use of pumps. Most of the time, at the state of the industry today, if you are wanting to get into fish culture, you do not have to go to a pumping system. If you have a certain piece of property that you want to develop into fish production, you may have to use pumps. If you are flexible in movement, you can select another site where you can use gravity flow. In another 10 to 20 years you may be unable to find such sites. Right now there are plenty of sites available where you can avoid the use of pumps.

JAMES F. MUIR: Getting back to this matter of criteria, I just wanted to make a few comments from our experiences in the United Kingdom. The first is that it is just as easy to make mistakes in expensive facilities as it is in inexpensive ones. The second is in regard to production of larger facilities such as those we visited yesterday. I made a quick estimate and concluded that on commercial terms you would either have to reduce costs about ten times or increase production ten times to make the operation competitive. That would suggest that you would get a lot more for your money by going to commercial sources.

Bio-Engineering Symposium for Fish Culture (FCS Publ. 1):201-209
© 1981 by the Fish Culture Section of the American Fisheries Society

Experiences with a Fouling-Resistant Modular Marine Fish Cage System

JOHN EDWARD HUGUENIN

Groton BioIndustries
P.O. Box 133
Woods Hole, Massachusetts 02543

SPENCER C. FULLER

Fox Islands Fisheries, Inc.
P.O. Box 427
Vinalhaven, Maine 04863

FRANK J. ANSUINI

Kennecott Copper Corporation
128 Spring Street
Lexington, Massachusetts 02173

WAYNE T. DODGE

Fox Islands Fisheries Inc.
Mason Power Station
Wiscasset, Maine 04578

ABSTRACT

The inservice experiences in a commercial context with an innovative prototype of a rigid marine fish cage system are described.[1] This modular design approach has inherent biofouling resistance, increased productivity due to better waterflow, easier maintenance and operations, and an anticipated life longer than more conventional fish cage systems. The basic design incorporates pultruded fiberglass structural frames and panels covered with 90/10 copper-nickel expanded-metal mesh of 76% open area. The prototype cage is deployed at the Fox Islands Fisheries, Inc., site at Wiscasset, Maine, where in early 1978 it successfully survived six major winter storms that were clearly the worst in 100 years. Techniques evolved for carrying out operational functions — such as grading, harvesting, cleaning, and maintenance — as well as modifications to the sytem's design are described. The overall system's performance and economics are evaluated in relationship to comparable conventional nylon-bag cage designs.

Introduction

In a previous study on the technology and economics of marine fish cages (Huguenin and Ansuini 1978), it was pointed out that while nylon-net bags are by far the most common type of fish cage, they tend to have short service lives, are particularly sensitive to biofouling, and often require secondary predator nets. On the other hand, rigid cages, despite higher initial costs, generally have lower maintenance requirements and longer service lives, and are more amenable to mechanized operations. Rigid cages also are more readily adaptable to a wide variety of materials for both mesh and structures, while conventional bags are limited primarily to nylon and other synthetic netting.

Biofouling is one of the severest operational and maintenance problems encountered in large-scale commercial floating cage culture. Biofouling on nets and floats can increase weight and drag,

[1]The design, construction, and evaluation of this system has been supported by the International Copper Research Association, Inc., under INCRA Grant No. 268.

reduce bouyancy, and seriously limit water circulation through the cages. Since it is the frequency and duration of the minimum water flow through a cage that usually determines the maximum stocking density, biofouling can significantly reduce the quantity of fish that can be maintained in a cage without inducing a stressed environment. Thus, fouling can have significant effects on unit productivity. Major net system failures, with fouling as a strong contributing factor, have already occurred in several large commercial marine aquaculture operations, including at least one large floating marine fish cage complex.

Many copper alloys have considerable biofouling resistance, very low corrosion rates and can, under many circumstances, be used in areas with only modest water flow without any detrimental effects on culture organisms (Huguenin and Ansuini 1975). Copper-based antifouling paints are already commonly applied to synthetic-fiber fish cages. Metallic copper alloy meshes can be used in rigid structures to give long life and low maintenance service and have previously been successfully used on a smaller scale for marine fish containment (Powell 1976). These meshes when fitted onto a suitably designed rigid cage structure provide the additional benefits of reduced probabilities of system structural failures, greatly enhanced protection against predators, and increased productivity from improved water flow. In short, fouling-resistant rigid cages appear to have considerable development potential for large-scale commercial marine cage systems. As marine culturing technology develops in the coming years, culturing activities will progress to more exposed and less accessible, but more available, marine sites where access and maintenance functions become increasingly difficult and expensive. This will place a greater premium on the use of materials and systems with long service lives, low maintenance requirements, and greater strength.

The general goal of the project discussed in this paper has been to develop a fish cage system that would use advanced design and materials technology to alleviate design-related problems confronting cage culture. The specific objective was to develop a design for a marine fish cage system that would virtually eliminate the need for short-term maintenance, be structurally sound so as to withstand a very severe storm and a 5-knot (2.6-m/s) current without permanent damage, increase resistance to predators, last in excess of 10 years, and, finally, be cost effective on a commercial basis. An additional objective was to make the proposed cage modular in design so that cages of other sizes and configurations — including floating, submersible, and bottom-mounted systems — could be assembled from the same standard components. In particular, the size of the prototype cage (57 m³) was recognized as being too small for most commercial uses (Huguenin and Ansuini 1978). Therefore, from the start of the project the objective has been the ultimate application of developments to a range of cage sizes up to 453 m³.

The panels in the prototype cage are of a standard size and design and are fitted into a space frame made of commercially available fiberglass sections that must be chosen according to the specific cage dimensions and anticipated maximum service conditions. In most cases the fiberglass frames, rather than the mesh panels, will fail first. By selecting appropriate space frame structurals, cages of any desired strength can be assembled. The ability to assemble these stock components into a variety of configurations would allow the culturist to design a cage system to meet the unique requirements of his operation without having to pay the higher costs that usually accompany a completely custom design. A prototype cage using this modular system was deployed during July 1977 and is currently undergoing commercial comparative testing with rainbow trout and coho salmon at Wiscasset, Maine, alongside conventional nylon cages.

Cage System Description

The basic structural element of the 57-m³ prototype cage is a 40- x 120-in (1.0- x 3.0-m) panel consisting of a fiberglass

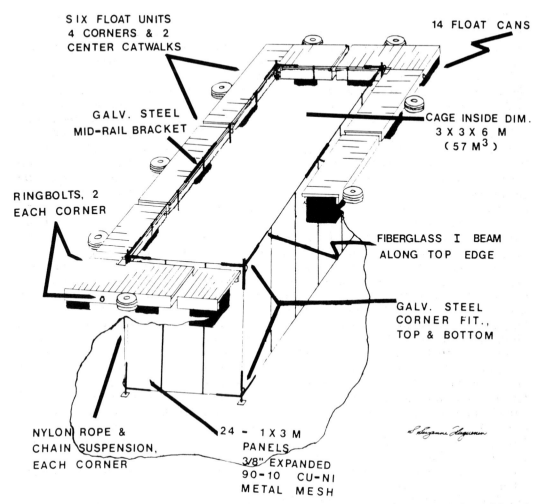

FIGURE 1. — This marine fish cage prototype system has been in use since July 1977. The cage sits within a well formed by the float collar. A five-panel rigid predator barrier on top of the cage and lines from the corner ringbolts to either adjacent cages or mooring fixtures (not shown) complete the system.

C-channel frame wrapped with 90-10 copper-nickel 3/8-inch nominal (7.9- x 12.7-mm hole size) expanded-metal mesh. Twenty-four panels are interbolted to form an open top rigid rectangular cage 10 x 10 x 20 ft (3 x 3 x 6 m). An artist's rendition of the complete system is shown in Figure 1. The design of the cage is such that all structurals and fasteners are located on the outside, where they are readily accessible for assembly and inspection. The inside of the cage is a smooth wall of copper-nickel mesh with no protrusions or sharp edges that might injure the fish. Expanded metal was selected for the mesh as it offered a large variety of mesh sizes, open areas, and strand gauges, coupled with the further advantages that the nodes were fixed, thus providing freedom from both fretting and crevice corrosion, and also enhancing mesh rigidity. The actual mesh selected was 3/8-in hex mesh with 76% open area and individual strands measuring 0.9 x 1.3 mm made of 90-10 copper-nickel (Copper Alloy 706).

In order to avoid problems with galvanic corrosion, the selection of materials for structural elements was limited to either similar alloys as the mesh or a nonmetallic. Pultruded fiberglass structurals (Tickle

1977) were selected, although they have no known prior uses involving long-term seawater exposure, on the basis of ready availability in a variety of section sizes as well as for their high strength and low weight. Hot-dipped galvanized steel weldments were used at the high stress locations, which include the eight corners and the midpoints of the long sides at the top. It is at these locations that loads are transferred from the cage to the floats. Care was taken to assure electrical isolation of the galvanized steel by use of nylon bushings to avoid bimetal contact with the expanded metal, which would seriously reduce the biofouling resistance of the mesh. It should be pointed out that this 90-10 Cu-Ni expanded-metal and pultruded-fiberglass modular construction approach also has considerable potential in broader-use seawater screening applications, such as for power plant cooling systems (Ansuini, Huguenin and Money 1978).

An articulated rectangular float collar surrounds and suspends the cage. The collar itself is about 0.8 m wide and consists of four L-shaped units at the corners and two straight units along the long walls of the cage. The center of bouyancy of the L-shaped units is located at the inside corner. These points show the least vertical motion in a sea and thus are the locations where the cage suspension lines are attached. The design of the float units, like the cage, is modular and can be readily adapted to larger cages by simply adding additional straight sections between the L's. For a nested cage installation, the L units are readily adapted to T's, or if necessary, X's. All mooring lines extend from the float collar so that, also considering the location of the suspension system, the amount of wave forces transmitted to the cage is minimized. Considerably more information on design criteria and details of system design can be found in a prior publication (Ansuini and Huguenin 1978).

The tidal range at the test site in Maine is about 3 m, so this was used to advantage in launching the cage. At a site just above the high-tide mark, panels, structurals, and fittings were assembled to form five major subassemblies. Farther down the beach, the elements of the flotation collar were interconnected to form two large C-shaped collars, which were floated free and moored to one side. At an outgoing tide, the cage subassemblies were moved to a position at the mid-tide level and final assembly of the cage was completed during one period of low tide by three people with simple hand tools. On completion of the final assembly, two halves of the float collar were towed into position, suspension lines connected, and short tethers used to hold the cage. After the rising tide completed the lift, the cage-float system was towed to its intended location and secured to the adjacent cages. The cage was then lowered to its full draft by gradually, and under perfect control, slacking off on the temporary tethers in each of the four corners. The tide-lift method of deployment is not limited to areas with very large tidal ranges and can be reversed to recover cages if this should become necessary. By using even shorter temporary tethers, the minimum water rise necessary to float a cage would be only the difference in draft of the floats with and without support of the cage; in this specific case, about 0.6 m. This makes it possible to use this deployment and recovery scheme at locations with even a relatively modest tidal range.

Design Experience and Modifications

The cage system, since its deployment in July 1977, has already withstood conditions exceeding its design limits. It saw six major storms during the winter of 1977-1978, one of which was extremely severe, causing wide damage along the coast. This winter was clearly the worst for storms in the deployment area for 100 years. Three more major storms struck during the winter of 1978-1979. During both years the copper-nickel cage was sited in the most exposed position and acted as a breakwater for the rest of the cages in the complex. Conventional cages in the facility were severely damaged by the storms and, in many cases, destroyed. At one point during a storm in December 1978, the rigid cage held the rest of the facility together. Observers report that in the two years the cage has been deployed it saw at least a five-knot tidal current, had three-foot waves breaking over it (triggered by a log boom a few feet away), was covered with considerable broken ice, about one foot thick, and debris, and was at the near center of an earthquake registering 4.2 on the Richter scale. In addition, the top of the cage was frozen into a solid piece of ice about three inches thick — all this without any damage to the cage itself or the loss of fish.

One of the problems encountered in designing the cage included the lack of design information on stresses and deflections in uniformly loaded expanded-metal panels. For this project an analytical model for predicting such stresses and deflections was developed (since none previously existed), which gave predictions that compared favorably to the results from an experimentally loaded cage panel (Gularte and Huguenin, manuscript). This work confirmed that the mesh would not yield with a five-knot current going through it. Efforts to refine further and generalize these design techniques are still in progress.

The cage is suspended from the floats by a tuned compound chain and nylon-line bottom-connected suspension system. Two years of rugged use has confirmed that this concept works well. While the floats respond to waves and the motion of people walking on the floats, the cage remains essentially stationary. In spite of the short distance between the floats and the cage, there are no collisions, even under extreme conditions and with damaged float hinge joints. The limiting conditions have been torsional loads in the float hinge joints caused by storm waves hitting the cage at an angle, which did cause some damage in the joints. However, this was easily fixed by using heavier hinge hardware.

One change made to the system at the request of the operator was to raise the top edge of the cage from the water line to about 0.4 m above it. Since the top part of the suspension system is chain, this change was easily accomplished. The long strip of

barrier netting between the cage and the float collar and the bird netting draped overhead, as originally supplied, did not provide any improvement over conventional cage top-side predator protection but did prove equally inconvenient. Raising the top edge of the cage enabled a five-panel rigid frame and chicken wire top to be placed on the cage. These panels not only proved to be more practical than conventional netting but also provided substantially improved protection against both birds and land mammals. Predation at this site has at times been as high as 20% over 3 months. Thus the ability to use light rigid-panel tops is a significant advantage.

An additional advantage of the suspension system is that the cage can be lifted clear of the water, if this should become necessary, without having to disconnect any of the main suspension lines. Four small A-frames were built, one for each inside corner of the float collar, and each provided with a small hand winch. With the ends of the winch cable attached to the strong point fixtures at each of the four bottom corners of the cage, the complete cage has been lifted about 0.5 m clear of the water surface. This technique is simple and quick and provides excellent control, and only four A-frames and winches are needed, even for a large number of cages. We believe this approach would work for considerably larger cages, probably as large as 6.25 x 12.5 m, but this size would require six A-frames and winches.

Experience to date indicates that while conceptually simple, the cage system requires a great deal of care and sophistication in choices and combinations of materials. Improper selections, such as in fastener materials, can easily lead to a nullification of the copper-alloy mesh's fouling resistance (Huguenin and Ansuini 1975) or even to rapid failures due to unfavorable area ratios between the surface area of mesh and fasteners. The best choices are not always obvious. The original fasteners on the cage were 304 stainless steel, which were electrically isolated from the mesh. Since this isolation was not perfect for all fasteners, the fasteners were switched to monel after one year of use. Monel is cathodic to the copper-alloy mesh and doesn't need to be electrically isolated. Small-scale tests are continuing to monitor long-term materials' performance and to find improved and cheaper materials. This includes fiberglass structurals, meshes, and fastener components, both singularly and in combinations.

Operational Experience

The cage after two years of in-service use has mesh that is almost completely free of biofouling. However, as was expected, the external fiberglass structurals and galvanized-steel fittings are covered with fouling, predominantly large mussels. However, this does not obstruct the water flow through the cages. After one year of service in July 1979, the cage was lifted clear of the water for inspection and modification, and as a result both the mesh and the structurals were cleaned. Other than this, not a single man-hour has been spent on biofouling control on this cage in two years of service. In late May 1979, the cage was inspected in situ and the mesh, including that on the bottom, was found to be as clean as when reintroduced into the water about a year previously. At the next inspection in mid-August 1979, although loosely attached, a few small clumps of mussels were seen on the mesh, particularly on the bottom panels. The biofouling resistance of the mesh is not absolute. When some biofouling does occur, it is usually very weakly attached. A quick brushing from the surface with a long-handled brush, possibly once or twice a year, will guarantee a completely unobstructed mesh. In comparison, smaller conventional nylon cages at this site are usually changed about four times a year and require at least 40 man-hours/year per cage for net removal, cleaning, repair, and replacement. Smaller meshes require considerably more man-hours.

It had been expected that this system, because of its greatly improved water circulation as a result of fouling reduction, would provide substantial benefits by enabling increased fish density and corresponding yields per unit volume compared to nylon cages (Ansuini and Huguenin 1978). Experience to date has tended to confirm that this benefit exists, but it has yet to be scientifically tested and quantified, because of a number of operational constraints unrelated to the cage system. The maximum fish density contained (September 1978) has been about 22.7 kg/m^3 (1.4 lbs/ft^3). This is only marginally higher than normal maximum values common to conventional cages at this site (16 kg/m^3). However, comparative density tests, using measured drops in dissolved oxygen concentrations are planned for next summer's high water-temperature stress period (1980).

One of the areas of operational concern with respect to rigid cages is the requirement to crowd fish for grading and harvesting. With synthetic mesh cages this is done simply by collapsing the net. Many schemes, including a variety of complex in situ adjustable graders, were considered in an iterative process with the user. The final choice was a simple 4.7 x 4.7 m of one-inch stretch-mesh nylon seine having a 2/5 basis of bunt (forming a partial pocket) and with weights on the bottom edge and floats on the top edge. Two long poles were tied to the sides of this seine.

The operational procedure for use of the seine involves chumming the fish to one end of the cage with feed, while the two poles and seine are lowered along the corners at the opposite end of the cage. One man on each side then walks the seine down the cage, being careful to maintain contact between the poles and the cage sides and bottom. The poles are pushed along the side of the cage at an angle, to make sure that the weighted foot-rope goes first. When the bottom ends of the poles reach the opposite side of the cage, the side poles are maneuvered so that the foot-rope is brought up along the opposite wall. At this point a third man, while not essential, is useful in raising the foot-rope to the surface by way of lines previously attached. The end result is the capture of most of the fish (90% on first pass in the only instance where a specific count was made) in a pocket formed by the bunt of the seine near the surface. This pocket can be collapsed to any degree desired or subdivided by passing poles underneath, and grading and/or harvesting can proceed as it would for a nylon cage. However, the mess and high weights associated with handling fouled nylon cages are avoided. This process for use with rigid cages has proven to be practical and requires only a few minutes.

Overall the management and operation of this rigid cage has been shown to be significantly easier than a comparatively sized conventional nylon bag-net cage. This is due to a variety of factors including fouling resistance, superior float design, improved system stability, greater predator resistance, and increased confidence in system integrity and strength. After two years of use, the fish in the prototype cage have growth rates and health, at least as good as, if not better than, those in nearby conventional cages. Mortalities in the prototype cage appear at times to be less than those in conventional units. The fish also appear to be more vigorous. The performers of the system will continue to be monitored for at least three more years so that the long-term economics may be evaluated.

Economics

A modified 57-m³ rigid-cage system, similar to the prototype but having less expensive 30-in (76-cm) wide panels, is projected to have a materials and component cost of about $4,200, based on a total purchase of around 3,800 m² of cage surface — including floats, moorings, and accessories. This cost can be expressed as $74/m³, which is approximately three times the volumetric cost of comparably sized conventional nylon-bag case systems ($26/m³). By increasing the volume of the cage eightfold (453 m³), to represent a larger commercial culturing cage, the volumetric cost of both systems decreases (10/m³ conventional; 33/m³ rigid) but the relative cost difference between the two systems remains approximately the same. It should be noted that a substantial portion of the cost increase is due solely to the added expense of a rigid cage. If a rigid cage was built to the design of the modified prototype, but used synthetic mesh rather than copper-alloy mesh, it would still cost twice as much as the flexible cage although offering none of the maintenance and performance advantages of the copper-alloy mesh. Put another way, switching from a flexible cage system to a modular rigid-cage system doubles the volumetric containment costs; upgrading the mesh to copper alloy represents a further 50% surcharge.

Despite the substantially higher acquisition cost of the copper-alloy rigid-cage system, the annual cost of the 57-m³ cage is projected to be less expensive than conventional cages. The largest savings comes from a reduction in maintenance costs and increased service life. The copper-alloy cage should last 10 years, while the nylon mesh on the conventional cages, only about 3 years. The annual cost of a conventional cage, including both maintenance and depreciation, is about $1,087 as compared to $619 for the rigid 57-m³ copper-alloy cage. The annual volumetric containment costs are estimated at $19.07/m³ for the conventional cage and $10.86/m³ for the rigid cage. As is the case for acquisition cost, these figures will vary with cage size but the ratios should remain approximately the same. At some installations, the frequency of net cleaning would further favor the cost advantage of copper-alloy cages at these sites. At other installations, the nets are periodically dipped in copper-based antifouling paints to reduce the frequency of cleaning. This approach reduces the total fouling control costs but it still does not completely eliminate the problem. Fouling and fouling-related costs are highly variable and are very dependent on specific site conditions, acceptable mesh blockage, and material, pretreatment, initial condition, and size of the mesh. While the economic numbers presented in this paper were conservatively developed over a year and a half ago, they are still valid estimates today. A more in-depth economic analysis comparing cage types at different sizes can be found in Ansuini and

Huguenin (1978) and Huguenin and Ansuini (1978).

The annual cost per cage is not necessarily the best measure of the true economic worth of this concept as it does not allow for differences in relative risks and in stocking density between the flexible and the rigid designs. The monetary value of fish contained in a cage can be high, often in the order of $85/m^3$. It should be noted that the value of the fish can be higher than the initial materials cost of even a relatively small 57-m³ copper-alloy cage. A possibility of a total loss of fish due to structural failures or rips in nets as well as associated hardware repair and replacement costs can assume some economic importance. As shown by the experiences of the prototype, this factor certainly works in favor of the copper-alloy cage; however, this economic advantage, while substantial, can not be quantified at the present time. A rigid cage would also have more usable volume than a comparable nylon cage, since it maintains its shape and does not collapse under the effect of a current. A fouling-resistant mesh allows a greater water flow into the interior of the cage, especially under limiting low-flow conditions. Both of these factors point to the use of a higher stocking density in a copper-alloy rigid cage than in a conventional cage. The maximum stocking density in the conventional cages at the test site is usually about 16 kg/m³, which gives an annual fish containment cost of about $1.19/kg. The prototype has already held 23 kg/m³ without any problems. If we make the reasonable assumption that the optimum stocking density of the rigid cage can be 50% higher (24 kg/m³), the annual fish containment cost would become $0.45/kg, or about 60% less than the conventional cage. However, this higher density expectation has not yet been scientifically proven by actual comparative tests and may in fact be conservative.

Summary

Two years of operational experience have verified the biofouling resistance of the copper-nickel mesh and proven the structural strength of the modular standard panel and application-specific space frame approach. The bottom-attached cage suspension system has been particularly successful, because of its general stability, flexibility to adjust draft, designed stiffness, and ability to raise the cage clear of the water without requiring disconnections. The entire system has clearly proven itself easier to manage and operate than similarly sized nylon cages and gives every indication of ultimately exceeding its 10-year design life.

There is currently considerable worldwide interest, particularly in Europe, in 90-10 Cu-Ni mesh rigid cages. The use of the design concept described in this paper appears at least as attractive for submersible bottom-mounted fish cages and seawater screening applications as it does for floating cage systems. If present trends continue, such systems or some derivations of them are very likely to find substantial commercial acceptance.

Acknowledgments

The authors wish to thank Mr. Brian Moreton, Ms. Pauline Riordan, Dr. Harold Webber, and Dr. George Cypher for their comments on the draft for this paper. Special thanks go to Gary Towle for his vital help early in the project.

References

ANSUINI, F.J., AND J.E. HUGUENIN. 1978. The design and development of a fouling resistant marine cage system. Proc. Ninth Annual Meeting of the World Mariculture Society 9:737-745.

ANSUINI, F.J., J.E. HUGUENIN, AND K.L. MONEY. 1978. Fouling resistant screens for OTEC plants. Proc. Fifth Ocean Thermal Energy Conference (OTEC), Miami Beach, Florida, February 20-22, 4:283-294.

GULARTE, R.C., AND J.E. HUGUENIN. An analytic method for determining stresses and deflection in thin gauge expanded metal panel. (ms.) 18 pp.

HUGUENIN, J.E., AND F.J. ANSUINI. 1975. The advantages and limitations of using copper materials in marine aquaculture. Pages 444-453 in Ocean '75 Conference Record, Marine Technology Society, Institute of Electrical and Electronic Engineers.

———. 1978. A review of the technology and economics of marine fish cage systems. Aquaculture 15(2):151-170.

HUNT, J.R., AND M.D. BELLWARE. 1967. Ocean engineering hardware requires copper-nickel alloys. Pages 243-275 *in* Proc. Third Annual Marine Technology Conference, Marine Technology Society.

POWELL, M.R. 1976. Resistance of copper-nickel expanded metal to fouling and corrosion in mariculture operations. Prog. Fish-Cult. 38(1):58-59.

TICKLE, J.D. 1977. Pultrusions step up the challenge to structural steels. Machine Design (October):163-167.

Bio-Engineering Symposium for Fish Culture (FCS Publ. 1):210-226

Solar Aquaculture: An Ecological Approach to Human Food Production

RONALD D. ZWEIG, JOHN R. WOLFE, JOHN H. TODD,
DAVID G. ENGSTROM, AND ALBERT M. DOOLITTLE

The New Alchemy Institute
237 Hatchville Road
East Falmouth, Massachusetts 02536

ABSTRACT

Solar-algae ponds (translucent fiberglass cylinders five feet in diameter and height) are efficient passive solar collectors and fish production components in greenhouse structures. Several outdoor reflective designs have also been evaluated for the maximization of light entering the ponds. The indoor thermal energetics of the ponds demonstrate that for every calorie used in their manufacture, five calories will be returned in the form of winter heat on Cape Cod during their expected twenty-year lifetime. The ponds' phytoplankton-based aquaculture has proven to be a stable productive ecosystem. The algal cells in the water column are useful (1) as a feed for phytophagous organisms, (2) in the oxygenation of the water through photosynthesis, (3) as micro-heat exchangers absorbing solar energy, and (4) in the purification of the water through directly metabolizing toxic fish wastes. Fish productivity exceeding ten times that previously documented in still water has been achieved with this aquaculture design.

The water chemistry anad phytoplankton dynamics have been monitored and recorded through several fish production trials. Different feeds have been evaluated regarding their impact upon the structure of these aquatic ecosystems and the growth of the tilapia, *Sarotheradon aureus,* within them. These data are being used toward the development of an ecological model to assist in optimizing the use of the solar-algae ponds and the development of a simple guide for fish culture.[1]

Design

Designs for aquatic ecosystems optimizing the entry of solar energy maximize the biological productivity within them. During the past three decades, aquatic biology has demonstrated that phytoplankton are the most efficient organisms to use solar energy. The potential of algal culture in sewage and waste treatment (Lincoln 1977), energy production (Oswald and Golueke 1960; McLarney and Todd 1972; Zweig 1978), and food production (McLarney and Todd 1972; Zweig 1977, 1978) is coming to be realized more and more. At The New Alchemy Institute, a synthesis of many of the characteristics of algae has been used in structuring a solar aquaculture design.

[1]This research has been supported in part by the Office of Problem Analysis, National Science Foundation Grant #OPA77-16790A02.

Dynamics

Highly translucent cylinders can create very productive algae cultures. The earliest reported design (Cook 1950) used a Pyrex glass cylinder 10.2 cm (4 in) in diameter and 1.83 m (6 ft) in height for the culture of the phytoplankton *Chlorella pyrenoidosa* Link. This flow-through, continuous-culture apparatus was tested both under laboratory conditions with artificial lighting and outdoors with natural sunlight. This study determined that 5% of sunlight reaching the water column or 2.5% of the total solar radiation was converted to organic matter. The most favorable conversion in agriculture has been 0.3% of the year's total radiation. Cook concluded that there is a great potential for solar-driven algal culture and that further study would help to realize its possibilities.

A literature survey (Tamiya 1957) reviewed mass culture of algae and com-

210

pared several culture techniques. The work of Myers et al.(1955) further supported the premise that light optimization improved productivity of algae, $g/m^2 \cdot d^{-1}$. In a deep tank with transparent side walls, a closed (no water exchanged), constantly stirred culture of *Chlorella* proved the best system of all systems described, including both open and closed culture schemes.

In 1974 The New Alchemy Institute took the idea of algal culture in transparent cylinders and combined it with the concept of raising herbivorous fish (Todd 1976, 1977; McLarney and Todd 1977). Solar-algae ponds, 2.38-kl (630-gal) translucent fiberglass silos, 1.52 m (5 ft) in diameter and height, mark the most successful design to date (Zweig 1977). These $2.4\text{-}m^3$ aquatic ecosystems are used primarily for the culture of fishes as human foods.

A key factor to consider with closed system fish culture is the management of toxic fish wastes. The most critical compound to control is un-ionized ammonia. Concentrations as low as 0.12 mg/l have been shown to inhibit channel catfish growth (Robinette 1976). Another beneficial aspect of an algae-based fish culture system is that some Chloroccocale, such as *Chlorella* and *Scenedesmus*, are capable of metabolizing ammonium directly, thereby helping to eradicate the problem (Syrett 1962).

One difficulty with intensive algal culture is that the photosynthetic activity is limited to the upper layers in an undisturbed, subsurface pond (Oswald and Golueke 1960). Oxygen concentrations, therefore, are usually high at the surface and negligible in the lower regions of the pond. These ponds require periodic stirring to prevent anaerobic conditions from developing. In a translucent cylinder the entire column of water is capable of photosynthesis, reducing the chance of development of anaerobic conditions. Stirring or aeration need then occur only nocturnally. Also, in a fish culture system, fish species can be chosen that will assist mixing through their swimming behavior.

The increased photosynthetic activity in the solar-algae ponds may have further significance beyond improved oxygenation of the water and toxic waste removal. In one study (McConnell 1965), it was demonstrated that fish growth *(Tilapia mossambica)* in natural bodies of water could be directly correlated to photosynthetic activity. A linear relationship was found to exist between photosynthesis and the cube root of the mass of growth per individual fish.

The phytoplankton in the water column, thereby, are useful (1) as a feed for phytophagous organisms, (2) in oxygenation of the water through photosynthesis, (3) as micro-heat exchangers absorbing solar energy, and (4) in the purification of the water through directly metabolizing toxic fish wastes.

Strategy

The most efficient strategy for the utilization of a pond for fish culture was first developed in Chinese polyculture (Fan Lee, 5th century; Chen 1934). In these culture designs, species of food fish were used that could exploit different trophic levels. These included plankton, bottom, and macrophytic plant feeders. This concept was integrated into greenhouse aquaculture design (McLarney and Todd 1972), where water volume was also used for heat collection and storage. In addition, the fertile fish water was used for the irrigation of vegetables grown within the structure. From this pioneering work, the solar-algae pond fish culture design evolved.

The polyculture strategy for the solar-algae ponds is defined by the nature of fish feed organisms growing within and the supplementary feeds that can be introduced from outside. The ponds present the best possibility for planktivorous fish species capable of utilizing the zooplankton and the phytoplankton that flourish in the water column. For tropical temperatures species of *Tilapia* are very efficient. The Sacramento River blackfish, *Orthodon microlepidotus*, has been cultured in solar-algae ponds and can be useful when cooler temperatures prevail (Murphy 1950). These species may also insure the turnover of the phytoplankton population, preventing a senescent condition from developing. A species such as the Israeli carp, *Cyprinus carpio* var. *specularis*, which can recycle sediments and feces, would also be advantageous. Sessile organisms growing on the side walls and bottom are another potential nutrient source that could be grazed by any of these fishes.

Productivity

In an early fish production experiment with the solar-algae ponds (Zweig 1977), a polyculture strategy was used including the tilapia, *Sarotheradon aureus*, the Israeli carp, and the Chinese bighead, and silver carps. The tilapia comprised a majority of the population. During this ninety-eight day trial, 6.98 kg/yr (57.3 lb/yr), 2.5 kg/m³, or 3.84 kg/m² surface area, was produced with an S conversion of 1.0 (Zweig 1977). The results of this experiment proved this solar-energy-based aquaculture system to be nearly ten times more productive than any other described standing body of water (Zweig 1977). Barry Pierce of Goddard College has documented the best growth rate so far for solar-algae ponds. In a single trial using a 100-fish polyculture of *Sarotheradon mossambicus*, *Cyprinus carpio* var. *specularis*, and hybrid sunfish, initially weighing 9.1 kg, a growth rate of 132 g/d over an 82-d trial was achieved. If extrapolated to a full year, a potential of 48.2 kg/yr (106 lb/yr) is possible for a single pond. The fish were fed 175 g/d dry weight of flying insects; the food conversion ratio was 1.2. Further, in using the more conservative figure of 22.7 kg/yr (50 lb/yr) per pond consistently achieved at present in The New Alchemy trials using commercial trout feeds, solar-algae pond culture can be calculated to require 140 kcal of energy to produce 1 g of dryweight protein (Wolfe, in press). This efficiency can be bracketed by egg production at 53 kcal/g and milk at 145 kcal/d protein. Feed-lot beef is of the order of 310 kcal/g.

Thermal Energetics

This aquaculture design has also proven to be economically viable as a passive solar collector in greenhouse structures (Zweig 1978). For every calorie used in their manufacture, five calories will be returned in the form of solar winter heat for indoor use with the Cape Cod climate over the expected twenty-year lifetime of the ponds (Wolfe, in press).

Aquatic Ecology: Phytoplankton Population Dynamics and Water Chemistry

In one experiment designed to test the impact of nitrogen input rates and types, phytoplankton population and written chemistry dynamics were examined in relation to their influence on fish growth. The experiment was performed in twelve solar-algae ponds (Ponds C-N) in the reflective courtyards adjacent to the Cape

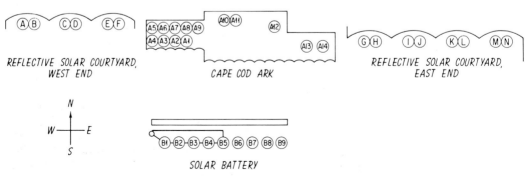

DIAGRAM 1.

212

Cod Ark at The New Alchemy Institute during the summer of 1978. (Diagram 1.) A monoculture of the tilapia *Sarotheradon aureus* was used with an initial stocking density of 2000 grams per pond (size classes between 35- and 115-g fish). Four different kinds of nutrient inputs were used:

(a) Fertilized ecosystem — Ponds C, D, E. (Pond D was intensively monitored.) Human urine was added to these ponds at a nitrogen input rate approximately equal to that of the rabbit feed described below. After three weeks, when water quality conditions deteriorated and measured fish growth was negligible, the urine inputs were terminated and later replaced with comfrey.

(b) Rabbit feed at 4% rate — Ponds I, J, and K. (Pond J was intensively monitored.) Rabbit feed (predominantly soy and alfalfa meal) with 20% protein and one-third the cost of commercial feeds was fed at a rate of 3% of fish body weight daily except Sunday.

(c) Trout developer feed at 3% rate — Ponds L, M, N. (Pond L was intensively monitored.) A 40% protein commercial preparation of trout feed was fed at a rate of 3% body weight daily except Sunday.

(d) Trout developer feed at 6% rate — Ponds F, G, and H. (Pond H was intensively monitored.) The same feed as above was fed at the high rate of 6% daily except Sunday. Feeding at this rate had a feedback component. When un-ionized ammonia levels reached 1 mg/l, feeding stopped until the ecosystem regained equilibrium and then was fed at the 3% rate.

Analytical Methods and Materials

Measurement methods and materials are described in detail in Progress Report 3 (Todd et al. 1979). They included the use of a linked computer-based data acquisition and analysis system comprised of a DEC PDP-11/03 and a Microlog (R) M6800 computer. The Microlog was used in conjunction with Yellow Springs Instruments Model 57 Dissolved Oxygen meters, temperature thermocouples,

Chemtrix Model 40 pH meters, and solar radiation with an Eppley Black and White Pyranometer. A series of data points was collected every fifteen minutes. Colorimetric analysis of orthophosphate, nitrogen compounds (ammonia, nitrite, and nitrate), sulfate, and alkalinity were measured every other day with a Bausch and Lomb Spectronic 20 spectrophotometer. Algae counts were made with the use of a Unitron (R) phase microscope using an Improved Neubauer Hemacytometer.

Algae

Species composition

Throughout the second experiment green algae — as opposed to blue-green algae or diatoms — dominated in every pond. As shown in Figures 1B, 1C, 2A, and 2B, the green algae *Ankistrodesmus brauni* and *A. falcatus,* prevalent initially, dropped out after the introduction of fish, possibly suggesting *Ankistrodesmus* sp. are susceptible to fish predation. Different species replaced *Ankistrodesmus* in each pond, as expected since each pond shown experienced different kinds and rates of nutrient loading (Fig. 1A). In Pond D, *Scenedesmus quadricauda* asserted dominance. *Chlorella* entered Pond J and persisted. In Pond L *Scenedesmus* proved the major species. In Pond H a variety of species played dominant roles at different times: *Scenedesmus, Micractinium, Radiococcus,* and a *Chlorella* species less than 3 microns in diameter.

Biovolumes

Looking at the overall biovolumes in Ponds J, L, and H (Fig. 1C, 2A, and 2B), one realizes that on an aggregate volume basis the ponds undergo remarkably similar transformations. All of the ponds with dry feed inputs experienced a sharp rise followed by collapse roughly a month after the experiment's inception, although the species responsible for each peak differed: *Sphaerocystis* in Pond J, *Scenedesmus* in Pond L, and *Micractinium* in Pond H. According to systems theory, overshoot and collapse occur where there is (1) exponential growth (2) toward an erodable limit

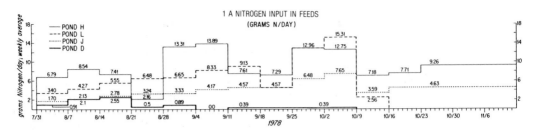

1 A NITROGEN INPUT IN FEEDS
(GRAMS N/DAY)

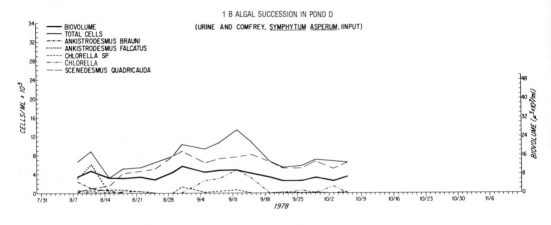

1 B ALGAL SUCCESSION IN POND D
(URINE AND COMFREY, SYMPHYTUM ASPERUM, IINPUT)

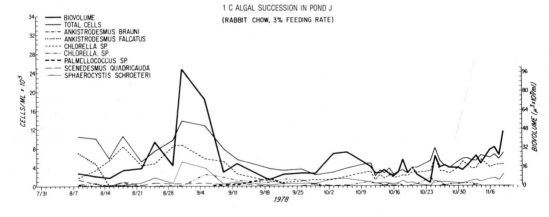

1 C ALGAL SUCCESSION IN POND J
(RABBIT CHOW, 3% FEEDING RATE)

(such as a depletable nutrient or an accumulative toxin), with (3) a delay in the response to the limit. The limit may be the buildup of detritus that blocks incoming light, or it may be the depletion of a trace micronutrient such as iron, magnesium, or a vitamin. In all ponds the macronutrients carbon, nitrogen, and phosphorus seem present in adequate amounts when the algae crashes occurred.

An interesting difference can be seen between ponds fed rabbit chow and those fed trout chow. The rabbit chow treatment represented by Pond J exhibited a biovolume peak, and in fact biovolume levels throughout, that were roughly double those for the two trout chow regimes, Ponds L and H. Pond J also avoided the slow algae recovery experienced by Ponds L and H after the initial decline in *Ankistrodesmus,* and biovolume in Pond J peaked earlier. Note that the exception, the early sharp spike in Pond L, was based on a few sitings of very large *Volvox* colonies; that spike may have been much less pronounced than shown. All this suggests

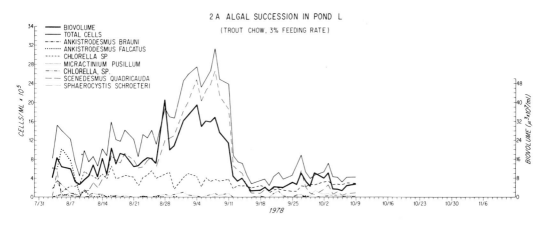

2 A ALGAL SUCCESSION IN POND L
(TROUT CHOW, 3% FEEDING RATE)

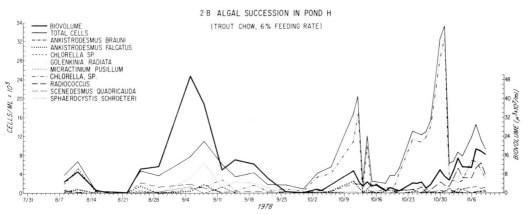

2 B ALGAL SUCCESSION IN POND H
(TROUT CHOW, 6% FEEDING RATE)

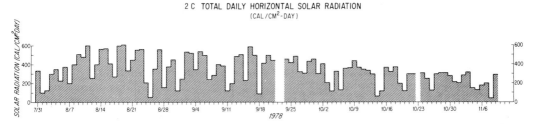

2 C TOTAL DAILY HORIZONTAL SOLAR RADIATION
(CAL/CM²-DAY)

that rabbit chow either causes less turbidity or contains a micronutrient absent in trout chow. In fact, rabbit chow contains large amounts of magnesium, the building block of chlorophyll, which trout chow does not.

Water Chemistry and Fish Growth

The nitrogen input rates from feeding and fertilization (shown in Fig. 3A) greatly influenced the concentrations of inorganic nitrogen compounds (Figs. 3C, 4A, 4B and 4C). The nitrogen in feeds had a direct effect on fish growth (Fig. 3B) as well as an indirect effect through water quality. In Pond D nitrogen entered the pond in the form of urine, a fertilizer, rather than in the form of feeds. In that pond (Fig. 3C), the rapid decomposition of urea to ammonia, combined with active algal photosynthesis that pushed pH above nine many afternoons, caused toxic un-ionized ammonia (NH^+_3) to climb above 3 mg/l. Such concentrations are undoubtedly deleterious to tilapia growth (they would

215

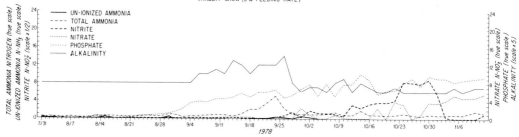

3 A   NITROGEN COMPOUNDS, ORTHOPHOSPHATES & ALKALINITY (mg/ℓ) IN POND J
(RABBIT CHOW, 3% FEEDING RATE)

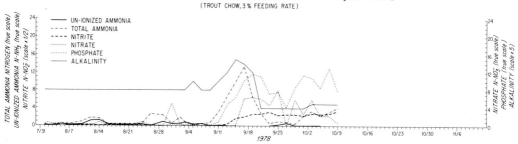

3 B   NITROGEN COMPOUNDS, ORTHOPHOSPHATES & ALKALINITY (mg/ℓ) IN POND L
(TROUT CHOW, 3% FEEDING RATE)

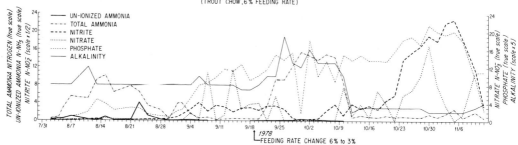

3 C   NITROGEN COMPOUNDS, ORTHOPHOSPHATES & ALKALINITY (mg/ℓ) IN POND H
(TROUT CHOW, 6% FEEDING RATE)

have killed trout or catfish). Urine inputs were discontinued after this time and replaced with comfrey leaves as a feed input.

Ponds J, L, and H (Figs. 4A, 4B, and 4C) had consecutively increasing nitrogen input rates. During the first month the consequential ammonia concentrations stayed below 1 mg/l nitrogen in Pond J, climbed to a moderate 2-3 mg/l in Pond L, and rose above 8 mg/l in Pond H. In the latter pond, nitrogen input clearly exceeded fish and algae assimilation rates, even though the algae, in a steady-growth phase, were undoubtedly incorporating significant amounts of ammonia. After this first peak in largely benign, ionized ammonia (NH_4^+) in Pond H, feeding in the 6%

ponds was discontinued temporarily, and then resumed at a 3% feeding rate.

The influence of the algae bloom and crash can be seen clearly in Figure 4. One to two weeks after the point in the first week of September when algae biovolumes dropped precipitously, ammonia concentrations climbed sharply. The 5-15-day delay was probably the time required for the sedimented cells to lyse, and for bacteria populations to increase sufficiently to decompose the dead cells. Phosphate rose simultaneously in the case of Pond L, or well beforehand, during the algae peaking, in Ponds J and H.

For all ponds, as ammonia increased, alkalinity also rose in an expected manner, since hydrogen ions are consumed as pro-

216

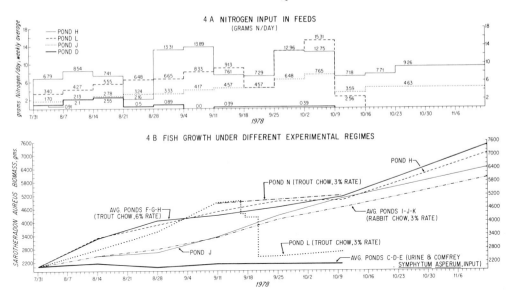

4 A NITROGEN INPUT IN FEEDS
(GRAMS N/DAY)

4 B FISH GROWTH UNDER DIFFERENT EXPERIMENTAL REGIMES

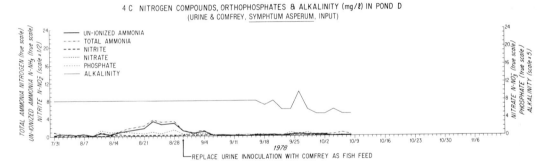

4 C NITROGEN COMPOUNDS, ORTHOPHOSPHATES & ALKALINITY (mg/ℓ) IN POND D
(URINE & COMFREY, SYMPHTUM ASPERUM, INPUT)

teins decompose. (Alkalinity is carbonate plus bicarbonate plus hydroxyl *minus* hydrogen ions. Brewer and Goldman, 1976, detail the process.) If the ammonia is reassimilated by algae, the alkalinity should return to initial levels; whereas, if the ammonia is oxidized by nitrifying bacteria into nitrite, twice as many hydrogen ions result and alkalinity should drop below initial levels. Because subsequently to the disappearance of ammonia, alkalinity returned to initial levels in Pond J, much of the ammonia nitrogen was probably reassimilated by the algal biomass. On the other hand, in Ponds L and H alkalinity dropped well below initial levels, suggesting that the ammonia had been nitrified (i.e., oxidized) to nitrite (NO_2) and then nitrate (NO_3). This hypothesis about different escape routes for ammonia in the ponds is supported by

the sharp increase in nitrite and nitrate in Ponds L and H as ammonia disappeared, compared to the muted increase in Pond J.

A shifting network of nitrogen pathways causes the complex nitrite and nitrate fluctuations seen in the latter part of the experiment for Ponds J, L, and H. Nitrite accumulates, denitrifies to nitrogen gas, or oxidizes to nitrate, depending on bacteria populations, oxygen levels, temperature, and pH. Nitrate accumulates, is assimilated by algae, or reduces back to nitrite (in low oxygen situations).

Figure 3B shows the resulting fish growth. Pond D grew negligibly because of the dominance of largely inedible algae during the fertilization phase, and because of low feeding rates of comfrey, apparently an incomplete protein source, after that. The algae bloom in Ponds L and H seemed to spur growth, and the aftereffects of the

TABLE 1. — Summary of growth efficiency, nutrient excess, nitrogen water quality, and physiochemical spot checks for Pond H (trout chow feed at 6% of fish body weight).

Growth period; dates	Growth g/d	N conver. eff. (%)	Sunny days % total	Weather	n	pH	DO mg/l	Temp. deg C	Secchi cm	Alkal. mg/l	CO_2 mg/l	N/d Nutr. excess mg/l	N Water quality mg/l
1 7/31- 8/14	93	49	33	Sunny	4	7.7	13.5	28.4	30.3				
				Cloudy	8	8.0	9.6	25.5	24.8	50	0.62	1.6	N-NH$_4$ to 9.0
				Total	12	7.9	10.9	26.5	28.5				
2 8/15- 8/28	40	33	58	Sunny	7	7.6	12.0	27.8	24.7				
				Cloudy	5	8.5	10.8	24.7	20.2	40	0.31	1.3	N-NH$_3$ to 4.0
				Total	12	8.1	11.5	26.5	22.8				
3 8/29- 9/11	42	12	50	Sunny	6	7.4'	14.3	26.0	17.9				
				Cloudy	6	6.6	7.8	23.5	15.3	42	4.13	4.9	N-NO$_2$ to 2.0
				Total	12	7.0	11.1	24.9	16.6				x̄ N-NO$_2$ = 1.0
4 9/12- 9/25	32	17	50	Sunny	6	6.7	9.6	28.5	18.2				
				Cloudy	6	6.4	6.4	23.5	12.8	42	10.39	2.8	N-NO$_2$ to 1.7
				Total	12	6.6	8.1	26.2	15.5				x̄ N-NO$_2$ = 1.5
5 9/26- 10/10	4	1	67	Sunny	8	6.8	2.3	30.1	22.9				
				Cloudy	4	6.9	3.8	24.1	17.8	70	10.92	4.0	N-NH$_4$ to 15.0
				Total	12	6.8	2.8	28.1	21.2				
6 10/11- 11/13	69	30	62	Sunny	18	6.0	11.3	25.7	14.4				
				Cloudy	11	5.8	8.1	25.8	12.9	15	18.60	1.7	N-NO$_2$ to 10.0
				Total	29	5.9	10.0	25.7	13.9				x̄ N-NO$_2$ = 5.4
Average	50												

TABLE 2. — Summary of growth efficiency, nutrient excess, nitrogen water quality, and physiochemical spot checks for pond L (trout chow feed at 3% fish body weight).

Growth period; dates	Growth g/d	N conver. eff. (%)	Sunny days % total	Weather	n	pH	DO mg/l	Temp. deg C	Secchi cm	Alkal. mg/l	CO_2 mg/l	N/d Nutr. excess mg/l	N Water quality mg/l
1 7/31- 8/14	55	57	33	Sunny	4	8.5	14.5	27.9	25.8				
				Cloudy	8	8.6	11.3	25.5	25.1	40	0.10	0.7	
				Total	12	8.6	12.4	26.3	25.3				
2 8/15- 8/28	56	38	58	Sunny	7	8.6	16.3	28.8	12.8				
				Cloudy	5	8.1	11.6	25.1	8.4	40	0.15	1.5	
				Total	12	8.4	14.3	27.3	11.0				
3 8/29- 9/11	93	50	50	Sunny	6	9.6	17.3	26.0	8.8				
				Cloudy	6	7.5	11.8	23.7	9.9	42	0.10	1.6	N-NH$_4$ to 1.9
				Total	12	8.6	13.6	25.0	9.3				
4 9/12- 9/25	*	*	50	Sunny	6	6.5	8.5	27.6	18.1				
				Cloudy	6	6.5	3.9	23.9	13.6	75/20	23.36/	*	*
				Total	12	6.5	6.4	25.9	15.9		6.23		
5 9/26- 10/10	*	*	67	Sunny	8	6.8	10.4	28.3	15.0				
				Cloudy	4	6.4	4.2	25.7	12.8	25	4.91	*	*
				Total	12	6.7	8.3	27.4	14.3				
Average	68												

*High mortality at the beginning of growth period 4 precluded additional analysis of growth efficiencies and nutrient load for periods 4 and 5.

algae crash seemed to harm growth. In fact, Pond L suffered extensive mortality and biomass reduction when its aeration supply was cut off inadvertently by a visitor during the critical period of intense decomposition after the algae crash. Pond N, a replicate of Pond L, continued with little mortality until the experiment ended on October 10. The ponds fed rabbit food (e.g., Pond J) did not experience the same decline in growth that the 3% and 6% trout chow ponds did. Surprisingly, growth seemed actually to accelerate, perhaps because of acceptable low ammonia and nitrite concentrations, combined with highly digestible bacteria thriving on dead algae detritus. In the end, the rabbit feed ponds grew about 4 kilograms, while growth in the 6% trout chow ponds (the adjusted rate was actually almost 3% overall) averaged 5.4 kilograms.

Fish Growth and Food Conversion Efficiencies

During good water quality conditions in the respective ponds fed the dry feed supplements, *Sarotheradon aureus* utilized only 49% of the nitrogen available in the feeds. Tables 1 and 2 review the fish growth efficiency and water quality and the impact of the buildup of nitrogen residue on feed conversion for two representative ponds. Because the fish in Ponds C, D, and E that depended solely on the algae as a nutrient source showed insignificant growth, we assume the thick-walled green algae present in the ponds were not a significant source of nitrogen (protein) for the fish. Analysis of the fish feces showed that the fish did not appear capable of efficiently lysing the algal cell walls at the pH ranges found in the ponds.

The duration of good water quality depends not upon the fish directly, but instead upon the excess nitrogen loading rate and its capacity to be assimilated by the microorganisms present in the ponds. It appears that the live algal biomass in the solar-algae ponds contains a relatively small percentage of nitrogen waste products. During intensive feeding of 2-kilogram populations of tilapia at a rate of 3% fish biomass, dry weight 40% protein trout developer feed, the standing crop of algae was found to have assimilated only up to 2-4 days of the nitrogen waste input, or approximately 2-3 grams nitrogen. (Todd et al. 1979). At higher nitrogen-input rates in relation to the fish population's growth, it seems necessary to crop the algal cells constantly to permit continuous algae growth and turnover for nitrogen removal. Tilapia filter-feeding has not been able to control completely these rates of increase within the feed input regimes tested. It is, therefore, necessary to remove the cells mechanically. This can be done through (1) periodic pond water dilutions replaced with a fresh supply, (2) recirculating techniques using a sand filter placed just above the ponds coupled with a simple air-lift water pump, or (3) circulation through a separate settling pond that is periodically drained and refilled with fresh water. All of these methods would remove algal cells and detritus but the latter two methods would also conserve water and retain heat. These methods should assist in the maintenance of good water quality.

Without them, the growth efficiencies of the fish in this phytoplankton-based aquaculture will be reduced as water quality deteriorates.

Fish Growth under Poorer Water Quality

Before examining the periods of lowered water quality, a review of factors affecting water chemistry and fish toxicity is needed. Table 3 summarizes the major oxygen, carbon, and nitrogen transformations in the ponds. Factors affecting nitrogen dynamics are listed in Table 4. Table 5 provides an overview for comparing the concentrations at which various compounds become stressful.

Pond H, with its high feeding rate and intense biological activity, provides an excellent context in which to discuss the toxins listed in Table 5. At different times in Pond H, each of the toxins may have had a significant impact on fish growth.

Pond H, Periods 1 and 2 (7/31 - 8/28)

Cloudy weather, an algae crash, and a high nutrient loading rate caused ionized ammonia (NH_4^+) to build up during the first growth period (Fig. 4C). Subsequent sunny weather pushed pH above 8.0 and shifted as much as 4 mg/l of ammonia-nitrogen to its toxic un-ionized form (NH_3). The reduction in growth efficiency from 49% to 33% was probably caused by un-ionized ammonia stress.

Once fish stress occurs, the resulting decrease in nitrogen assimilation accelerates ammonia formation (positive feedback). Though algae may initially outcompete nitrifying bacteria for ammonia, the intense ammonia production during periods 2 and 3 outstripped algal assimilation rates, and nitrifying bacteria had time to establish themselves.

Pond H, Periods 3 and 4 (8/29-9/25)

Here the system becomes more complex. Nitrite levels appeared high enough to stress fish, reducing growth efficiencies to 12% and 17% in periods 3 and 4. Sharp increases and decreases of nitrate (NO_3^-) concentrations during this time were

TABLE 3. — Transformations in the nitrogen and oxygen/carbon dioxide systems.

Organism	Pathway process	Simplified biological reaction[a]	Effect on pH, alkal.	Theoretical net gain or loss per mg N, unless noted otherwise	
				O_2 mg/l	CO_2 mg/l
Fish	assimilation	Dissolved Org N $\rightarrow$ Org N	none		
Fish	excretion from gills (ammonification)	Amino acid + 1.5 O_2 $\rightarrow$ 2 CO_2 + H_2O + NH_3 (gills)	increase	-3.4	+5.4
Aerobic decomposers	decomposition (ammonification)	NH_4^+ + OH^- + 2 CO_2			
Green algae	assimilation	NH_4^+ + 6.6 CO_2 $\rightarrow$ Org N + 6.6 O_2 + H^+	decrease	+15.1	-20.7
Green algae	assimilation	NO_3^- + 6.6 CO_2 $\rightarrow$ Org N + OH^- + 8.6 O_2	increase	+19.7	-20.7
Nitrosomonas bacteria	oxidation (nitrification)	NH_3 + HCO_3^- + 1.5 O_2 $\rightarrow$ 2H^+ + NO_2^- + H_2O + CO_2 + .5 O_2	decrease	-3.4	-4.3 mg HCO_3 +3.1
Nitrobacter bacteria	oxidation (nitrification)	2H^+ + NO_2^- + .5 O_2 $\rightarrow$ 2H^+ + NO_3^-	decrease	-1.1	0
All	respiration	Glucose + 6 O_2 $\rightarrow$ 6 CO_2 + 6 H_2O	decrease	-1/mg O_2	+1.4/mg O_2

[a]Brewer and Goldman 1976; Delwiche 1970.

sometimes inversely correlated with algae densities (which in turn were affected by daily solar radiation). At other times it appeared that nitrate and nitrite (NO_2^-) cycled out of phase, suggesting (1) an oscillation between nitrite oxidation (nitrification) and nitrate reduction (denitrification); or (2) inhibition of nitrite oxidation by high nitrate concentrations, followed by nitrate assimilation by algae; or (3) pulsing ammonia inputs causing spurts of nitrite, in turn causing influxes of nitrates. At any rate, the increased rates of change suggested that an unsteady balance now characterizes the ecosystem.

Pond H, Period 5 (9/26 - 10/10)

Between September 22 and 24, an algae population crash drastically upset Pond H's water chemistry. Oxygen concentrations plummeted, indicating intense aerobic decomposition, and almost no fish growth occurred for two weeks following the crash. Ammonia accumulated in solution, either because a sufficient nitrifying bacteria population had not established

itself, or because oxygen levels were low enough to inhibit nitrifying bacteria activity. Luckily, the low pH caused by rapid carbon dioxide creation forced the high levels of ammonia into its non-toxic, ionized form. Orthophosphate concentrations also climbed, since large quantities were liberated from the decomposing algae cells, and neither nitrifying bacteria nor algae populations were active enough to reabsorb the phosphorous.

Pond H, Period 6 (10/11 - 11/13)

During the last growth period the nitrogen assimilation efficiency rebounded to 30%, despite nitrite concentrations that climbed to 10 mg/l. This is five times the nitrite concentrations suspected of causing the low assimilation efficiencies of 17% and 12% during periods 3 and 4. The conversion efficiency for Pond H in period 6 was also twice that for Pond J, having half the nitrite concentrations. The only other measured parameter setting Pond H in period 6 apart from the other ponds cited is a lower afternoon pH (5.9 vs 6.6-7.0).

TABLE 4. — Factors affecting microbial nitrogen transformations.[a]

Organism	Relationship
Aerobic decomposers	Release ammonia. Decomposition can occur in suspension in water column. High decomposition increases the mineralization of CA^+ and Na^+.
Green algae	Assimilate ammonia and nitrate. Populations grow and decline rapidly according to availability of solar energy and nutrients.
Nitrosomonas	Oxidize ammonia to nitrite (first stage of nitrification). Generation time is 8-36 hours. Growth is limited at less than 2 mg O_2/l. No growth occurs at less than 0.8 mg O_2/l. Postulated: NH_3 inhibits nitrification. NH_4^+ at 60-110 mg/l does not affect nitrification. Temporary accumulation of NO_2^- can occur from slower growth rate of *Nitrobacter*. Optimum pH is 7.0-7.5. Acclimation to pH 5.5-6.0 takes ten days to attain previous nitrification rate.
Nitrobacter	Oxidize nitrite to nitrate (second step in nitrification). Generation time is 8-36 hours. Growth is limited at less than 2 mg O_2/l. No growth occurs at less than 0.8 mg O_2/l. HNO_2 inhibits nitrification. 0.1-1.0 mg NH_3/l inhibits nitrification. At low O_2 concentrations NO_3 can reduce to NO_2 and then N_2 (denitrification). Optimum pH is 7.0-7.5. At 30 °C, *Nitrobacter* deactivates at pH greater than 8.5 or less than 6.5.

[a]From Konikoff 1975; Ruttner 1953; Wong-Chong and Loehr 1975; and Sharma and Ahlert 1977.

Pond Conditions That Clearly Inhibit Fish Growth

Though the relationship between nitrite and fish growth remains complex and replete with anomalies, some simple statements can be made about the growth-inhibiting effects of high un-ionized ammonia, high carbon dioxide, and low oxygen concentrations:

(1) When low initial algae densities are combined with a high nutrient loading rate (e.g., early periods in Ponds D and H),

TABLE 5. — Factors related to stress and reduced growth in fish.[a]

Stress	Relationships
High un-ionized ammonia, NH_3	Interferes with respiration and excretion in the gills. Toxicity varies with fish species. Affects some species at less than 1.0 mg/l.
Low oxygen, O_2	Stresses respiration by reducing the passive hemoglobin transport of O_2 (water) → O_2 (tissue) $Hb + O_2 \longleftrightarrow HbO_2$ (reversible)
High carbon dioxide, CO_2	Low O_2 and high CO_2 causes high CO_2 partial pressure and low O_2 partial pressure at gill interface. Resulting diminished hemoglobin-oxygen and excessive hemoglobin-CO_2 can produce anoxia in tissues.
High nitrite, NO_2^-	Oxidizes hemoglobin to methemoglobin (Fe^{++} → Fe^{+++}). MetHb cannot release O_2 and CO_2 after binding them. MetHb increases affinity of normal Hb for O_2, further impairing O_2 supply to tissues. Active Hb reductose system in fish reduces MetHb. Bound oxyhemoglobin (Hb-O_2) is not easily oxidized to MetHb; thus high O_2 may reduce nitrite toxicity. Affects some species at less than 1.0 mg/l. Toxicity varies with fish species. At pH 5.6-5.8, H_2S + MetHb → (anaerobic) → → MetHbS → (auto-reductive) → Hb. Cl^- at concentrations 15 times nitrite interferes with nitrite toxicity. High Ca^{++} interferes with nitrite toxicity. Low pH may enhance nitrite toxicity in clean water.
High nitrate, NO_3^-	Can be reduced to NO_2^- in digestion, leading to MetHb toxicity. Toxic at concentrations at least 30 times that of NO_2^-. KNO_3 four times as toxic as $NaNO_3$ for bluegills.

[a]From Baird et al. 1979; Bodansky 1951; Crawford and Allen 1977; Harrison 1977; and Perrone and Meade 1977.

sudden shifts from cloudy weather (when ionized ammonia accumulates) to sunny weather (when pH rises dramatically) can result in high concentrations (over 1 mg/l) of toxic un-ionized ammonia. This appears to harm fish growth in all circumstances where it occurs.

(2) In mature ponds pH drops near or below neutral, and ammonia occurs only in its relatively non-toxic, ionized form (NH_4^+).

(3) Rapid decomposition from algae population die-offs produces high carbon dioxide, alkalinity, and ionized ammonia levels, and depletes oxygen concentrations. The carbon dioxide and oxygen conditions half growth, and in extreme circumstances (e.g., a failed aeration system) can even kill tilapia. Afternoon oxygen concentrations below 3 mg/l indicates serious night-time depletion.

(4) It is difficult to improve water quality after it has declined because of an algal crash (water quality worsens 5-15 days after a crash). Replacing large fractions of the water mass did not remedy the situation. A period of recuperation must occur before fish and algae growth revives. Fore-seeing and preventing algae crashes, or siphoning immediately after a crash, may prove the most effective management strategy.

Ecosystem Modeling: A Tool for Optimizing the Use of an Aquaculture Design

Mathematical modeling of aquatic ecosystems can be useful toward gaining a detailed understanding of the ecological dynamics within an aquaculture strategy. With this it is possible to predict what the impact of alterations in a design may be. This knowledge can be used toward improving the economic efficiency and increased productivity of a design through the manipulation of management strategies such as feeding rates and aeration. These methods are presently being employed to optimize the use of the solar-algae pond design.

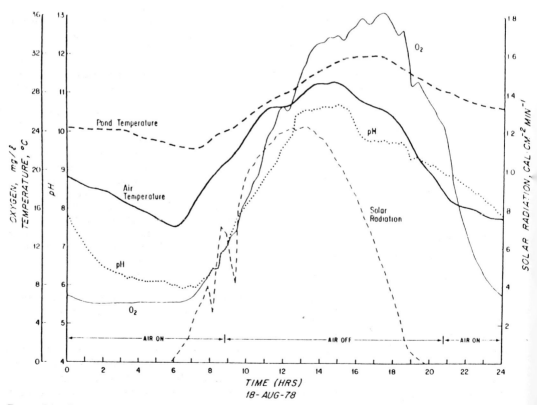

FIGURE 5. — Computer record of solar-algae pond L on August 18, 1978.

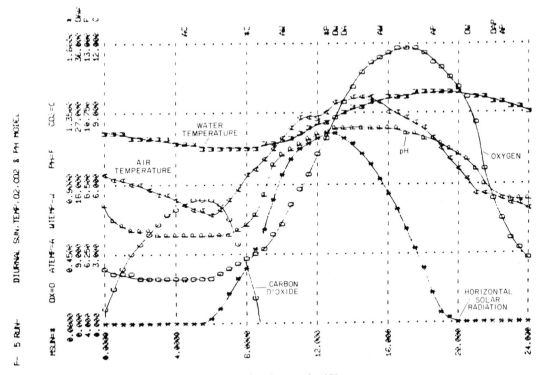

FIGURE 6. — Computer simulation of solar-algae pond on August 18, 1978.

The following figures illustrate how a computer simulation of real time data can be used to limit the amount of energy used for aeration. (Aeration is half the energetic costs of New Alchemy's fish culture systems: feeds and the manufacturing energy costs of the fiberglass are the other half.) Figure 5 is a graphic representation of data points collected during one twenty-four hour monitoring period of Pond L (3% Trout Developer Feed) on 18 August 1978. Solar radiation, dissolved oxygen, pH, and temperature were measured every fifteen minutes.

Figure 6 is a computer simulation of the same information through the use of System Dynamics modeling techniques. The programming notes and equations can be found in Progress Report 3 (Todd et al. 1979). Figure 7 is a manipulation of the data used in building the model. It shows that one can conserve energy through minimizing the use of mechanical aeration at times when not needed. During super-saturated conditions, aeration tends to

perturb the water so as to drive these excess quantities out. It would therefore be better to aerate only at times when the oxygen level in the ponds drops to levels stressful to the fish rather than on an inflexible basis where it may work against the efficiency of a design.

The dissolved oxygen levels and the pH of the water have direct impacts upon the rates of bacterial nitrification in the ponds. therefore, the importance of these factors has another direct correlation to the fish productivity in the ponds beyond respiration. All factors interrelate. Because of the efficiency of nitrogen conversion from feed into fish and the resulting wastes that ultimately feed back to inhibit fish growth, the research thus far has defined the necessity to focus upon the nitrogen flows through the system. The research is now working toward constructing a mathematical ecological model of these factors. Figure 8 illustrates the biochemical pathways through which the feeds put into the ponds are shunted. The

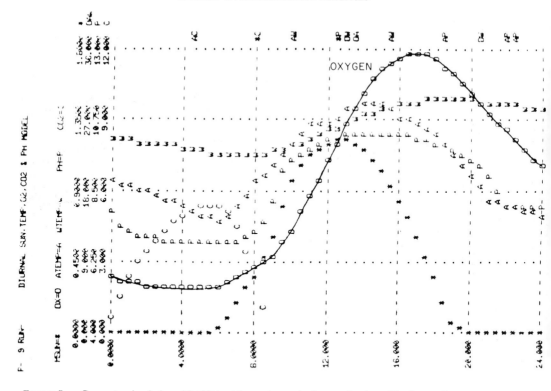

FIGURE 7. — Computer simulation of DAY2A with aeration on in the morning but off in the evening.

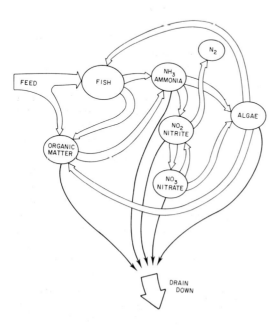

FIGURE 8. — Physical flow of nitrogen through the solar-algae pond ecosystem.

resulting model describing the rates of nitrogen flow into and through the various pathways will allow us to optimize feeding rates through understanding the capacity of the aquatic ecosystem to use the nitrogen put into the ponds and to be able to anticipate the evolution of fish growth-inhibiting wastes. This capability can lead to redesign considerations and management strategies to increase efficiency.

Potential Management Strategies

It may seem to be a rather circuitous route to attempt to achieve simple management tactics using this rather sophisticated approach to aquaculture. In essence, the focus of the work is an attempt to develop our ability to raise fish in a manner no more elaborate than what may currently be used by present-day fish farmers. The idea here is to increase the efficiency of raising fish with an in-depth understanding of the phenomena occurring in contained aquatic ecosystems. The work at The New Alchemy Institute is using current scientific techniques to achieve this expeditiously. At the same time a parallel monitoring program using simple tools and observation techniques is being done. This includes documentation of fish behavior, color of the water, and smell of the ponds. The kinds of tools used are items such as thermometers and secchi disks. This is added to observations of solar radiation — clear versus cloudy days — which can help to establish the most efficient feeding strategies, since sunlight is directly correlated to the amount of energy available to the phytoplankton for assimilation of dissolved fish wastes.

The modular nature of the solar-algae ponds allows the scaling up of an aquaculture operation through the potential of increasing the number of ponds. Experiments are also presently underway testing the ponds in linked arrangements with water rapidly recirculating through them combined with slow fresh water dilutions replacing sediment-rich water. Depending on the size of the operation, automated electronic sensors and controls may be useful in operations large enough to warrant their need for assistance in system monitoring and maintenance.

Summary

Many subtleties of aquatic ecology and fish productivity in solar-algae ponds have been revealed by the routine application of a battery of chemical tests combined with continuous electronic monitoring. The resulting time series data provides the base upon which to build a model of the flows of energy and nutrients through the ponds. Five basic points can be made:

1. Fish feeding rates and types affect fish growth.

Under good water conditions tilapia incorporates half of the nitrogen in the commercial feeds tested. With adequate feed, growth rates exceed 50 pounds per pond per year. Algae in a pond, supplied with fertilizer, can supply the maintenance energy needed by two kilograms of tilapia, but cannot foster significant growth.

2. The ecosystem absorbs nutrients potentially toxic to fish.

This year's research has documented a central premise of solar-algae pond fish culture: Algae absorb large amounts of toxic fish wastes during active algal growth.

3. Feeding rates affect water quality.

At moderate feeding rates the fish and plankton can assimilate most nitrogen, preventing the accumulation of dissolved inorganic nitrogen compounds (ammonia, nitrite, and nitrate). To maintain good water quality, the feeding rate must be within the ecosystem's capacity to absorb residual toxic nutrients.

4. Water quality affects fish growth.

Two water quality conditions have been observed that harm fish growth. (a) High pH during strong algae growth combined with high feeding rates can cause ammonia to accumulate in its toxic un-ionized form. (b) Heavy decomposition from high feeding rates and/or the accumulation of dead algal cells leads to low pH and oxygen, and high ionized ammonia, nitrite, nitrate, phosphate, and carbon dioxide.

5. Solar-algae ponds conserve energy.

The energy costs of producing fish protein in solar-algal ponds compare favorably to the most efficient methods of providing animal protein (i.e., Israeli pond fish polyculture and egg production). In solar-heated buildings, the ponds ultimately capture five times more energy than was used in manufacturing the fiberglass silos.

References

BAIRD, R., J. BOTTOMLEY, AND H. TAITZ. 1979. Ammonia toxicity and pH control of fish toxicity bioassays of treated wastewaters. Water Res. 13(2):181-184.

BODANSKY, O. 1951. Methemoglobin and methemoglobin-producing compounds. Pharm. Rev. 3:144-196.

BREWER, P.G., AND J.C. GOLDMAN. 1976. Alkalinity changes generated by phytoplankton growth. Limnol. Oceanogr. 21:108-117.

CHEN, T.P. 1934. A preliminary study of association of species in Kwantung fish ponds. Langman Sci. J. 13(27):275-283.

COOK, P.M. 1950. Large-scale culture of Chlorella. Pages 53-75 in J. Brunel, G.W. Prescott, and L.H. Tiffany, eds. The culturing of algae, a symposium. The Charles F. Kettering Foundation, Yellow Springs, Ohio.

CRAWFORD, R.E., AND G.H. ALLEN. 1977. Seawater inhibition of nitrite toxicity to chinook salmon. Trans. Am. Fish. Soc. 106:105-109.

DELWICHE, C.C. 1970. The nitrogen cycle. Sci. Am. 223(3):137-146.

FAN LEE. 5th Century B.C. The Chinese fish culture classic. Tr. by Ted S.Y. Koo. Contribution No. 489. Chesapeake Biological Laboratory, Univ. Maryland, Solomons, Maryland. Am. Fish Farm. July 1972:10.

HARRISON, T.R. 1977. Harrison's principles of internal medicine. G.W. Thorn, R.D. Adams, E. Braunwald, K.J. Isselbacher, and R.G. Petersdorf, eds. 8th edition. McGraw-Hill Book Company, New York. 2088 pp.

KONIKOFF, M. 1975. Toxicity of nitrite to channel catfish. Prog. Fish-Cult. 37(2):96-98.

LINCOLN, E.P., D.J. HILL, AND R.A. NORDSTEDT. 1977. Microalgae as a means of recycling animal wastes. Am. Soc. Agric. Eng. Pap. No. 77-5026 (June 26-29).

MCCONNELL, W.J. 1965. Relationship of herbivore growth to rate of gross photosynthesis in microcosms. Limnol. Oceanogr. 10:539-543.

MCLARNEY, W.O., AND J.H. TODD. 1974. Walton two: A compleat guide to backyard fish farming. J. New Alchem. 2:79-117.

_____. 1977. Aquaculture on a reduced scale. Comm. Fish Farm. Aquacult. News 3(6):10-17.

MURPHY, G.I. 1950. The life history of the greaser blackfish, *Orthodon microlepidotus*, of Clear Lake, Lake County, California. Calif. Fish Game 36:119-133.

MYERS, J., A. EISENBERG, AND M. EVENARI. 1955. Studies on the deep mass culture of algae in Israel. Pages 227-230 *in* Conference on solar energy: The scientific basis, Tucson, Arizona (Oct. 31-Nov. 1).

OSWALD, W.J., AND C.G. GOLUEKE. 1960. Biological transformation of solar energy. Adv. Appl. Microbiol. 2:223-262.

PERRONE, S.J., AND T.L. MEADE. 1977. Protective effect of chloride on nitrite toxicity to coho salmon *(Oncorhynchus kisutch)*. J. Fish. Res. Board Can. 34(4):486-492.

ROBINETTE, H.R. 1976. Effect of sublethal levels of ammonia on the growth of channel catfish *(Ictalurus punctatus)*. Prog. Fish-Cult. 38:26-28.

RUTTNER, F. 1953. Fundamentals of limnology. Univ. Toronto Press, Toronto, Ontario. 295 pp.

SHARMA, B., AND R.C. AHLERT. 1977. Nitrification and nitrogen removal. Water Res. 11:897-925.

SYRETT, P.J. 1962. Nitrogen assimilation. Pages 171-188 *in* R.A. Lewin, ed. Physiology and biochemistry of algae. Academic Press, New York.

TAMIYA, H. 1957. Mass culture of algae. Annu. Rev. Plant Physiol. 8:309-329.

TODD, J.H. 1976. The new alchemists. Co-Evolution Q. Spring 1976:54-65.

_____. 1977. A search for sustainable futures. Pages 45-57 *in* D. Meadows, ed. Alternatives to growth. Ballinger Publishing Company, Cambridge.

TODD, J.H., R.D. ZWEIG, A.M. DOOLITTLE, D.G. ENGSTROM, AND J.R. WOLFE. 1979. Assessments of a semi-closed renewable resource-based aquaculture system. Prog. Rep. 3, Office of Problem Analysis, Applied Science and Research Applications, Nat. Sci. Found., Washington, D.C. 126 pp.

WOLFE, J.R. The energetics of solar-algae pond aquaculture. In press.

WONG-CHONG, G.M., AND R.C. LOEHR. 1975. The kinetics of microbial nitrification. Water Res. 9:1099-1106.

ZWEIG, R.D. 1977. The saga of the solar-algae ponds. J. New Alchem. 4:63-68.

_____. 1978. Observations of limiting factors in semi closed aquatic systems. J. New Alchem. 5:93-104.

Bio-Engineering Symposium for Fish Culture (FCS Publ. 1):227-234
© 1981 by the Fish Culture Section of the American Fisheries Society

An Approach to Functional, Economical, and Practical Fish Culture Through Better Bio-Engineering

KEEN W. BUSS

Mansfield State College
Mansfield, Pennsylvania 16933

ABSTRACT

The cost of construction and maintenance of modern hatcheries is a major constraint in the development of commercial fish culture and, in many cases, the improvement of State and Federal hatchery programs. The dearth of research on hatchery construction and the requirement of the species has generally kept aquaculture in a holding pattern instead of reaching the heights so often predicted for it.

All fish cultural installations must be designed to conform to the tolerances of the species for water quality, water quantity, and space factor. Unfortunately, these tolerances are not well known but with the small amount of available knowledge, great progress can be made in new hatchery designs. Even better designs can come from intensive research on the requirements of species at different size.

The quality of water is of utmost importance since the change in pH of one unit can mean 10 times more or 10 times less production in a given water flow. An equally important criterion is kilograms of fish that can be raised in a cubic meter of water. Presently hatcheries are designed around 32-48 kg/m^3; however, there is much evidence to support the raising of salmonids in densities to 128 kg/m^3. If fish can be raised at these densities, there could be a commensurate reduction in hatchery size. Savings such as this would drastically reduce the cost of hatchery construction and operation.

Other savings could be accomplished by eliminating expensive rearing units in buildings, changing feeding techniques to increase growth rate, spawning brood fish twice a year through light control, developing labor-free incubation units, installing more practical, trouble-free instrumentation, designing an efficient cleaning system specifically adapted to fish culture units and EPA regulations, and utilizing disease-resistant strains of fish.

Introduction

One of the major constraints to the progress of aquaculture in this country has been the enormous initial cost of hatchery construction. Not only are the large, impressive physical facilities expensive, but also the space-age electronic instrumentation brings up the cost. In reality, the new hatcheries have not been so productive as the price would indicate nor so efficient as the instrumentation would lead one to believe.

The State and Federal governments have been responsible for these expensive installations because, in part, they have been working with "unlimited" funds. Instead of setting examples by creating show places of economy and function, they have discouraged private investment and fish and game commissions with lesser funds from entering into such a commitment because of the overdesigned installations.

Design engineers and biologists cannot be entirely to blame for these overdesigned and expensive establishments. Academia has either consistently treated fish culture as an art, or proliferated the art with modifications on the old systems. In many cases, fish culture research is not even a consideration in fisheries study in the academic environment. To slow down essential progress further, Federal and State research facilities in many instances have been busier trying to find materials to keep the fish alive in a poor facility than they have been in developing the parameters for a good facility.

It may be a surprise to many, but the basic requirements of any species of cultured fish are still unknown in fish culture. For instance, no one can say with certainty what the densities of any species of fish could be under a specific circumstance. James and Sandra Clary in

New Mexico and Harry Westers in Michigan have been attempting to alert fish culturists to new concepts of basic fish requirements that would greatly reduce the size of the new installations. In general, their publications have had little effect.

Improving traditional fish hatchery methods and research have had little impetus in the last ten years and are still more labor- and energy-intensive than they should be.

The purpose of this paper is to discuss the possible changes that can be made in design, construction, and maintenance that will take into consideration the fundamental requirements of fish, particularly salmonids, and thus reduce the initial and continuing costs. Suggestions will also be made to improve the stock, restructure the research, and simplify rearing methods.

Hatchery Design

The Construction Plan

It may sound trite to say that a hatchery should be designed around the fundamental requirements of species for which it is designed to cultivate. However, there are very few species of fish, if any, whose basic biological and behaviorial requirements are known. Although the principal thrust of biological research has been with salmonids, the ultimate kilograms of these species that can be reared per cubic meter is still questionable.

If a hatchery is constructed properly, fish can be reared at much greater densities than they are now. Indications are that they can be reared at 2.5 to 4 times the densities conceived under present design policies. The savings in structures alone would be immense.

The hatchery design, if based on greater densities of the fish as noted above, will be a far more compact unit. In addition, the rearing units must be large and contiguous. All separate units are units of work and increase labor costs. Circulating ponds almost invariably are small and comprise a unit of work and, according to Westers (1976), do not provide so good an environment as a flow-through type ex-

emplified by the raceway. The raceway also has the advantage of rearing large numbers of fish in essentially one work unit.

The Water Requirement
Kilograms of Fish Produced
per Liter of Flow

The supply and quality of water in the future hatchery should be given more consideration. The water flow performs two major functions for the well-being of the fish. First, it transports oxygen, which is consumed by the fish to convert food into muscle and energy; and, secondly, it carries away the metabolites that, at a certain level, will stress or kill the fish. Therefore, future hatcheries should be designed around the kilograms of oxygen, which are naturally available or can be made available through aeration, and the pH and temperature of the water, which in turn dictate the ultimate limiting factors — the metabolites. The toxic metabolites are usually in the form of un-ionized ammonia, possibly pheromones, and perhaps other substances that we are not aware of at this time.

It would seem that the above statements are not very profound and are known by everyone of the fishery biology or fish culture profession. Yet, almost without exception, when hatcheries are designed, preconceived formulas based on previous construction are implemented to design the hatchery with a set number of kilograms of fish per cubic meter or kilograms of fish per liter flow. Little, if any, consideration is given to the water quality in terms of temperature, pH, and available oxygen.

Almost invariably, when more fish production is desired at an old installation, the first, and sometimes the only, consideration is more space. Rarely are water flow and quality assessed to estimate if there is enough oxygen available and if the quality of water flow warrants an increase in production and hatchery size. Usually, an increase in flow and exchange rates without an increase in structural size will accomplish the same purposes.

Hatcheries have been built with enough

space to produce over 290,000 kilograms of fish at 64 kg/m³ of fish and a water quality and quantity that, even with intensive aeration, couldn't possibly support over 113,400 kilograms of fish.

When too much space is provided for the flow characteristics in old or new installations, it starts a never-ending process of chemical warfare just to keep the fish alive.

The water quality (pH, temperature, and available oxygen) and the flow (exchange rate) must be the basis for hatchery construction. Very simply, the kilograms of fish produced per liter per minute is based on the individual water quality and the flow, and every new hatchery must have its primary design tailored to that individual quality and flow.

The pH and temperature have almost as much influence on rearing potential as flow. A difference in pH from 7 to 8 can produce 10 times more or fewer fish in a given flow. Exchange rates to carry out the metabolites and bring in oxygen can probably be increased from 3 times to as high as 8 times in a 90-m raceway. The rearing capacity per cubic meter can also change in the same proportion. The available oxygen before aeration and the cost of supplementary aeration before metabolites build up above stress levels are additional factors that will dictate hatchery size. These considerations can make hatchery construction more economical.

The energy crisis has limited reuse systems because of the pumping that is necessary. Looming in the future are two potential breakthroughs that again make reuse systems feasible. First is a commercial development of a small air reduction machine that can be purchased by private enterprise to produce oxygen (Speece, personal communication). Secondly, the promise of removing metabolites by naturally occuring zeolites or batchwise ion exchange methods offers new hope for reuse systems.

The Space Requirement
of Fish Species
Kilograms per Cubic Meter

The fish hatchery of the future will be designed around the fundamental requirements of the fish species to be reared. One of these considerations is the space factor (*lebensraum*) or kilograms per cubic meter that the species will tolerate when oxygen, metabolites, and exchange rates are not limiting factors.

Raceways must be designed so that water flows are not split up into many units. Such splitting has been disastrous, not only in the initial cost and subsequent upkeep but in attempting to maintain the fish in good physical condition. In addition to these problems, densities must be kept at low levels and great amounts of space are wasted when flows are split. When water flows are concentrated through relatively few units, exchange rates are increased, fish are healthier, densities are greater, no space is wasted, and cost of construction and maintenance is greatly reduced.

Historically, when space factors have been a consideration, salmonid hatcheries have been designed around 32 to 48 kg/m³. Westers (1976) indicates that 64 kg/m³ is a reasonable density. Clary (1978) had excellent growth and health with rainbow trout at 123 kg/m³. Buss, Graff, and Miller (1970) grew rainbow trout at a density of 544 kg/m³ although growth was limited because of the physical limitation of space. Westers (1976) apparently limits his design to 64 kg/m³ because exchange rates over 0.03 m/s will not allow solids to settle. If a dropout zone were placed in the raceways, then flows and exchange rates could be increased with even greater savings. In silo systems, 112-128 kg/m³ are common. There is no reason why densities such as this would not be possible in horizontal-flowing units. The savings in hatchery construction would be enormous. If hatcheries were designed around the flow, exchange rates, and water quality to produce 128 kg/m³ instead of 32 kg/m³, the savings in construction alone would be almost 400 percent in addition to the reduction of future costs in labor and maintenance.

This aspect of fish culture must be researched immediately for all species and sizes of fish if hatchery costs are going to become reasonable.

Hatchery Building

The Design

Hatchery buildings are usually over-designed because the intent is to hold many young fish at low densities. S.D. Clary (J. Clary 1969) held small fish at $186.24 \, kg/m^3$ and had a normal growth rate. This compares to a normal design of much fewer kilograms per cubic meter.

The extreme was practiced in California where a few hatchery buildings were constructed with only a cleaning trough and no holding tanks. Upon swimming up, young fish were immediately moved outside into raceways and fed at very short intervals. According to Macklin (1968, personal communication), this was a very satisfactory method of starting and rearing fingerlings.

These leads to building small hatchery buildings should be pursued. If the large indoor rearing units can be eliminated, the cost of hatchery buildings can be drastically reduced.

Incubation Units for Salmonids

Eggs can be incubated in several different types of incubators. Picking of dead eggs is not necessary under any conditions. In tray-type incubators, the fry can be allowed to remain in the incubators until they swim up as long as water flows are increased and fungicide treatments are continued. The fry can then be removed to starting and growing units.

The eggs can also be incubated in a drum with a vertical flow; and when the fry hatch, they can swim up and out if the drum is not screened. The fry can be collected in a second screened drum and feeding can start. The fry can then be transferred to any growing unit. This technique eliminates all labor except fungicide treatment. Advantages of this method are that all dead eggs and dead or crippled fry are left in the incubating drum and the fish in the catch drum are similar in age.

This latter method can eliminate large buildings, expensive incubators, cleaning troughs, and sometimes the extra labor required during the spawning season.

Instrumentation

Instrumentation is a controversial subject because it involves safety factors built into every modern hatchery. Unfortunately, even though these hatcheries have modern instrumentation, much of it is wasted since it is never utilized. The reasons for this waste are relatively simple: (1) much of the instrumentation was superfluous; (2) the personnel did not understand its operation; (3) the facility wasn't staffed with enough trained people to operate the sophisticated electronics or analyze the data; and (4) the instrumentation had a history of breakdown, which made it difficult to justify its existence.

It would seem that, within reason, the less complex the instrumentation, the cheaper and more efficient is its operation. A reservoir tank at the head of the water supply with a simple, inexpensive water alarm can sound any kind of alarm if the water level drops. Flows can be read from permanently installed gauges, and oxygen depletion and metabolite buildup can be anticipated in most instances before they occur.

The above are essentially the parameters that have to be measured, and they are certainly never so complex as the instrumentation sometimes leads us to believe.

Genetics

Every time one thinks of genetics, one thinks of long-term, expensive development to produce strains that are adequate for commercial production, or to develop fish for put-and-take fisheries, or to breed strains that survive in the wild.

First, if the need is for strains or species that survive for long periods of time in the wild, the problem is solved. Nature has been working for eons to produce such fish and these fish are available for the catching. Any manipulation for hatchery use may destroy the natural selection of thousands of years.

If a requirement is to produce a fish to grow faster, this too has already been accomplished. Many strains have been developed throughout the country and these strains, or even species, have different characteristics that can be utilized. The only requirements are that these characteristics be measured. Growth can be measured by determining the temperature units (T.U.) required to grow a given amount. By measuring the growth at different T.U.'s for different strains or species, the best hatchery salmonids for temperatures ranging from 5 to 18 °C can be identified.

If disease resistance is a criterion for genetic improvement, then maintaining a source of broodstock at each hatchery is of primary importance. Nothing has been more detrimental than transferring eggs, or grow-out fish, throughout the country. Not only has disease been spread in this fashion, but the introduced fish may lack resistance to local strains of pathogens. Maintaining good water quality by proper hatchery construction and management and by selecting and maintaining broodstock on the site where the progeny are going to be reared will eliminate years of genetic selection to produce the disease-resistant strains. It is not possible or probable that any genetic station can develop a strain of fish that will be adaptable to the varied conditions found in the United States. Such a station can, however, measure the capabilities of the various strains and recommend them for the different environments.

One of the major necessities for the commercial trout farmer is to be able to obtain eggs every month of the year so that he has a salable stock for each month of the year. Many, many years have been spent to develop strains of trout that spawn in all the months of the year. Rainbow trout have been manipulated to spawn in volume for about nine months of the year; the other three months are a hiatus as far as eggs are concerned. The selected strains do not usually spawn over a short period, but spawning usually stretches out over a period of weeks or months, although a fair proportion of fish may mature within a few weeks. Obviously, this long-time process of selection can be a never-ending program.

The answer to all these problems is photoperiod control by which fish can be induced to spawn at a given time, although the spawning is usually over a short period. Best of all, eggs can be obtained from the same broodstock every six or seven months instead of waiting for a year. Although this system is functional now, selection and research is needed to attempt to produce larger eggs, manipulate time periods and study their effects, and determine how long a stock can be spawned every six months before it is exhausted.

In summary, every new hatchery does not need a long-term genetic program to develop broodstock. An inexpensive, well-designed photoperiod control building incorporated into the spawning program can produce eggs to meet a variety of requirements.

Feeding Techniques and Equipment

Feeding techniques for new hatcheries are usually based on the amount of money available to purchase equipment. The more money available, the more sophisticated the equipment and the more complex the method of calculating the required amount of food. Expensive feeding equipment does not necessarily mean faster growth or better conversions, but it does mean more expense, upkeep, and, in some cases, problems that require expensive professional repairs. There appears to be a more simple, inexpensive method to feed fish with less expense and few breakdowns — once the technique is learned.

The amount of food fed has been based on temperature, fish size, and sometimes the species, but never has it been based on appetites of the fish. Any angler will tell you that fish just aren't hungry all the time. In fact, they may go for days without feeding and then suddenly go on a feeding spree that seems to be insatiable. Fishery managers will verify the fact that salmonids, in particular, will gorge themselves in the spring when the insect population is at its peak and, consequently, gain the largest proportion of their annual growth. Any observing hatchery manager will tell you this same phenomenon occurs during this same period in a hatchery when the days start to lengthen in January. During these two or three months much growth on the small fishes may be made up by feeding above "book value." Yet with all these facts that have been so obvious for so long, trout and other species are being fed by programs worked out by expensive computers, slide rules, charts, and tables. The obvious is not to program the computer, nor decide the amount of food by a slide rule or chart, but to allow the fish to decide when and how much it wishes to eat. This covers every variable and, potentially (given good water quality), optimum growth can be obtained. This is exemplified by the almost unbelievable growth obtained when rainbow trout, such as the kamloop rainbow, were introduced into Lake Pend Oreille in Idaho. With an unlimited source of food in the kokanee, one rainbow set a world record that has not been equaled. Good water quality? Space? Food quality? Very definitely, but adequate food at the feeding period was an absolute necessity. The introduction of rainbow trout into South American lakes produced some unprecedented growth, yet these same strains, including kamloop, do not obtain this growth in a hatchery. Again, it might be water quality, space, or food quality which deters this growth, but it also could be limited feed, the "weight-watchers" diet calculated by a computer.

The answer to the restriction in trout growth is the European-developed demand feeders where fish feed themselves when they are hungry and satiate themselves when inherent traits demand it. These demand feeders take a little more imagination than the demand catfish feeder developed in the United States. Because of the high density of fish, the "trigger" must be out of water and the feeding pan regulated so that a limited amount of feed falls with each activation of the trigger. Limited use in this country indicates that this relatively cheap feeding mechanism is more efficient and produces better conversions than the accepted techniques of feeding from charts or tables with other types of mechanical feeders.

A feeding system such as this would drastically reduce the cost of the complex feeding mechanisms so often used in the most modern hatcheries.

Effluent Control

In some new hatcheries, techniques to clean up the effluent have been adapted from municipal sewage plants. These units are very expensive to construct and to operate. Such overdesigned units are not so efficient as one would suppose since there are vast differences between the concentration of the material not only in the effluent from municipal sewage and fish hatcheries (Mudrak 1979) but in the consistency of the organic matter that is being settled out.

Clarifiers, the activated sludge process, and sophisticated sedimentation procedures are not required or necessary to be included in the hatchery design. The only excuse for such development is the lack of land to install an adequate settling basin. Settling basins with a sludge clean-out trap are relatively inexpensive to construct and very cheap to operate, and little can go wrong to require repairs.

The most dangerous concept of the Federal and State developments where municipal sewage systems are used and where money is virtually unlimited is that it sets a nationwide precedent for effluent cleanup. If commercial hatcheries were required to install expensive municipal sewage units, commercial fish production in the United States could conceivably be wiped out.

Pathology and Diseases

It is just as important for pathologists to know the fundamental requirements and biological tolerances of a fish species as it is for a design engineer to know these factors for design and construction. When these requirements and tolerances are known, a stress-free environment can be created that will greatly reduce the incidences of diseases and the future large expenditures for diagnosis and treatment.

Many new hatcheries have been stocked with disease-free fish with little, or no history of previous diseases. This appears to be a progressive move until one considers that there are few, if any, continuing disease-free environments. In many cases, a year or two of production have been lost because these ''disease-free' fish with no resistance have encountered an introduced or indigenous pathogen. Usually the mortality is complete and very costly, and the first few years of production are lost.

Wouldn't it be better to exploit disease-resistant strains that have had a cycle or two in isolation and test environments to eliminate disease carriers? If disease does strike the fish, the losses are usually minimal and can be controlled. To complement this approach, there should be a massive effort to clear legally the effective drugs that are now available and reduce the dependence on new drugs that might be effective.

Disease control requires a properly designed, low-stress environment that should be stocked with a good strain of disease-resistant fish. This is an added bonus to start-up and operational costs.

Transportation

Many different types of transportation tanks have been developed over the years, and the trend seems to be that the more expensive they are, the more efficient they are. Four events take place which limit the hauling capacity of a transportation truck: (1) the consumption of oxygen; (2) the production of CO_2; (3) the production of other metabolites — notably ammonia; and (4) the change in the temperature of the water.

The simplest method to keep oxygen levels high is to introduce pure oxygen. Carbon dioxide can be eliminated with the inexpensive, small electrical agitators. Incidentally, the agitators are also a backup unit for the oxygen supply if something should go wrong or clog up. Marine plywood, which is strong and relatively cheap, is a good insulator for tank construction and helps to maintain the temperature that controls metabolites production. Fish should be transported in cool water; and to provide this, it is cheaper to have an electrically precooled water source than it is to haul about a breakdown-prone

refrigeration unit. Ice is usually available if more cooling is required. A simplified unit as described is adequate and cheap for daily shipping trips. For longer trips a commercial ammonia filter is available (Clary 1977).

Research

Fish culture research to improve hatchery design and function must change its concepts if we are to keep up with dwindling energy and increased costs.

Academia should reassess its views of fish culture and recognize it not as a traditional art that drains most fish and game departments of the major portion of their budgets, but as a dynamic science that, when properly investigated, can economically improve angling and provide impetus to commercial ventures to produce protein for the world food source.

Research in all installations and institutions must be restructured to study the fundamental requirements of fish species that are to be cultured. First, the tolerance of all species and sizes of fish to metabolite buildup should be determined. This, in effect, would indicate the modification needed on an Atlantic salmon hatchery as compared to a rainbow trout hatchery. Secondly, the lowest oxygen level before fish are stressed should be established, and most important in this series of research is the relationship between oxygen and metabolites; i.e., if oxygen is maintained at a high level, how well will it overcome the effects of the metabolites?

How much flow can the sizes and species withstand before food conversions start to increase because of the increased energy requirements? The more flow or increased exchange rate that can be provided, the heavier the density of trout that can be reared.

These are basic questions that must be answered before an economical and practical hatchery can be built. Since hatcheries are the cornerstone of the entire culture program, wouldn't it be wise to have a research unit, or an installation with auxiliary help from colleges and universities devoted to ''How to Build a Hatchery''?

233

Summary

Modern fish cultural hatcheries and related activities must be designed not only to meet water quantity but also to meet the quality of water. The change in pH of one unit can mean 10 times less or 10 times more production in a given water flow. Therefore, it is obvious that water quality is one of the more important criteria for hatchery design.

An equally important criterion is kilograms of fish that can be reared in a cubic meter of water. Presently, hatcheries are designed around 32-48 kg/m^3; however, there is much evidence to support the rearing of salmonids in densities to 128 kg/m^3. If fish could be reared at these densities, there could be a commensurate reduction in hatchery size. Savings such as this would drastically reduce the cost of hatchery construction and operation.

Other savings could be accomplished by eliminating expensive rearing units in hatchery buildings, changing feeding techniques to increase growth rate, spawning brood trout twice a year through light control, developing labor-free incubation units, installing more practical, trouble-free instrumentation, designing an effluent cleanup specifically adapted to fish culture units and EPA regulations, and utilizing disease-resistant strains of fish.

Acknowledgments

The author would like to thank Gordon L. Trembley and Dr. Roger Herman for reviewing this manuscript.

References

BUSS, K.W., D.R. GRAFF, AND E.R. MILLER. 1970. Trout culture in vertical units. Prog. Fish-Cult. 32(4): 187-191.

CLARY, J.R. 1977. An improved distribution tank for long distance hauls. Salmonid, Sept./Oct.:16-18.

_____. 1978. High density trout culture. Salmonid, Jan./Feb.:8-9.

MUDRAK, V.A. 1979. Guidelines for economical commercial fish hatchery wastewater treatment systems. Proj. 3-242-R. U.S. Dep. Comm., NOAA, and the Pa. Fish Comm. 18 pp.

WESTERS, H. 1976. Some principles of salmonid hatchery design. Michigan Dep. Nat. Resour. Tech. Rep. No. 76-1. 21 pp.

QUESTIONER: I would like to ask Keen Buss, in reference to his attempts to reduce hatchery size, how he would reduce the size of incubation facilities such as the Heath incubator? How would you reduce the size and expense of this incubation?

KEEN BUSS: Through the years we have washed our eggs, we have picked our eggs, we have cleaned our eggs, we have cleaned our fry, and we have put in rather expensive units in our system at Benner Springs. I gave my superintendent an alternative, one so simple and inexpensive that it is unbelievable. We use a 55-gallon drum, set up with gravel so that the eggs do not roll. We put in the eggs, treating them as we do every day. That is the secret. We treat them with fungicide and allow them to stay there until the fry swim up. You do not touch anything. You allow them to swim to the next barrel, the same as if you are hatching walleye fry. You clean them up and feed them. Fifty-five-gallon drums are cheap and easy to use. You must remember when you put the water through you need enough oxygen, not for the eggs, but for the fry that will swim up. There is only one thing bad about it — cleaning the drum is an odoriferous job and the boys complain. I do not recommend this system for a quart or two of eggs — only for a big egg take. It is cheap and easy with very little work except for the daily treatment.

(UNIDENTIFIED): In Idaho, it is traditional to use the up-pulling system. Just to expound on how simple it is, Keen, last year we incubated 55 million trout eggs in up-pulling PBC incubators. You can take a 10- or 12-in PBC and make one of these in a matter of minutes.

BUSS: It is a very simple technique. Various designs are used, I chose the 55-gallon drum because it holds more eggs, but any cylinder will work.

(UNIDENTIFIED): The type of incubation that Keen described has been used for over twelve years in Italy.

Bio-Engineering Symposium for Fish Culture (FCS Publ. 1):235-237
© 1981 by the Fish Culture Section of the American Fisheries Society

Intensive Eel Farming
in a Nuclear Power Plant Effluent

PHILIPPE LEMERCIER

Aqua Service
Centre Commercial du Château Vert
34200 Sète, France

Introduction

Whatever can be the selected fish farming technology — intensive or extensive — the following constraints are always present:

1. Technical feasibility of the process, the know-how of the farming process; and the reliability of the results.

2. High commercial value of the product to permit a short pay-back-time and profit on the investments.

The Aqua Service Company is presenting an intensive eel farming example that uses hot water effluents of the Saint Laurent Nuclear Power Plant. The farm has been designed and is run by Aqua Service.

The Selection of Species and the Market

The present and potential market of the eel presents large advantages:

1. The demand is very constant because of traditional eating customs in Europe and in the Far East. It will reach 80,000-100,000 tonnes in 1985-1987, with 30,000-35,000 tonnes in Europe.

2. The supply is limited by overfishing. In 1985-1987, the deficit between the supply and the demand will reach the value of nearly 30,000 tonnes for the world and 13,000 tonnes for Europe.

3. The present value of the total wholesale market in Europe is similar to that of the trout market (US$140-200 million) with wholesale prices close to US$4.80-5.50/kg for eels and US$3.10/kg for trout. In comparison, the eel market is much more valuable after processing as it can reach the value of US$19-24/kg, smoked (Table 1).

Selected Processes on the Farm

Different technologies are available:

1. Utilization of wide areas for extensive farming:

The main characteristics are the following: low investment cost per surface unit; slow and natural growth of the animals; limited production per surface unit; stagnant water (without renewing); small or unexisting control of the farming factors. This technology is adapted to regional valorization as utilization of natural ponds.

2. Intensive farming:

This process is characterized by important investment costs per surface

TABLE 1. — Contrasts of the importance of the market after processing (retail price).

Species	1979 demand in tonnes	Value of wholesale market (US$)		Value of retail market and processing (US$)	
		(unit price per kilogram)	(Total)	(unit price per kilogram)	(Total)
Eels	25,000 to 35,000	$5.48	$150 million	$19-24	$405 million to $524 million
Trout	70,000	$3.10	$217 million	$4.80 to $6.00	$381 million

FIGURE 1. — Inside eel production facilities.

unit; necessity of small surface; reliable process by increasing control of the production factors (waterflow, O_2, NH_3, food supply, etc.); high production level per surface unit; necessity of quick growth of the animals to reduce the pay-back-time and increase the profit on the investment.

An Intensive Eel Farming Process

In relation to the absence of breeding control, the post larvae, or elvers, are conditioned to a meal and selected on their growth potentiality. The selected are ongrown to the commercial size, 200 g. During this period the eels are rationally fed and graded.

The conditioning hall includes 12 reinforced 16-m² fiberglass tanks. Well water with a temperature of 20-27 °C is supplied to these tanks with a waterflow of 40-70 m³/h. These installations are placed under an agricultural plastic film (greenhouse) (Fig. 1). The power plant effluents heat the well water through a plate heat exchanger. This permits a complete and reliable thermoregulation of the rearing water temperature.

Seven ponds, each 75 m², and eight ponds, 300 m² each, permit an industrial ongrowing of the animals (Fig. 2). All of these ponds are concrete. Two hundred cubic meters of hot water per hour is pumped from the effluent and supplied directly to the ponds. During the year the temperature of the effluent will slowly vary from 15 °C in winter to 30 °C in summer.

Built close to the ponds, the food storage and food preparation hall permits a rational feeding of the animals after the food has been prepared (hydration of the meal and oil supply).

The farm has been designed to permit control of the rearing parameters and provide maximal mechanization. Maximum control of the rearing parameters is gained through waterflow control, oxygen requirements and supply control, accurate knowledge of the biomass, and optimal feeding in relation to the former controls.

The production experience has shown that ongrowing includes two main problems, feeding and fish manipulation (grading etc.). As for control of the rearing parameters, the design of the farm permits an important reduction of the necessary

236

FIGURE 2. — Outside eel production facilities.

time and a rationalization of these two operations.

Economy and Profitability

It is obviously necessary to be very careful on the still preliminary elements of profitability of this farm. Nevertheless, the initial investment costs including earth-moving, concrete, building, and all diverse equipment were exactly US$693,534 in 1978 (date of construction of the farm). For a production of 80 tonnes of fish per year, the cash flow after taxes (50% in France) is expected to be roughly US$161,500 and the pay-back-time would be between 3.5 and 5 years.

Conclusion

If the technology is obviously interesting for the utilization of hot effluents from a nuclear power plant, it must not be limited to this aspect. The interest is to extend it to the utilization of all hot water, including natural hot water (tropical areas) and industrial effluents.

It appears to us that this technology could be developed in the United States. The reasons are that fish feed and energy are available. In the farm presented here, the food cost is 64% of the total costs (including amortization) for a food price of US$0.77/kg, transportation and taxes included. In the United States this food could be supplied for US$0.50/kg, so the total operation costs in the United States could be reduced to 77% of the operation costs in France.

Such a difference in the operation costs permits a direct action and sale on the European and Japanese markets. Furthermore, there is a local market in the United States that must be considered. It is based on the different Asian ethnics on the Pacific coast.

A final important factor is the large and natural American elvers supply, *Anguilla rostrata*. It appears clear that this process could be applied to the American elvers, keeping the main factors of mechanization, rationalization of the process, and reliability of the process.

Bio-Engineering Symposium for Fish Culture (FCS Publ. 1):238-239

Considerations for Selecting Fish Production Facilities

NICK C. PARKER

Southeastern Fish Cultural Laboratory
U.S. Fish and Wildlife Service
Marion, Alabama 36756

Extensive and intensive fish production facilities have been differentiated by the degree of man's involvement to control production (Stickney 1979). Extensive systems — natural aquatic systems in which fishery managers manipulate one or more factors to exert some positive influence on favorable species — produce relatively few kilograms of fish per hectare when compared to intensive systems. Intensive production systems — closely managed by aquaculturists to maximize production of selected species by exclusion of competition and sources of mortality — may annually produce harvestable fish in the range of 3,000 to 10^6 kg/ha.

Intensive culture systems include ponds, cages, tanks, raceways, semiclosed recirculating systems and closed recirculating systems, but not all production facilities are suitable at any given location. A system that is optimum for one site may be most inappropriate for another site because of relative cost of land, water, and energy (Parker 1976). For example, in the urban environment land costs are usually high; thus, pond culture systems that require large tracts of land might be cost prohibitive. Raceways require large volumes of continuously flowing water of high quality and are cost prohibitive in areas where water is limited or highly contaminated with industrial, agricultural, or municipal pollutants. Recirculating systems not only require input to maintain water quality but also require intensive management and operator expertise. Complex water reuse systems can have only limited success in underdeveloped nations where electrical service may be erratic and operator expertise is in short supply.

In the past, aquaculture has been more of an art than a science. Based on the combined research of fish culturists, engineers, economists, chemists, physiologists, nutritionists, and others, aquacultural practices are now being analyzed and defined to provide data bases necessary to make performance projections and to establish aquaculture as a science. An indication of our nationwide interest in aquaculture was provided by the National Research Council's 1978 report, which considered the constraints and opportunities for aquaculture in the United States. Recent contributions to the body of knowledge such as Wheaton's (1977) engineering text, Boyd's (1979) water quality text, and papers presented previously in this symposium provide guidance for future developments in the aquaculture area.

The economic potential of the aquacultural industry has generated investments that in some cases have failed because not all factors were carefully considered. Components of the system were not appropriate for the site. Considerations necessary before selecting a particular type of intensive fish production facility include, but are not restricted to, availability of the following: (1) species, (2) land, (3) water, (4) energy, (5) equipment, (6) labor, (7) technical expertise, (8) diagnostic services, (9) maintenance services, (10) emergency backup systems, (11) feeds, (12) markets, (13) storage, (14) processing, (15) transportation, (16) investment capital, and (17) economics.

In this last session of the Bio-Engineering Symposium, operational and design characteristics of several different fish production facilities reflecting state-of-the-art will be presented. Each system, regardless of size, complexity, or expense, was selected to meet design criteria at a given site. They include systems designed for underdeveloped nations, for backyard aquaculture, for commercial ventures with marine cages, for waste heat utilization in electric power plant effluents, and for the production of salmonids, cyprinids, and aquatic macrophytes. Evaluate these systems for the seventeen points listed above to assess their application potential in other sites. Remember, that as aquaculturists we must first consider the species' requirements, and, secondly, regardless of the design, if a system is not economical, it will not be practical.

References

BOYD, C.E. 1979. Water quality in warmwater fish ponds. Agric. Exp. Sta., Auburn, Alabama. 359 pp.

NATIONAL RESEARCH COUNCIL. 1978. Aquaculture in the United States: Constraints and opportunities. Nat. Acad. Sci., Washington, D.C. 123 pp.

PARKER, N.C. 1976. A comparison of intensive culture systems. Pages 110-120 *in* Proc. Fish Farm. Conf., Annu. Conv. Catfish Farmers Texas. Texas A&M Univ., College Station, Texas.

STICKNEY, R.R. 1979. Principles of warmwater aquaculture. John Wiley & Sons, New York. 375 pp.

WHEATON, F.W. 1977. Aquacultural engineering. John Wiley & Sons, New York. 708 pp.

Bio-Engineering Symposium for Fish Culture (FCS Publ. 1):240-248

The Transition and Scale-Up of Research Results to Commercial Marine Aquaculture Systems

JOHN E. HUGUENIN AND HAROLD H. WEBBER

Groton BioIndustries
P.O. Box 133
Woods Hole, Massachusetts 02543

ABSTRACT

Scaling up of research results to larger-size commercial systems appears to be a major problem in aqua-cultural development. It involves both technical and managerial considerations. The nature of the development process and the magnitude of problems, including dramatic changes in objectives and criteria, involved with progressing from laboratory tests to pilot plant to full-scale commercial applications have not always been understood. Determining optimum scale for individual production units within a full-scale farm and the total size of the farm involve many complex considerations. There are economies of scale and economic breakpoints for vertical integration. Common misunderstandings, sources of error, and factors bearing on the aquacultural development process are also discussed.

Introduction

It is essential to define the objectives and limitations of this subject. We direct our attention to marine culture systems that are required to be feasible in a technical, economic, marketing, social, and political environment, basically similar to that which currently exists. We are talking about commercial systems where all inputs and outputs, including labor and land, have market values and the purpose is to produce at a profit substantial net aqua-food to support urban/suburban populations. Our interest in size is not for the sake of bigness but rather as a variable design parameter, which is determined by a project's objectives, constraints, and system requirements. Large-scale means large with respect to prior research efforts and can be attained either by having big individual units or by replicating many smaller units, which in some way work together as a system.

We do not intend to discuss the basic systems decisions that have been, at least partially, addressed elsewhere. These include site selection (Webber 1971), species selection (Webber and Riordan 1976a), and choices of basic technical approaches. Additionally these decisions often include considerations of intensive/extensive, sub-sistance/commercial, closed/open cycle, monoculture/polyculture, and labor/capital intensity. These decisions are certainly *not* independent and are very complex in their interactions. Most disagreements about the best choices are argued on the basis of technical characteristics, while the basic disagreements in fact involve objectives, criteria, and the assumed environments (legal, social, economic, etc.) in which the systems are expected to operate. This is a very common mistake, which can waste considerable time and effort. Many apparent disagreements can be quickly resolved if the system requirements can be clearly defined. It is in this context that the assumptions, objectives, and perspectives for the rest of this paper must be considered.

Very little information is available about the design, construction, and operation of commercial marine aquaculture systems. What does exist is generally not published and is often considered to be proprietary information. Most of this prior experience involves unsuccessful commercial attempts and current operations that have not yet been proven as commercial ventures.

This situation is not quite so serious for freshwater aquaculture, because of many more numerous precedents and experiences along with the firmly established status of commercial catfish and trout culture. However, the use of seawater combined with biological restrictions greatly increases the engineering problems while reducing the choices of acceptable materials and available equipment. This is further compounded by the necessity of having to construct and operate most facilities at the edge of the sea, a location that is always expensive and risky, partly because of the destructive effects of wind, waves and currents, and legal/political complications. Thus, the proper design, engineering, and management of large marine aquaculture operations can be expected to be both difficult and critical.

The importance of systems problems is substantiated by experience with commercial culture such as with pompano (Berry and Iversen 1967), where the breaching of dikes repeatedly produced failures in several different operations. Excessive freshwater inflow and loss of seawater supply due to biofouling have been the causes of other disasters. Net and barrier problems have been experienced by Marifarms Inc. of Panama City, Florida (Anderson 1972), and more recently by DØMSEA Farms, Inc., of Seattle, Washington (Lindbergh 1976). Scaling up too soon without a pilot plant phase was significant in the demise of Seapool Fisheries, Ltd., Clam Bay, Nova Scotia (Helliwell and Hutcheson 1976). Many of the serious problems and failures of commercial marine aquaculture facilities have been due to poor site selection, poor planning and design, inadequate equipment, economically inefficient methods, extreme weather conditions, erroneous assumptions concerning markets, and inadequate financing rather than to a lack of basic biological knowledge. Unfortunately, there are no studies available that precisely evaluate the successes and failures of commercial marine aquaculture ventures or assess their actual research needs. We believe that the limiting factors are frequently not science, or even technology, but rather management capabilities, when management is defined in its broadest context as the ability to make all the diverse

elements work together. Management limitations are highlighted by the often inconsistent performance of large-scale systems. While this is not surprising for an emerging technology, consistently good performance is essential for achieving projected returns and for stimulating needed investment.

Considerable effort has been expended in investigating the biological and biochemical aspects of marine aquaculture and as a consequence much progress has been made in recent years. This scientific knowledge outstrips the development of equipment and management systems to exploit it in commercial culturing operations. On a national basis the percentage of the total effort being applied to the development of marine aquaculture systems is small. The greater part of this effort involves the construction of experimental equipment rather than the development of coherent plans, systems, equipment, and management techniques that will be necessary for practical applications. The reason for the small effort is simple. The problems inherent with the transition from scientific experiments to commercial operation are just not perceived as being critical. Most of the marine aquacultural systems development is being done by dedicated entrepreneurs, usually with inadequate resources, who often fail in the process because of avoidable mistakes and undercapitalization. These entrepreneurs usually do not have the resources to carry out an orderly development sequence and to repeatedly make mistakes, learn from these mistakes, and try again. A few large companies with adequate resources are also involved. Communications between the academic research community and these practitioners, at least on the marine side, regrettably are usually quite poor. There is often very little similarity in objectives, perspectives, nature of pressing problems, sources of funding, professional affiliations and contacts, and backgrounds of individuals, even though both may be working with the same species.

Some scientists view the development of marine aquaculture as strictly a biological and biochemical problem. The problems most commonly mentioned as limiting are disease control, nutrition, and larval rearing. The role of the technologist is usually viewed basically as providing support in the development of experimental equipment. Large-scale systems management aspects are generally not even considered. The hope is to carry out sufficient experimentation in the laboratory so as to answer all the critical questions limiting large-scale practical applications. Only then is it judged necessary to factor-in seriously the engineering, marketing, and economic aspects. Many of the failures to date can be partially attributed to this neglect of the basic problems involved. Probably the biggest misunderstanding is the assumption that the scale-up of laboratory experiments to commercial size is a low-risk, straightforward linear bio-engineering process. Thus, it is considered that there is little need for engineering or systems management input to experimental design or in determining priorities required to guide research efforts into productive channels.

Nature of the R&D Process

The answer to some of the criticisms advocating early technological and management inputs is that there are so many scientific unknowns and uncertainties that the planning and design of large-scale applications is premature. This is in large part a valid argument, as a history of failures can attest. However, this argument neglects the fact that to achieve successful developments, information feedbacks and iterations of design and performance are required. A common but fundamental misconception is that the research and development process is sequential, as shown in Figure 1, when it is in fact more realistically an iterative process (Fig. 2). Without evaluating what is already known when applied to large applications, how

can the critical data voids, functional requirements, problem areas, and sources of inefficiencies be identified and the research effort intelligently directed? At a minimum, in the absence of access to actual full-scale experience, careful, but not necessarily expensive or time-consuming, systematic preliminary design studies are required to provide this type of feedback information. (Examples: Huguenin 1976; Huguenin and Rothwell 1979.) Pilot plants if directed towards practical applications can also provide valuable information essential for proper analyses and planning. However, it sometimes is necessary in a more general developmental context for complex systems to go through several inefficient generations in actual operations, usually in highly specialized applications that maximize the relative advantages of the new technology, before developing to generally more broadly usable forms. This is demonstrated in aquaculture by the commercial emphasis on very high-priced premium aquafoods for limited markets rather than on developing more broadly usable food sources. Since the risks in commercial ventures are many and diverse (Webber 1973), high potential returns are needed to justify the sizable risks involved. As the risks are reduced and production economics improved through experience, the markets can be expected to broaden and the costs drop.

It is unrealistic to expect great success or substantial beneficial impacts on our society or food supplies from initial marine aquacultural attempts, because of the inherent risks and hazards. The expectations of many people are that the first attempts will produce operations with the scale, performance, and efficiency more realistically characteristic of third generation systems. The breakthrough myth, involving great technological advances achieved instantly

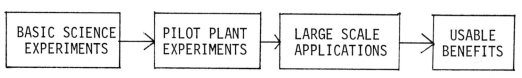

FIGURE 1. — Simplistic development sequence and information flow.

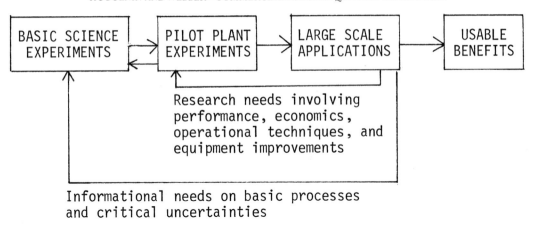

FIGURE 2. — More realistic development process and information flow.

with little effort by individual geniuses and immediately producing tangible benefits, is unfortunately part of the general misunderstanding of the development process. If marine aquaculture is to realize its very substantial potential, it will require a great deal of conscious effort and the dedication of considerable resources.

The first question usually asked about an experimental marine aquaculture system concerns its economics at commercial sizes. Only with careful and thorough design and operations studies can even a good estimate be attempted and this estimate is very inferior to actual experience. In fact, without careful study it is impossible even to determine the size range for potentially viable commercial systems. Without a preliminary design involving *quantified* judgments about size, productivity, location, processes, materials, specified subsystems, dimensions of units, labor requirements, types of construction, and assumed performance, it is not possible to evaluate costs. Since an adequate bio-engineering data base on which to make these types of engineering judgments may not exist, it is in many cases very difficult to carry out realistic analysis without including the impacts of possible numerical variations on important systems parameters. Possibly for these reasons, few evaluations of proposed large-scale marine aquaculture systems are being published. Some projections of future systems and evaluations are found in the

many business plans in circulation but they are often incomplete and of poor quality.

Scale-Up Problems

There are many basic differences existing in objectives, criteria, and figures of merit between experimental systems and the much larger applications. Some of these differences are quite fundamental and are shown on Table 1. The table lists an intermediate pilot plant or semi-works stage, which is often defined quite differently by scientists and engineers. In order to be useful in systems development a pilot plant should be a production-oriented test facility, and not just a more expensive research project. These fundamental differences among the development stages, and the fact that the implications of these differences are generally not appreciated make it difficult to apply directly small-scale, short-term research experiences to large-scale, long-term commercially oriented operations.

The approaches and techniques that are practical on a small experimental basis may not be rational or even feasible at much larger sizes, and the converse is also true. These scale-up problems can arise in bulk handling of materials such as animals, water, and feeds, which in laboratory situations are easily transported and held in small containers. Performing necessary life support functions can also

TABLE 1. — Differences among typical aquacultural system development stages.

	Initial experiments	Pilot plant	Large-scale application
Size	Desk top or equivalent	Highly variable; maybe full-scale module	1-10,000 ha
Objectives	Research	Production research	Economical production
Primary outputs	Research data	Management and scale-up information	Tons of seafood/ financial profit
Environment	Indoor laboratory	Indoors or outdoors	Usually exposed year round
Period of operations	Short periods of testing	Intermittent	Continuous year round, if possible
Staff skill and manning levels	Very high	High	Low
Efficiency of processes	Unimportant	Important	Critical
Economics of operations	Unimportant	Relatively unimportant	Critical
Magnitude of consequences of failures	Very low	Moderate	Very high

become complicated, since individually monitoring the distribution of water, providing food, and checking general health for large numbers of animals each day quickly become prohibitive. Even the routine maintenance and cleaning of culture units, while trivial in the laboratory, can create major problems with increased scale. More fundamental is the fact that there may be differences in the basic process between the research and commercial approaches. An example might be that all available test data, for practical reasons, were developed in short-term batch cultures while the larger-scale applications, for equally valid practical reasons, will be a long-term culture with continuous inputs and outputs. The biochemistry and system dynamics of batch and continuous culture systems are different. These problems are further compounded by the many environmental and test conditions in a laboratory that can be quite different from those of probable large-scale applications. These differences can also include the pretest environmental history of the experimental animals as well as their initial state of health and genetic background. Classical scientific experimentation in the laboratory frequently holds all but one or two test parameters constant. There is often an implicit assumption that the fixed parameters are independent of the test parameters. We all know that we are, in fact, dealing with very complex processes having a great deal of parametric interdependence, and all superimposed on time-dependent seasonal and biological cycles. In short, there can be substantial risks in extrapolating results gained under laboratory test conditions to commercial field conditions, where a large number of parameters are very different and/or are continuously varying. Therefore many of the available scientific data are of somewhat limited value in the design of production systems.

Unfortunately, it appears that most of the experimental marine aquaculture systems are not being developed to provide easy scale-up and optimal production efficiency at large sizes, but rather are being optimized as basic research tools. Thus, they often produce results that are readily acceptable for publication in the scientific literature but do not tend directly to increase our capability for efficient large-scale production. Limits on publication space and different perspectives on experimental design often result in important system parameters' being ignored or inadequately presented. In short, the questions of the greatest scientific interest are not necessarily the most important for

244

systems design and scale-up and, even if important, the research may be carried out under conditions so as to make the data of questionable value for practical applications. While a system built for research objectives might be capable of being scaled up to an efficient system for mass production, the two quite different objectives are not necessarily compatible (Huguenin 1975).

Scaling up systems by several orders of magnitude is not at all a straightforward planning process, especially in the absence of available prior full-scale experience. The biological components of a system much larger than any precedent may alone introduce major uncertainties. While it is generally accepted that biological and behavioral processes can be scale dependent, we are not aware of any literature dealing primarily with the biological effect of scale and the importance of this effect. In addition, the engineering problems and attendant risks in any first-of-a-kind system, if the design or the conditions of use are new, are significant. The performance and economics are particularly hard to scale and the physical parameters generally do not increase linearly.

Determining the optimum sizes for application is itself a major issue, which is not so straightforward as it may seem. An early decision is the determination of how many phases the culture process is to have; in other words, how many times during the total culture period the culture units are to be changed or altered. As an example, this may be a two-step process of hatchery and grow-out, a three-step process of hatchery-nursery-grow out, or even a four- or five-step process. This is primarily an operations research decision made by management. The major scale decisions then involve the size and number of the basic culture containers in each of the phases and the size and design of the total system including support subsystems.

The size of the culture confinement is usually a trade-off between decreased purchase cost per unit volume or unit surface area versus diseconomies resulting from the greater difficulties and decreased efficiencies often accompanying larger units.

Increasing container size may also enable the efficient and advantageous use of expensive equipment, if the equipment in fact works and is practical. These containers may be tanks, raceways, ponds, silos, or cages. In fact, monitoring and control limitations and related management and operational problems often preclude economies that might otherwise be realizable with larger units. This certainly is true in marine cage culture (Huguenin and Ansuini 1978) and in the pond culturing of shrimp. Unit size may at times also be limited by tasks that a few men can physically handle at one time or accomplish in one work period. Another set of important considerations in determining basic culture unit sizing involves operational advantages of having multiple independent units. These advantages may result from increased flexibility to meet marketing requirements and decreased consequences of individual failures. This is another important area for management judgment. Lastly, under some circumstances, keeping size below some logistical size limit may enable cost reductions through centralized high volume production of specific equipment. Modular construction approaches may provide this advantage even though the assembled equipment may be quite large. Example would be prefabricated buildings and some large tanks. In short, diseconomies of increased size sometimes exist at sizes where on-site construction becomes necessary because of logistical constraints. Since culture units sizing is dependent on the state-of-the-art and management capabilities, such decisions can be expected to be subjective, non-deterministic, and time dependent.

The other size determination involves overall system size in each of the phases as well as in the ancillary functions. When dealing with economic criteria, there are minimum scales for economic viability (Webber and Riordan 1976b). This is due to economies of scale inherent in many culturing functions such as acquisition of seed stocks, bulk feed purchases, regulatory requirements, affordable access to specific technical expertise when needed, marketing arrangements, and monitoring/control systems, including required analytical capabilities. In fact, some of these have relatively fixed minimum requirements and are somewhat independent of system size. However, it is important to realize that large systems need not mean physically larger production facilities or co-location, if several smaller units, through collective agreement or governmental support services, can acquire the benefits of larger scale.

While there may be minimum scales for economic viability, as the scales increase, functions and services, which have previously been purchased, become economically justifiable as integral parts of the system (Webber and Riordan 1976b). These economic break points for inclusion in the system may involve such things as a dedicated hatchery, processing plant, feed preparation facility, cold storage capability, and improved marketing arrangements. At the appropriate scales, these self-contained functions can prove to be very profitable, both within the system and as an additional source of revenue through the sale of excess capacity. Decisions about appropriate scales can be time dependent. A system may start with a scale just sufficiently large to be viable but also possess significant growth capability. In such a way a boot-strapped development sequence towards a vertically integrated system can be initiated.

Matching the overall sizes and production capacity for each of the culture phases involves many management decisions. It must include provisions for expected mortalities, anticipated utilization/efficiency, desirability of selling excess production, and expected learning curve benefits for each phase of the system. For systems where significant process development is expected to continue for some time, these applied research needs must also be considered. Since major facilities are expected to have long service lives, it is essential to project their needs in time, and establish a compromise between short- and long-term requirements. This is rather difficult to do in areas of developing technology. Short-term requirements are often quantifiable and usually subject to frequent change, and are often of peripheral importance. In contrast, long-term requirements are very difficult to define quantitatively but are usually of a more fundamental nature.

Current Needs

It is obvious that to develop anything as complicated as profitable marine aquaculture systems it is necessary to combine the expertise of a great many scientific and technical disciplines. Unfortunately, interdisciplinary research teams of more than a few professional members in this field are rare, especially in academia, and most of these are heavily dominated by a single discipline, usually biology. This is particularly interesting since practical experience has shown that legal, political, and technical constraints affecting overall management capabilities may be more limiting than biological unknowns. There are several factors in the working patterns and sociology of the marine science community that appear to work against the formation of stable interdisciplinary teams. However, organizational structures and the narrow areas of interest of many of the state and federal funding agencies are contributing factors in precluding more balanced efforts. A broader perspective is desirable to achieve truly interdisciplinary development teams capable of providing tangible and useful systems. The academic and industrial communities involved with aquaculture have quite different strengths and weaknesses. While considerable benefit might result from joint development efforts, many of these efforts may have to overcome serious attitudinal and institutional obstacles. If these obstacles can be overcome, true collaboration would be most useful at the transitional or pilot plant stage. It is important that pilot plants be approached with commercial development rather than basic science objectives.

It is important to remember that innovative engineering developments in agriculture have not been left solely to private funding, but have, to a great extent, been promoted with public funds. Thus, there is a precedent for funding research on the more practical aspects of aquaculture. Since the risks are often high and potential rewards may be distant, there is little incentive for privately funding the development of major new aquaculture systems. Investors, by and large, have adopted a watch-and-wait policy. Information on past successes and failures is not generally published or available in any usable form so that new development efforts do not necessarily benefit from prior experiences and knowledge. If it is desirable to promote aquaculture in a manner similar to that done previously with agriculture, then basic technological developments must be at least partially funded with public funds so that the results and benefits, including information on "failures," can be more widely disseminated.

While a great deal of valuable data, experience, and equipment can undoubtedly be transferred from the more traditional engineering disciplines, this is a selection and modification process. The requirements are different and engineers, by and large, do not fully comprehend the biological constraints imposed by marine aquaculture on equipment and system design. They generally do not appreciate the complexity, variability, occasional unpredictability, and control problems of biological systems. Additional communications problems result from radically different jargons in the biology and engineering communities. Even when common terminology is used the definitions may differ widely. Thus, there is a need for a specialized type of technologist who is familiar with basic marine biological processes. In fact, what is needed is a cadre of technologists with a role analogous to that of the agricultural engineer. It is clear that all the progress in plant industry and animal husbandry would be severely limited if farming equipment and operating methods were still those of 100 years ago. Thus, the agricultural engineer has an important and recognized role in American agriculture.

The lack of capability to apply the present scientific knowledge in aquaculture is substantiated by the very limited literature on large-scale applications and on technological and management aspects. There are few design studies published and many of the articles technically describing systems involve small research projects and are of limited value for practical applications. As an example, a literature survey on large seawater culture systems (Huguenin 1975) indicates that a flow rate of 100 gallons per minute is judged "large." However, most commercial applications would have rates that are orders of magnitude higher. Useful publications describing the iterative development sequence of an aquaculture system over several design generations are virtually nonexistent, with at least one notable exception (Lindbergh 1976). Most available comments on this subject are overly qualitative with little substantiating information and are therefore of little value. Equally rare are descriptions and assessments of commercial attempts that "failed." This is a serious loss since many were in some ways very successful and innovative. The experience and knowledge gained, sometimes at great cost, are lost except by word of mouth and in the memories of the participants. It appears that repeating mistakes and reinventing the wheel are not uncommon occurrences in marine aquacultural development.

What little published information currently exists is widely scattered in journals concerned primarily with biological aspects or in discipline-oriented engineering journals generally unconcerned with

TABLE 2. — Technological factors requiring more effort with respect to large-scale aquaculture.

- Culture unit hydraulics on large-scales
- Construction approaches with economies of scale
- Monitoring and control methodologies and systems
- Cheaper and more efficient water processing (pumping, heating and cooling, filtering, sterilizing, aerating, etc.)
- Feed management methodologies and systems
- Bulk handling and harvesting
- Cleaning, waste handling, and hygiene control
- Definition of markets
- Evaluation of alternate operating policies
- Economic analysis to identify high leverage R&D

biological systems. What is needed is a central source of professional discussions on such important practical subject areas as suitability of materials, monitoring and control systems, tank and hydraulic design, feed preparation and distribution systems, cheap but effective construction methods, filtration systems, automated handling/processing systems, economic aspects and operations management. These important areas identified as requiring more effort for large-scale applications are listed on Table 2.

Summary

Some segments of the marine research community do not comprehend nor appreciate the magnitude of the problems involved in scaling up small research experiments to economically viable large-scale production facilities. A better understanding of the basic differences between research and commercial operations and of the nature of the developmental process may significantly improve the value of future research.

Current research does not benefit significantly from feedback information on the needs and problems of existing or future production systems that would result from parallel systems evaluations and technology development programs. If the hope is to develop marine aquaculture for use in practical applications, there is a need for broader perspectives and a more balanced division of development efforts among basic science, applied technology, and management techniques. It is vitally important to remember that biology, engineering, economics, and operating policies are highly interdependent. Ignoring this principle would invite disaster in any development program.

Acknowledgment

The authors wish to thank Ms. Pauline Riordan for her comments on the draft for this paper.

References

ANDERSON, A. 1972. Florida shrimp farm loses part of crop to high tide. Maine Coast Fisherman (April), p. 10A.

BERRY, F., AND E.S. IVERSEN. 1966. Pompano: Biology, fisheries, and farming potential. Proc. Gulf Caribb. Fish. Inst. 19:116-128.

HELLIWELL, J.R., AND J.E. HUTCHESON. 1976. The industrial engineering role in salmonid aquaculture. Pages 205-211 *in* Proc. Bras d'Or Lakes Aquaculture Conf., Cape Breton Island (June 1975).

HUGUENIN, J.E. 1975. Development of a marine aquaculture research complex. Aquaculture 5(2):135-150.

_____. 1976. An examination of problems and potentials for future large-scale intensive seaweed culture systems. Aquaculture 9(4):313-342.

HUGUENIN, J.E., AND F.J. ANSUINI. 1978. A review of the technology and economics of marine fish cage systems. Aquaculture 15(2):151-170.

HUGUENIN, J.E., AND G.N. ROTHWELL. 1979. The problems, economic potentials and system design of large future tropical marine fish cage systems. Proc. World Maricult. Soc. 10:162-181.

LINDBERG, J.M. 1976. The development of a commercial Pacific salmon culture business. FAO Tech. Conf. Aquaculture, Kyoto, Japan (May 20 to June 2). FIR:AQ/Conf./76/E.19. 12 pp.

WEBBER, H.H. 1971. The design of an aquaculture enterprise. Proc. Gulf Caribb. Fish. Inst. 24:117-125.

_____. 1973. Risks to the aquaculture enterprise. Aquaculture 2(2):157-172.

WEBBER, H.H., AND P.F. RIORDAN. 1976a. Criteria for candidate species for aquaculture. Aquaculture 7(2):103-123.

_____. 1976b. Problems of large-scale vertically integrated aquaculture. FAO Tech. Cong. Aquaculture, Kyoto, Japan (May 20 to June 2). FIR:AQ/Conf./76/R.4. 13 pp.

Bio-Engineering Symposium for Fish Culture (FCS Publ. 1):249-258

An Ecological Approach
to a Water Recirculating System for Salmonids:
Preliminary Experience

KENNETH T. MACKAY AND WAYNE VAN TOEVER

The Ark Project
The Institute of Man and Resources
R.R. #4, Souris, Prince Edward Island
Canada C0A 2B0

ABSTRACT

A water recirculation system for salmonids has been designed using ecological principles. The system conserves energy, recycles wastes by using algae and hydroponic crops to remove nutrients, and is an integral part of a solar greenhouse.

The system consists of linked translucent silos coupled to hydroponic troughs in which a gravel rooting media also serves as substrate for the nitrifying bacteria. Water is circulated and aerated by air-lift pumps.

The system of 8,500 liters was stocked with 18 kilograms of rainbow trout without allowing time for preconditioning of the bacterial filters. Fish growth and water quality were monitored for 43 days.

Oxygen was the limiting factor resulting in 42% mortality. Oxygen conditions can be improved by design changes. Water chemistry remained within acceptable limits for rainbow trout with the exception of nitrite; however, no mortalities were attributable to nitrite toxicity.

The system operated on close to 100% recirculation, was energy efficient, and required little maintenance. It appears to have considerable potential for salmonid hatchery and production systems in northern temperate locations and is simple enough to offer potential for small-farm-scale hatcheries.

Introduction

Recent predictions of energy supply indicate potential shortfalls and inevitable price increases. These have been highlighted by recent political events in the Mid-East and the resulting U.S. shortages of diesel fuel and gasoline. It is apparent that energy-intensive systems are vulnerable to these events.

North American fish production systems can be very energy intensive. Cage, pond (fed), raceways, and closed systems utilize more energy than is contained in the fish they produce. The energy ratios (input/output) were from 0.4-3.5 for cage culture; 1.3-18.4 for fed and fertilized ponds; and 59 for closed systems (Edwardson 1976). Taking into account the energy required for manufacturing the materials, for construction, for heating, and for pumping the water, 7 calories were needed for every calorie of fish produced in a recirculating salmonid hatchery in Manitoba (MacKay and Ayles 1979). Plans for new hatcheries often call for even more energy-intensive systems.

The use of recirculating water systems offers considerable advantages for salmonid hatcheries and rearing facilities (Burrows and Combs 1968; Meade 1974). These systems drastically reduce water requirements and decrease the energy required for heating, thus enabling economical acceleration of hatching and rearing (Henderson and Saunders 1978). Water reuse systems are also advantageous where water is in short supply and where pollution control of hatchery effluent is required.

Traditional recirculation methods using biological filters have a number of drawbacks. Organic material often accumulates on the filter requiring frequent backflushing, and oxygen demand of the filter is often equivalent to the demand of the cultured animals (Hirayama 1974). Accumulation of nitrate and decrease in alkalinity lower the pH in old filters (op.

cit.). Low pH reduces the nitrification rate and may require reestablishment of the filter. Establishment of the appropriate bacteria can take up to a month (Collins et al. 1975; Hirayama 1974); thus nitrification could be lost for considerable periods.

We have taken an alternative approach to biological filtration. We use a diverse mix of algae and hydroponic horticultural crops to assist in removing dissolved nutrients. Photosynthetic organisms such as phytoplankton and marine algae have been used to assist in purifying freshwater and marine recirculating systems (Kinne 1976) and to treat sewage (Ryther 1975), while hydroponically grown tomatoes have maintained water quality in a recirculating system growing channel catfish (*Ictalurus punctatus*) (Lewis et al. 1978). Our system is unique in that it combines a bacterial filter with algae and higher plants and is located in a solar greenhouse. This linking of fish and plant production in our system is similar to that occurring in natural water bodies and to techniques discovered by Chinese fish culturists over 2,000 years ago (Bardach et al. 1973).

Natural systems appear to be ideal models to follow in times of scarcity as they have survived the evolutionary test of time, being selected for their ability to process scarce energy and material resources. Ecosystem theory suggests that diverse systems have a greater resilience and stability and greater efficiency in utilizing energy and in recycling nutrients than simpler systems (Odum 1971). Our system has been deliberately designed using these ecological principles. A diversity of species and subsystems have been incorporated and the system makes maximum use of solar energy.

Our systems have been used to hatch, rear, and maintain broodstock of brook and rainbow trout and to grow Atlantic salmon fingerlings. A variety of vegetable and flowering plants have been cultured hydroponically. The actual design of the system is continually evolving. This paper reports on the design, performance, and monitoring of an unconditioned system operating from April 30 to June 11, 1979, which was initially stocked with 18 kg (7g each) of rainbow trout fingerlings.

Systems Design and Methods

Design

The basic units of the system are circular translucent 1,700-l tanks (1.2 m wide and 1.5 m high) constructed of fiberglass reinforced plastic (KALWALL). These solar algal ponds have been used for intensive Tilapia culture (Zweig 1977). At the P.E.I. Ark these tanks are located within an energy-efficient solar greenhouse (MacKay 1979). The growth of algae in the translucent tanks enhances absorption of solar radiation so that the aquaculture system acts as an efficient passive system for collection, storage, and distribution of solar energy. Additionally, the solar radiation enables the use of photosynthesis of the algae and plants cultured in the system to remove excess dissolved nutrients.

Four of these solar algal ponds filled with 1,500 l of water are linked together with 5-cm A.B.S. pipe to form a solar river, which in turn is linked to a settling tank (1,000 litres), an additional reservoir (1,250 litres) and three hydroponic troughs (one, 6 m long by 15 cm diameter; two, 4 m long by 15 cm diameter) (Fig. 1). Aeration and circulation between tanks and to the hydroponic troughs are provided by air-lift pumps powered by an electric air pump. The pumps consist simply of an air line with an attached air stone inserted into a vertical 5-cm plastic pipe (Fig. 1). The air-water mixture created in the pipe is less dense than the water outside the pipe so that it is displaced upwards and out the top, resulting in the pumping action (Spotte 1970). Flow rates of 8-10 l/min are routinely maintained.

The air-lift pumps are nonclogging, maintenance free, and self-regulating. The pumping efficiency depends on the percent submergence of the vertical pipe; therefore, the pumping rates of all pumps in a system automatically equilibrate at a certain head of water. Operating costs are also low: electrical consumption for this experimental system was 6 kWh or $0.42/d.

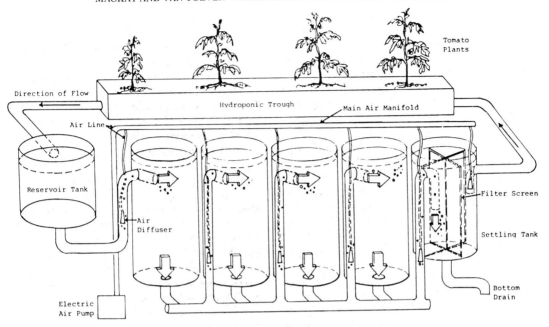

FIGURE 1. — Solar river aquaculture system.

The circulation pattern in the solar rivers is enhanced by the cylindrical shape of the solar ponds and the configuration of the piping used to interconnect the ponds. Water is pumped from the bottom center of each tank and directed to the surface of the next tank in the series. A 90° angle fitting at the point of entry to the tank directs the stream of water around the perimeter of the tank, thereby setting up a swirling circulation pattern. The vortex created draws sediments across the bottom to exit at the center. Suspended solids are removed in the final tank, which has been modified to serve as a settling tank. Water enters at the top of the tank and is piped directly to the bottom; thus the decrease in the circulation velocity of the water enhances the settling of suspended matter. Subsequently, the water must pass through coarse- and fine-mesh screens before exiting at the top of the tank into a hydroponic trough. A bottom drain with a shut-off valve in the settling tank simplifies removal of accumulated sediment, and the effluent is used to irrigate and fertilize greenhouse plants.

Nutrient Removal

Some freshwater algae can uptake ammonia directly (Healey 1977); thus the solar algal ponds may act directly as biological filters. However, increased nutrient loading will result in lower species diversity and unstable plankton populations. These conditions have been observed in solar algal ponds stocked with Tilapia at high densities (Todd et al. 1979).

We have added a hydroponic system to strip excess nitrogen and other nutrients from the recirculating system and supply an additional salable crop. Initial trials were with plants floating on styrofoam blocks with the roots extending into the tanks. This method of hydroponic culture requires no maintenance other than normal care of plants. Preliminary tissue analyses indicated the hydroponic crops are obtaining a full complement of nutrients from the fish wastes with the exception of the trace element iron. This is remedied by the routine addition of steel wool. However, this method, while simple and productive, is limited by available surface area and interferes with circulation, feeding, and fish removal. Tomatoes *(Lycopericon esculentum),* comfrey *(Symphytum peregrinum),* and leaf lettuce *(Lactuca sativa)* have been grown successfully.

To increase the area available for plant culture, an improved system employs modules of 15-cm-diameter P.V.C. plastic water pipes with rectangular slots cut in the top and wooden troughs lined with plastic sheeting. Coarse gravel and oyster shells serve as support for seedlings. This provides the added advantages of supplying a suitable substrate and surface area for the establishment of nitrifying bacteria, and increasing buffering ability. The soil attached to the roots of plants transplanted from pots to the hydroponic trough supplies the initial bacterial innoculant. Our recirculating system thus employs photosynthetic algae, higher plants and bacteria to remove nitrogen and other excess nutrients.

Method

In this experimental system the 6-m hydroponic wooden trough was planted with four tomato varieties (Sweet 100, Michigan Ohio Forcing, 107, and Jumbo). Celery and an assortment of flowering plants were transplanted to the two 4-m P.V.C. troughs. The four solar algae ponds were each stocked with 4.5 kg of rainbow trout at 7 g each. These fry had been hatched previously in another recirculating system within the P.E.I. Ark, from eyed eggs obtained from White Sulfur Springs Hatchery, West Virginia. The fish were fed 3-5% of body weight/day on a dry pellet food (Co-op Atlantic Trout Food).

Water quality was monitored for a number of physical and chemical parameters. The parameters, the method of measurement, and the sampling interval are outlined in Table 1. Dissolved oxygen (D.O.) and temperature were monitored automatically and data recorded on cassette tape for later analysis; however, difficulty was encountered with the reliability of the D.O. readings. Samples for chemical analysis were obtained from the center of each tank, while the hydroponic samples were taken at the entry and the exit of the 6-m trough. Chemical samples were filtered through a 0.8μ filter and collected in acid-cleaned glassware. Colorimetric analysis was done with a Bausch and Lomb Spectronic 20 spectrophotometer with precalibrated Bausch and Lomb Spectrokits (R). Gridded 0.45μ nitrocellulose filters were used to obtain samples for identification of planktonic organisms.

Results

Temperature

Temperatures during this experiment were within the optimal range for salmonid growth. Initial water temperature was $17.8\,°C$; a minimum of $14.7\,°C$ occurred on day 8 and a maximum of $19.4\,°C$ occurred on day 43. A more crucial period for hatcheries in northern temperate latitudes is December through March. Figure 2 illustrates representative temperatures for the Ark solar hatchery during 1977-78. In spite of sub-zero December-January outside temperatures, the solar algal ponds were over a degree warmer than groundwater normally used

TABLE 1. — Methods of measuring the physical and chemical parameters of water quality in a salmonid recirculating system at the P.E.I. Ark.

Parameter	Method	Sampling interval
Dissolved oxygen	Specific ion probe, electronic (YSI)	2-3 days
pH	Specific probe, electronic (Corning)	2-3 days
Temperature	Thermocouple electronic	hourly
Alkalinity	Titration	2-3 days
Nitrogen NH_4	Nesslerization, spectrophotometric	2-3 days
Nitrogen NO_2	Diazotization, spectrophotometric	2-3 days
Nitrogen NO_3	Brucine sulfate spectrophotometric	2-3 days
Phosphorus	Ascorbic acid, spectrophotometric	2-3 days

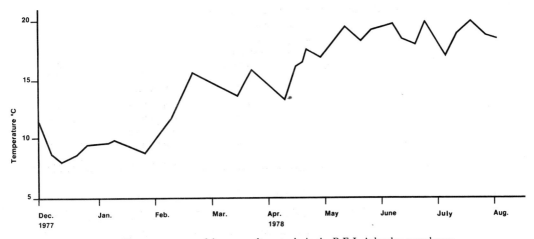

FIGURE 2. — Average weekly temperatures of the aquaculture tanks in the P.E.I. Ark solar greenhouse.

by hatcheries in Maritime Canada. Warming occurred rapidly during February. Temperatures of 13-18 °C were maintained from February to June, July-to-August temperatures ranged from 18 to 23 °C, and September-to-December temperatures ranged from 20 to 13 °C (not shown in Figure 2).

Dissolved Oxygen

Only manual D.O. readings were available as the computerized data system was not reliable. The manual readings were taken between 10:00 and 11:00 a.m. Diurnal variations of D.O. occur with daylight levels higher than night levels because of algal photosynthesis. D.O. concentrations for the tanks containing fish are plotted in Figure 3. Oxygen declined during the experiment from 9 ppm on day 1 to a low of 4.9 ppm on day 42. Mean oxygen levels for the entire experiment are shown in Table 2 and show variation between tanks. Oxygen declined from the reservoir tank through tanks 1-3 and then increased in tank 4 and the settling tanks. Oxygen became limiting during the experiment on two occasions. A crimped air line stopped circulation and supplemental air to tank 2 on day 2 and resulted in 100% mortality in that tank. In addition, on day 24, inadequate oxygen levels (less than 2 ppm) in tank 3 also resulted in 100% mortality in that tank.

Nutrient Chemistry, pH, and Alkalinity

The water chemistry data indicate a number of interesting trends. Figure 3 indicates the changes in ammonia, nitrite, nitrate, and alkalinity over the 43 days of the experiment. Unfortunately, because of a shortage of a number of chemicals, nitrite, ammonia, and alkalinity were not measured for the entire experiment. Ammonia increased from 0.3 ppm on day 1 to a high of 11.4 ppm on day 12; and decreased to 9.5 on day 15. No further measurements were taken until day 29, when no ammonia was present. Ammonia concentration for the rest of the experiment was less than 2 ppm.

Nitrate levels remained around 1 ppm for the first 12 days of the experiment, increased rapidly to 19.5 ppm on day 27, then decreased to 6.0 ppm on day 38 and increased to 12 ppm on the last day.

Nitrite concentrations of less than 0.5 ppm were found during the first 12 days; concentrations increased to 1.2 ppm on day 22, then decreased slightly on day 26 after which no further measurements were made.

Alkalinity was high for the first 12 days of the experiment (100-160 mg Ca CO_3/l). It dropped dramatically to 0 on day 15 and then climbed abruptly to 140 on day 23, dropping to 60 on day 29. This dramatic change was not reflected in the pH, which showed a gradual decline during the ex-

TABLE 2. — Dissolved oxygen, pH, nutrient chemistry for various components of a recirculating system at the P.E.I. Ark. (Confidence limits for students (*t*) at *p* <0.05 in parentheses.)

| | Dissolved oxygen | | | |
	Maximum	Minimum	Mean	pH
Reservoir	10.4	5.4	7.9 (±0.88)	7.0 (±0.15)
Tank # 1	10.1	4.5	7.5 (±0.76)	6.7 (±0.13)
#2	9.2	4.3	6.4 (±0.93)	6.7 (±0.13)
#3	9.0	4.1	6.3 (±0.90)	6.7 (±0.13)
#4	9.4	4.6	6.8 (±0.72)	6.7 (±0.13)
Settling tank	9.6	5.4	7.5 (±0.35)	6.8 (±0.19)

	PO_4^{-3} ppm	NO_3^{-1} ppm — N	NO_2^{-1} ppm — N	NH_4^+ ppm — N	Total N
Reservoir	4.7 (±1.29)	6.4 (±3.28)	0.22 (±0.22)	2.3 (±2.23)	8.92
4 solar algal ponds	5.0 (±0.66)	6.5 (±1.52)	0.23 (±0.10)	2.4 (±0.96)	9.13
Settling tank	4.9 (±1.21)	6.6 (±3.05)	0.25 (±0.26)	2.4 (±2.45)	9.25
Hydroponic troughs	4.6 (±1.15)	5.9 (±3.26)	0.22 (±0.24)	1.5 (±1.52)	7.62

periment. Initial pH's were 7.3 while final pH's were 6.2.

Phosphate concentrations gradually increased from 0 on day 1 to 5 ppm on day 12, stabilized until day 22, then gradually increased to 8.8 ppm on the final day.

The water chemistry can also be portrayed by examining each component. This is illustrated in Table 2. While there are few significant differences in the parameters, there are consistent changes within the system that suggest some of the system's dynamics. The pH decreased from the reservoir to the fish ponds while PO_4^{-3} and Nitrogen (NO_3^{-1}, NO_2^{-1}, and NH_3) increased slightly. The settling tank is quite similar to the fish ponds with a slight increase in nitrogen. The hydroponic system had lower PO_4^{-3} and nitrogen concentrations.

Fish Growth and Survival

During the 43-day period, the fish grew from the initial average weight of 7.0 grams to a final weight of 13.3 grams. This growth rate (<2%/d) is much lower than we have regularly achieved in the past by using similar systems. Rapid growth rates have consistently been achieved in the greenhouse aquaculture system during winter and spring. During the winter and spring of 1977-78, growth rates for speckled trout and rainbow trout, in a similar system, were twice that for this experiment. The survival was also poor at 42% — primarily because of occasions during which two entire tanks (total of 1,300 fish) were lost to low oxygen levels.

Hydroponics

Some problems were encountered with the hydroponics. The tomatoes were initially rooted too deeply in the gravel-oyster shell substrate and did not receive adequate oxygen. Low nutrient levels and shading from an overhead walkway further stressed the plants. The other hydroponic crops grew better but the length of this experiment did not allow the plants to mature so no yield data were obtained. Previous results (unpublished data) have indicated higher yields for these crops grown hydroponically than in adjacent greenhouse beds.

Other Species

In addition to the fish, hydroponic plants, and nitrifying bacteria, a number of other organisms have become established in the system. Five genera of algae and 11 groups of animals have been identified from the recirculating system. Of importance to nutrient uptake are the algae populations. While we have not undertaken a detailed survey of their population dynamics, we did notice the following changes. From days 6-8 a dramatic shift occurred from the green algae *Chlorella* to the diatoms *Navicula* and *Ankistrodesmus;* and by day 16 the green algae *Scenedesmus* predominated. Dense populations of *Oscillatoria* were present on the surface of the hydroponic troughs by day 30.

General Comments

The system operated at close to 100% recirculation. During the 43 days of the experiment, water was added only to replace that lost by evaporation and to replace the 1,200 litres drawn off during the periodic removal of sludge. Apart from repair to a crimped air line, no other maintenance was required.

Discussion

Dissolved oxygen was the major limiting factor in the recirculating system. Mortalities occurred twice related to low D.O. The first event was related to systems design and the crimped air line was repaired. More seriously, D.O. levels were low from day 19 onward. This resulted in mortalities in one pond and probably resulted in reduced growth rates in others. During sunny days under similar stocking rates, we normally experience high D.O. conditions (7-8 ppm) because of algal photosynthesis. We attribute the low D.O. in this experiment to low light levels partly due to a shading from a dense canopy of tomatoes growing in the greenhouse soil beds in front of the solar algal ponds. There is some indication that the slightly higher D.O. levels in pond 4 and the settling tank (Table 2) are related to higher light levels as these tanks were not shaded so much by the tomatoes. The high angle of the sun during May and June also contributed to low light levels as direct sunlight did not reach the back of the greenhouse during this period.

Daytime oxygen levels can be increased by relocating the fish tanks within the greenhouse and we have recently redesigned the system to do this. However, increased light will increase the solar warming of the water and some shading is required to prevent overheating.

Critical parameters of water quality have performed much better than D.O. In spite of the difficulty experienced in the hydroponics, nitrogen did not accumulate in the system after day 26. While ammonia concentration peaked at 11.4 ppm, the concentration of toxic un-ionized ammonia was only 0.049 ppm (Trussel 1972). This was twice the 0.026 ppm recommended as the maximum level fish can tolerate for long periods (European Inland Fisheries Advisory Commission 1973). However, after establishment of the nitrification process in our system, steady state levels of un-ionized ammonia were less than 0.002 ppm — levels that Henderson and Saunders (1978) found suitable for growth in rainbow trout.

Nitrate is toxic only at high concentrations. Meade (1974), for example, maintained levels below 200 ppm in a unit culturing chinook salmon. Nitrate during this experiment did not exceed 20 ppm.

Nitrite is toxic at very low levels. The 96-hour LD_{50} for 12 g rainbow trout was 0.22 ppm NO_2-N in bioassays performed by Russo et al. (1974). Our measured levels of NO_2 were much higher. Concentrations increased dramatically after day 12, and on days 18-26 were above the 24-hour LD_{50}. These levels were lower, however, than Collins et al. (1975) found in their unconditioned recirculating system with similar stocking rates of channel catfish. As their system became conditioned, nitrite dropped to very low concentrations. We assume a similar condition occurred in our system, but unfortunately we lack measurements of nitrite after day 26. As nitrite interferes with oxygen transport (Perrone and Meade 1977), the high temperatures and low oxygens of our experiment should have produced severe mortalities but did not. This may be due to the interference of some dissolved ions, as nitrite toxicity can be lowered by calcium and chloride ions (Crawford and Allen 1977; Perrone and Meade 1977).

Dramatic changes occurred in the water chemistry from day 12 onward (Fig. 3).

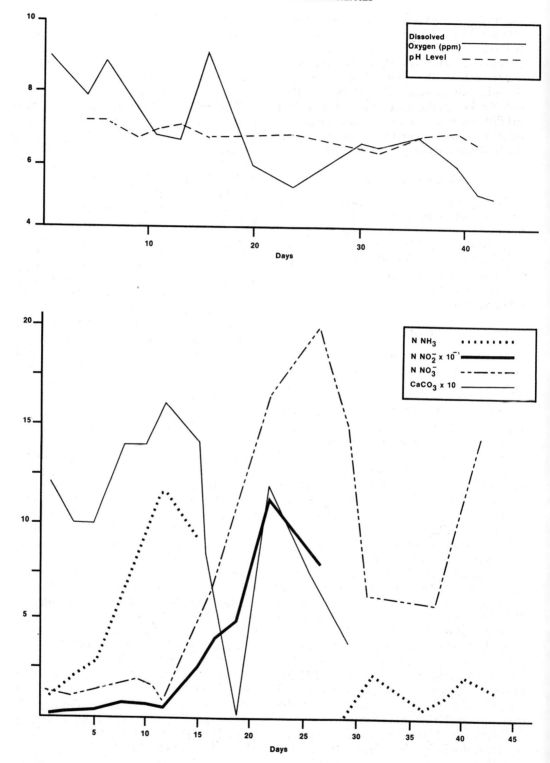

FIGURE 3. — Dissolved oxygen, pH, alkalinity, and nitrogen concentrations in a recirculating system at the P.E.I. Ark.

Ammonia decreased while nitrite and nitrate increased followed by a crash in alkalinity and then a decrease in pH. These are all associated with the onset of nitrification. Collins et al. (1975) found similar changes in the nitrogen concentrations as a freshly established recirculating system became conditioned. However, in their system, nitrite reached much higher concentrations (15.5 ppm), while nitrate increased much more slowly. The simultaneous increase in nitrate and nitrite in our system suggests that both the *Nitrosomonas* and the *Nitrobacter* bacteria became established simultaneously. Collins et al. (op. cit.) indicate a considerable lag in the establishment of the *Nitrobacter* group.

The nitrification of ammonia generates hydrogen ions, thus will decrease pH and alkalinity. Our system was initially well buffered and pH decreased slowly; however, alkalinity crashed to 0 on day 19. This decrease in alkalinity may also be due to direct uptake of ammonia by algae and hydroponic plants, which yield hydrogen ions (Brewer and Goldman 1976). The increase in alkalinity after day 19 is probably due to increased $CaCO_3$ released from the oyster shell hydroponic substrate; however, increases in alkalinity are also caused by plant uptake of NO_3 ions during photosynthesis (op. cit.).

It was not possible from the monitoring to assess the relative importance of the various biological filters. The system contained algae, grazing zooplankton, and decomposing bacteria in the solar algal ponds; and algae, higher plants, and nitrifying bacteria in the hydroponic filter. In addition sludge was periodically removed from the system. However, the end result of such a diverse mix was a system which conditioned quickly, maintained ammonia below harmful levels, and did not accumulate nitrate, and in spite of high nitrite levels there were no nitrite-related mortalities. This allowed us to grow fish under almost 100% recirculation. The low daytime D.O. levels were related to excessive shading. The low D.O.'s and initial high ammonia and nitrite levels probably accounted for the low growth rates

— lower than we achieved in other experiments with such a system. Slight modifications in the location of the tanks within the greenhouse, improvements to the aeration system, and pre-conditioning of the filter should solve these problems.

This system offers considerable potential for salmonid hatchery and rearing facilities, as it allows for economical accelerated rearing during winter. As our system has no effluent, it is ideal for quarantine facilities. It is also simple enough to be operated by small-scale decentralized hatcheries.

Acknowledgments

The P.E.I. Ark Project is funded by the Federal Departments of Environment and Energy, Mines and Resources through the Canada-P.E.I. Agreement on Renewable Energy Research and Development. Alan McLennan assisted in chemical monitoring and the P.E.I. Department of Fisheries assisted in supplying monitoring equipment.

References

BARDACH, J.E., J.H. RYTHER AND W.O. MCLARNEY. 1972. Pages 1-3 in Aquaculture. Wiley-Interscience. Toronto.

BREWER, P.G., AND K.C. GOLDMAN. 1976. Alkalinity changes generated by phytoplankton growth. Limnol. Oceanogr. 21: 108-117.

BURROWS, R.E., AND B.D. COMBS. 1968. Controlled environments for salmon propagation. Prog. Fish-Cult. 30:123-136.

CRAWFORD, R.E., AND G.H. ALLEN. 1977. Seawater inhibition of nitrite toxicity to chinook salmon. Trans. Am. Fish. Soc. 106:105-109.

COLLINS, M.T., J.B. GRATZEK, E.B. SHOTTS, JR., D.L. DAWE, L.M. CAMPBELL, AND D.R. SENN. 1975. Nitrification in an aquatic recirculating system. J. Fish. Res. Board Can. 32:2025-2031.

EUROPEAN INLAND FISHERIES ADVISORY COMMISSION. 1973. Water quality criteria for European freshwater fish. Report on ammonia and inland fisheries. Water Res. 7:1011-1022.

HEALEY, F.P. 1977. Ammonium and urea uptake by some freshwater algae. Can. J. Bot. 55:61-69.

HENDERSON, E.B., AND R.L. SAUNDERS. 1978. Comparative growth of rainbow trout reared in recycle and single pass system. I.C.E.S. 1978. 10 pp.

HIRAYAMA, K. 1974. Water control by filtration in closed culture systems. Aquaculture 4:369-385.

KINNE, O., ed. 1976. Marine ecology. Volume III Cultivation. Pages 129-134 *in* Part 1. John Wiley and Sons, London.

LEWIS, W.M., J.H. YOPP, H.L. SCHRAMM, JR., AND A.M. BRANDENBURG. 1978. Use of hydroponics to maintain quality of recirculated water in a fish culture system. Trans. Am. Fish. Soc. 107:92-99.

MACKAY, K.T. 1979. Solar heating at the P.E.I. Ark. Proc. SESCI Annual Meeting, Charlottetown, P.E.I. 8 pp.

MACKAY, K.T., AND B. AYLES. 1979. Aquaculture in an ecological agriculture: Background paper Science Council of Canada study, Canadian agriculture in the year 2001.

MEADE, T.L. 1974. The technology of closed system culture of salmonids. Univ. Rhode Island, Marine Tech. Rep. 30. 30 pp.

ODUM, E.P. 1971. Fundamentals of ecology. 3rd ed. W.B. Saunders Co. Philadelphia. 574 pp.

PERRONE, S.J., AND T.L. MEADE. 1977. Protective effect of chloride on nitrite toxicity to coho salmon *(Oncorhynchus kisutch)*. J. Fish. Res. Board Can. 34:486-492.

RUSSO, R.C., C.E. SMITH, AND R.V. THURSTON. 1974. Acute toxicity of nitrite to rainbow trout *(Salmo gairdneri)*. J. Fish. Res. Board Can. 31:1653-1655.

RYTHER, J.H. 1975. Preliminary results with a pilot plant waste recycling-marine aquaculture system. International Conference on the Renovation and Reuse of Wastewater Through Aquatic and Terrestrial Systems at the Rockefeller Foundation Study and Conference Center, Bellagio, Italy, July 15-21, 1975.

SPOTTE, S.H. 1970. Fish and invertebrate culture. Wiley-Interscience. New York. 145 pp.

TODD, J.H., R.D. ZWEIG, A.M. DOOLITTLE, JR., D.G. ENGSTROM, AND J.R. WOLFE. 1979. Assessment of a semi-closed renewable resource-based aquaculture system. Progress report 3 to Office of Problem Analysis, Applied Science and Research Applications, National Science Foundation, Washington, D.C.

TRUSSEL, R.P. 1972. The percent un-ionized ammonia in aqueous ammonia solutions at different pH levels and temperatures. J. Fish. Res. Board Can. 29:1505-1507.

ZWEIG, R. 1977. The sage of the solar algal ponds. Journal of the New Alchemists 4:63-68.

Bio-Engineering Symposium for Fish Culture (FCS Publ. 1):259-265
© 1981 by the Fish Culture Section of the American Fisheries Society

Evolution of the Design and Operation
of a Freshwater Waste Heat Aquaculture Facility
at an Electric Generating Plant

BRUCE L. GODFRIAUX

Public Service Research Corporation
Newark, New Jersey 07101

NILS E. STOLPE

New Jersey Department of Agriculture
Division of Rural Resources
Trenton, New Jersey 08608

ABSTRACT

Since start-up in 1973 the waste heat aquaculture facility at the Public Service Electric and Gas Company (PSE&G) Mercer Generating Station on the Delaware River south of Trenton, New Jersey, has undergone a series of evolutionary changes culminating in the current proof-of-concept facility. Initially, the program involved the grow-out of postlarval size shrimp *Macrobrachium rosenbergii* (de Man) in a single-lined excavated pond supplied with the heated effluent from the power plant. Presently, rainbow trout *Salmo gairdneri* (Richardson), channel catfish *Ictalurus punctatus* (Rafinesque), and the American eel *Anguilla rostrata* (Lesueur) are cultured in the heated effluent. High intensity techniques utilizing concrete raceways, partial recirculation, and oxygen injection have been used for these species.

Systems and procedures are discussed that allow the interfacing of a multispecies aquaculture program with the operations of a coal-fired electric generating station. The major topics addressed concern problems encountered in dealing with condensor chlorination, seasonal water temperature fluctuations, and diurnal water temperature fluctuations from the power plant. The paper concludes with a description of the proposed 20-raceway commercial aquaculture facility at the Mercer Generating Station.

Introduction

Since August 1973, Public Service Electric and Gas Company (PSE&G) with Trenton State College, Rutgers University, Long Island Oyster Farms, and more recently the New Jersey Department of Agriculture has been conducting research to determine whether an economically viable and intensive aquaculture industry can be established using thermal effluents from the Mercer Generating Station.

The research program has been supported by funds provided by the National Science Foundation (NSF) and PSE&G. In addition PSE&G contributed the aquaculture site as well as initial capital to build the pilot aquaculture facility. Beginning in November 1976, NSF and PSE&G provided additional research grants to build and operate a proof-of-concept aquaculture facility (prototype design for a proposed commercial facility) for a further three-year period.

The existing Mercer Proof-of-Concept Aquaculture Facility and proposed 4.5-ha commercial facility are located at the PSE&G Mercer Generating Station, 6 km south of Trenton (Fig. 1). The coal-fired station generates 600 MW of power and discharges approximately 1,700,000 liters per minute of heated discharge water at a maximum of 6 °C above the ambient Delaware River water temperature, giving the discharge water a temperature range of 6-35 °C.

Mercer Pilot Aquaculture Facility

The original Mercer Pilot Aquaculture Facility was built during the summer of 1973 and became fully operational by early 1974 and operated until April 1978. Pilot facilities are shown hatched in Figure 2. It consisted of two 3,000-liters-per-minute intake pumps located at the Mercer Station discharge canal. This pumped the heated condenser cooling discharge water

259

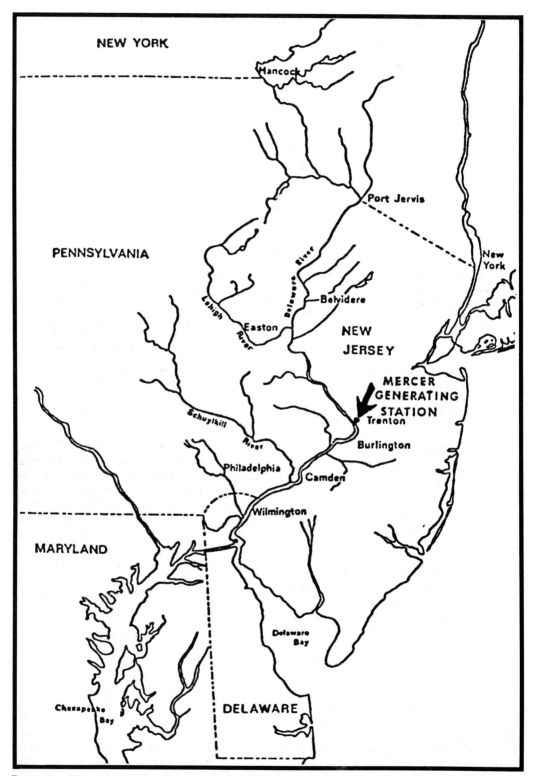

FIGURE 1. — Map of the Delaware River drainage basin, showing the location of the Mercer Generating Station.

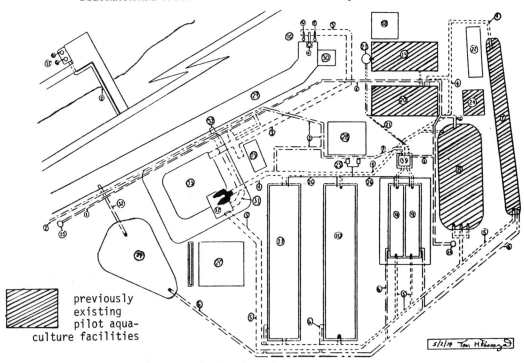

previously
existing
pilot aqua-
culture facilities

1. Lines from Well 1 and Well 2 to laboratories and raceways
2. Line from Delaware River to aquaculture system
3. Line from Mercer Station discharge canal to temperature-moderation pond
4. Line from recirculation chamber to raceways and laboratories
5. Line from raceways to recirculation chamber
6. Line from raceways to waste-settling pond
7. Raceway I
8. Pond I
9. Enclosed raceways A, B
10. Raceway II
11. Raceway III
12. Recirculation chamber
13. Temperature-moderation pond
14. Well No. 1
15. Well No. 2
16. Pumps for Mercer Station discharge canal to aquaculture system
   a 1,500-gpm pump (5.7 m³/min)
   b 800-gpm pump (3.0 m³/min)
   c 800-gpm pump (3.0 m³/min)

17. Delaware River Pump Station
   a 500-gpm pump (1.9 m³/min)
   b 500-gpm pump (1.9 m³/min)
18. Maintenance building (houses alarm system)
19. Laboratory I
20. Laboratory II
21. Line from Laboratory I to mixing chamber for enclosed raceways
22. Office-laboratory trailer
23. Office trailer
24. Food-storage shed
25. Compressed-air blowers
26. Air lines for compressed air
27. Greenhouse
28. Mercer Station transformer
29. Mercer Station discharge canal
30. Mercer Station Fire House
31. Drainage line from recirculation chamber to Mercer Station discharge canal
32. Line from waste-settling pond to Mercer Station discharge canal
33. Mixing chamber for enclosed raceways
34. Waste-settling pond

FIGURE 2. — Diagram of the Mercer Aquaculture Facility, showing the old pilot facility and the new proof-of-concept facility.

directly to the various facilities within the pilot complex, which included one vinyl-lined pond 30 m × 8 m, one vinyl-lined V-shaped raceway 50 m × 2.4 m × 1 m

deep and two fiberglass greenhouses 13 m × 6.7 m.

Several problems were encountered with this facility that at certain times made fish rearing more difficult than usual.

1. The generating station chlorinates its condenser cooling water for three one-hour periods daily.

2. During station chlorination periods, the discharge canal intake pumps were shut down completely and eliminated process flow within the pilot aquaculture facility.

3. There is a 35 °C change in water temperature between coldest winter conditions and warmest summer conditions.

4. There was no way to blend these extreme water temperature conditions higher or lower with ambient Delaware River or well water.

5. There could be large daily station discharge water temperature fluctuations.

Certain of the above mentioned problems could be rectified by instituting certain operational procedures at the pilot facility. To avoid any rainbow trout and shrimp mortalities due to residual chlorine levels in the station discharge water, all aquaculture intake pumps were shut down during periods of station chlorination. The method for automatically shutting down the intake pumps is described by Godfriaux et al. (1975). As long as the fish density was low — under 2 lb/ft³ (907 g/28 dm³) — oxygen levels remained sufficiently high and ammonia levels sufficiently low that no mortalities occurred during these "no flow" periods. However, if higher fish densities were to be reared, some method was needed to continue water flow during generating station chlorination periods either by using another chlorine-free water source and/or by recirculating the aquaculture facility process water during these periods.

Since there was such a large seasonal change in station discharge water temperature, a two-species aquaculture crop was initiated. A coldwater species rainbow trout was alternated with a warmwater species of freshwater shrimp. This method of alternating a warmwater species with a coldwater species was named "diseasonal aquaculture." Using the two species, reasonable growth rates could be maintained all year round at the aquaculture facility.

As the trout and shrimp crops moved closer to the corresponding warmer and colder rearing seasons, the species were more susceptible to temperature stress. If other different temperature water sources such as ambient Delaware River water or well water were available to moderate discharge canal temperatures, water temperature could be kept closer to the optimum growth temperature.

Mercer Proof-of-Concept Aquaculture Facility

Figure 2 shows the present aquaculture facility now in operation at Mercer Generating Station. The previous pilot aquaculture facility was integrated with the new proof-of-concept facility. The proof-of-concept facility was designed to take care of some of the problems mentioned earlier with the pilot facility.

The proof-of-concept facility (Fig. 2) consists of a tempering pond where water from three sources (ambient Delaware River water, well water, and generating station discharge water) is pumped and blended together. This pond gives aquaculture personnel some control over the temperature of process water used in the aquaculture facility. During normal operation of the aquaculture facility, process water then flows by gravity from the tempering pond to the rest of the aquaculture facility.

Other main features of the aquaculture facility are the prototype commercial grow-out concrete raceways. Both are 30.5 m long × 3.6 m wide. One raceway is 1.8 m deep and the other is 1.2 m deep. Also, enclosed in a double plastic layer air-inflated greenhouse, are two concrete nursery raceways 18.3 m × 1.8 m × 1.2 m deep. The remaining facilities were part of the original pilot program and were described earlier. The other new facility that takes the raceway cleanings from the concrete raceways is the vinyl-lined waste-settling basin. Water laden with wastes after cleaning is retained in the basin for 24 to 48 hours for the settling of particulates before being discharged.

In order to provide the new aquaculture

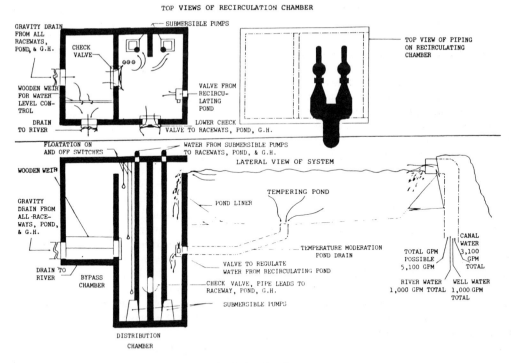

FIGURE 3. — Operation of the tempering pond recirculation chamber system.

facility with a cooler water source, two 2,000 l/min Delaware River intake pumps were installed and two 2,000 l/min wells were drilled. They were to provide for a cool water supply to rear rainbow trout fingerlings during the summer months. Unfortunately, the wells could not be used since their water contained a high iron content (17 to 23 ppm).

Another problem, the loss of process flow during station chlorination periods, was rectified by the installation of two 5,700 l/min recirculation pumps. When the chlorination cycle begins, the discharge canal intake pumps are automatically shut down. The water level in the tempering pond and consequently the recirculation chamber lowers (Fig. 3), and the float valve in the recirculation chamber turns on the two recirculation pumps. The water in the distribution chamber (Fig. 3) is pumped lower than the used water from the aquaculture facility in the bypass chamber. A check valve opens, which allows the water in the bypass chamber to flow into the distribution chamber for

recirculation to the aquaculture facility. When the chlorination cycle ends, the discharge canal pumps resume operation. The tempering water level rises, and the water level within the recirculation chamber increases to the level where the recirculation pumps are turned off. Water then flows from the tempering pond to the rest of the aquaculture facility by gravity. For further details on the design and operation of this system, see Godfriaux and Shafer (1979).

The tempering pond also served as a buffer against diurnal temperature changes with the Mercer Station discharge water. These temperature changes were caused by changes in electrical demand over a 24-hour period. Lightest loads occurred during evening hours, when the generating units were run at reduced load, creating a reduction in heat energy dissipated in the discharge water. This reduced discharge water temperatures.

The one other addition made to the Mercer Proof-of-Concept Aquaculture Facility was the addition of a side-stream oxygen injection system. Pure oxygen gas is vaporized from a cryogenic oxygen tank and is placed in contact with aquaculture facility process water at 60 PSIG pressure from the discharge end of the raceway. The oxygen/water mixture travels down a 90-ft (27.4-m) long contact loop and then is discharged through four spargers located at 20-ft (6.1-m) intervals along the length of the concrete raceways. This system always keeps dissolved oxygen readings near saturation.

Biological Program

It became apparent by the end of the shrimp outdoor grow-out cycle in October 1977 that they could not be reared at sufficiently high densities, 15-30 g shrimp/ft² (929 cm²), to be economically viable. The major problem was cannibalism. At high densities shrimp suffered mortalities until their density was reduced to one 30 g shrimp/ft² (929 cm²) of surface area.

Because of these facts, channel catfish were substituted for *Macrobrachium rosenbergii* as the main summer crop during 1978 and 1979. Harvest catfish density has reached 2 lbs/ft³ (907 g/28 dm³). In subsequent grow-outs, it is hoped to increase catfish harvest densities to 11 lbs/ft³ (5 kg/28 dm³).

Another species that is being reared on a limited scale is the American eel. Success to date has been mixed. Eels in the elver stage to approximately 30 g in size were found to be very susceptible to *Saprolegnia* infections. After reaching 30 g in weight, the eels are very disease resistant. Generally, the eels can add 150 to 200 g per growing season. During the colder months (November through April) aquaculture facility process water temperatures are generally below 60 °F (15.6 °C) and little growth occurs.

Rainbow trout have been the only species reared during the colder month period. The trout have adapted easily to generating station discharge water condi-

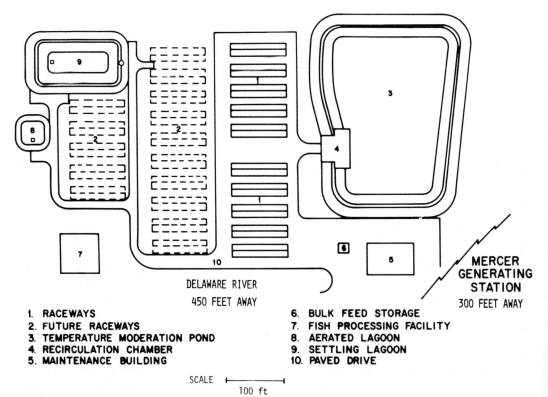

1. RACEWAYS
2. FUTURE RACEWAYS
3. TEMPERATURE MODERATION POND
4. RECIRCULATION CHAMBER
5. MAINTENANCE BUILDING

6. BULK FEED STORAGE
7. FISH PROCESSING FACILITY
8. AERATED LAGOON
9. SETTLING LAGOON
10. PAVED DRIVE

DELAWARE RIVER
450 FEET AWAY

MERCER GENERATING STATION
300 FEET AWAY

SCALE

100 ft

FIGURE 4. — Schematic of the proposed commercial aquaculture facility at Mercer Generating Station.

tions. During the 1978-79 grow-out, trout densities reached 6 lbs/ft³ (2.7 kg/28 dm³) just before harvest. With this next grow-out season harvest densities approaching 11 lbs/ft³ (5 kg/28 dm³) are expected. This has been made possible through the use of a side-stream oxygen injection system to provide the additional oxygen required to sustain this high trout density.

Mercer Commercial Aquaculture Facility

Figure 4 shows the proposed design for an initial 20 raceways. Each raceway is made of concrete and is 30.5 m long × 3.6 m wide × 1.8 m deep. The proposed commercial aquaculture facility would work in a similar manner to the proof-of-concept aquaculture facility including the use of a waste-settling lagoon and aeration lagoon. Other facilities include a fish-processing facility and a bulk-fish-feed storage facility.

Process flow of this facility is 136,000 l/min. This facility has the capability to produce 727,000 kg of rainbow trout during the colder grow-out period from November through May and 727,000 kg of channel catfish during the warmer grow-out period (June through October). Harvest periods for the two species are April-May for rainbow trout and September-October for channel catfish.

Associated with the proposed Mercer Commercial Aquaculture Facility are two off-site trout hatcheries to provide 17.5- to 20-cm-long trout fingerlings to Mercer each November for grow-out to commercial size. A feed mill may also be built or purchased to provide food pellets for the trout and catfish grow-outs to commercial size.

Acknowledgments

This research was funded by the National Science Foundation Research Applied to National Needs Grants ENV 76-19854 A01, A02, A03, and PSE&G authorization RO-443. Other contributions to the project have been made by Trenton State College.

Dr. A.F. Eble and personnel of the Mercer Aquaculture Facility provided comments on various topics discussed in this paper. We would also like to acknowledge the help provided in various stages of this program by Dr. E.H. Bryan, Program Manager, NSF, Applied Science and Research Applications; Mr. R.A. Huse, General Manager, R&D, PSE&G; Dr. C.R. Guerra, Manager, Advanced Systems, R&D; Ms. C. Stephens-La Brie, Research Assistant; and Mr. J. Morrison, Manager, Mercer Station, PSE&G.

References

GODFRIAUX, B.L., AND R.R. SHAFER. 1979. A method of maintaining aquaculture process flow during chlorination of electric generating station cooling water. Pages 202-212 *in* B. Godfriaux et al., eds. Power plant waste heat utilization in aquaculture, Workshop II. Allenheld, Osmun and Co., Montclair, N.J.

GODFRIAUX, B.L., H.J. VALKENBURG, A. VAN RIPER, C.R. GUERRA, A.F. EBLE, AND P. CAMPBELL. 1975. Power plant heated water use in aquaculture. Pages 233-250 *in* V. Langworthy, ed. Proc. Third Annu. Pollut. Control Conf. Water Wastewater Equipment Manufacturers Association.

Bio-Engineering Symposium for Fish Culture (FCS Publ. 1):266-273
© 1981 by the Fish Culture Section of the American Fisheries Society

Integration of Hatchery, Cage, and Pond Culture of Common Carp *(Cyprinus carpio* L.) and Grass Carp *(Ctenopharyngodon idella* Val.) in The Netherlands

E.A. HUISMAN

Organization for Improvement of Inland Fisheries
Buxtehudelaan 1, 3438 EA Nieuwegein, The Netherlands
and
Department of Fish Culture and Inland Fisheries
State Agriculture University
P.O. Box 338, 6709 PG Wageningen, The Netherlands

ABSTRACT

An outline and a technical characterization is given of three different types of fish husbandry facilities used in The Netherlands; e.g., stagnant ponds, an environmentally controlled hatchery, and cages, the latter being situated in the cooling water discharge canal of a conventional electrical power station.

The combined and integrated use of these facilities in the propagation of the warmwater cyprinid fishes carp *(Cyprinus carpio* L.) and grass carp *(Ctenopharyngodon idella* Val.) is illustrated.

Data relevant for fish culture management — such as growth, feed conversion, standing crop densities, and yields — are reported to stress the outstanding feasibilities for carp and grass carp culture in cool climatological regions using a combination of different types of fish husbandry systems.

Introduction

Fish culture in The Netherlands started in 1861, when the first salmon *(Salmo salar)* eggs were hatched in the facilities of the Royal Zoological Society "Artis Natura Magistra" at Amsterdam. It is a matter of striking coincidence that already in these early days fish culture was carried out for (re-)stocking purposes and this still is the main objective of fish culture in The Netherlands.

However, in the meantime recreational fisheries have become more important than the professional fisheries in the inland waters. Also, because of the deterioration of the water quality, (re-)stocking of warm- and coolwater fishes such as carp *(Cyprinus carpio)* and pike *(Esox lucius)* has come more in demand in fisheries management than stocking of salmonid fishes, of which only the rainbow trout *(Salmo gairdneri)* still has some importance.

Apart from carp, which has been cultured from the beginning of this century, the interest in grass carp *(Ctenopharyngodon idella),* introduced since 1966 from Hungary and Taiwan, is increasing for both recreational fisheries and aquatic weed control.

Specific propagation and production methods have been developed for these two warmwater cyprinid fishes, using a combination of different fish husbandry systems. This is carried out at the fish culture center "Lelystad" of the Organization for Improvement of Inland Fisheries and will be described below.

Technical Characterization of the Fish Culture Facilities

Pond Facilities

The pond farm is situated in a recently reclaimed area (Flevopolder). The level of the pond farm is about 4 meters below the water surface level of the neighboring largest Dutch freshwater lake (Lake IJssel, ± 200,000 ha). Water is supplied from the lake by means of a siphon with a capacity of 1,500 l/s. By gravity flow the ponds can be separately filled and drained, the latter by discharging the water into the drainage

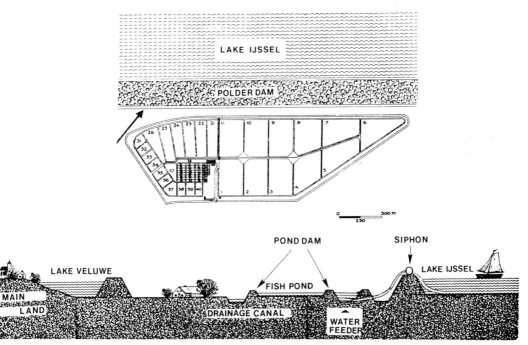

FIGURE 1. — Ground plan and schematic cross section of the pond farm "Lelystad" situated in the Flevopolder.

canal surrounding the pond farm. The layout of the farm and a schematic cross section indicating the direction of the water flow are given in Figure 1.

The total surface of the pond farm amounts to 217 ha. The lay-out of the different pond categories originated from a carp-farming management in stagnant water with a three-year production cycle for carp with a final body weight of 1,000-1,500 g. The ponds have an adjustable water level, but on the average the water depth is 1.2 m. The following types and series of ponds are available:

20 Dubish spawning ponds of 50 m² each

10 nursery ponds of 1.5 ha each

6 fattening ponds of 5 ha each

11 fattening ponds of 11 ha each

24 experimental ponds of 0.2 ha each

8 experimental ponds of 0.1 ha each

8 ponds of 160 m², each with running water (6 l/s) for trout production.

The carp farming management can be characterized by the following data:

The growing season outdoors is climatologically limited to about 4 months.

On the average the natural pond productivity amounts to about 150 kg/ha.

Fertilization with P and N increases natural productivity by 100 percent or slightly more.

Additional feeding with low-protein, high-energy feeds, which is usually practiced, results in a production of 1,000 to 1,500 kg/ha.

Farm management results in a 1-year-old carp of 25 g, and a 2- and 3-year-old carp of ± 350 and ± 1250 g, respectively.

In general, pond farming of carp in The Netherlands is marginal because of the climatological conditions. This can be illustrated by the fact that the weight of a 1-year-old carp may vary from 5 to 50 g in different years, depending on prevailing summer conditions.

Hatchery Facilities

The hatchery, situated in the center of the pond farm, is multipurposely used for routine reproduction, incubation, and fry rearing of rainbow trout, pike, carp, and grass carp, while on an experimental scale, the propagation of rudd (*Scardinius erythro-*

phthalmus), European catfish *(Silurus glanis),* and pike perch *(Stizostedion lucioperca)* is carried out. For these purposes the hatchery is in operation from November until July-August.

The hatchery has a surface of 2000 m² and is constructed as a greenhouse for the following reasons, valid for the Dutch circumstances:

Water evaporation and condensation related problems solved technically by the greenhouse construction;

Full and low cost advantage of daylight;

Low construction costs per unit of surface area.

These advantages outweigh the increase in heating costs during the winter because of the relatively poor insulation.

For incubation of eggs, 24 so-called improved 17-liter elf jars are available. This type of jar is used successfully for eggs of pike, carp, and grass carp. For eggs of trout and European catfish, 20 Californian incubation troughs and 3 incubation cases of the Swiss Veco type are in use.

Sac fry of trout, European catfish, and pike are reared in Californian troughs, but the sac fry of cyprinid fishes are reared in small stainless-steel wire cages floating in rectangular basins with running water.

For nursing the fry up to the fingerling stage of a few grams, 12 aquaria of 160 litres and 134 fiberglass reinforced polyester basins — 110 rectangular and 24 circular ones of 3.5 and 2.5 m³, respectively — are present.

The water supply to the hatchery comes from a sedimentation pond, which in turn receives its water from the Lake IJssel by means of the siphon. Before used in the fish-rearing units, the water is processed in 5 parallel units (spreading risks). In each unit the water is pumped through a high-rate filter, with both a surface diameter and a column height of 1.5 m and a capacity of 80 m³/h. After filtration, aeration takes place by means of a U-tube oxygenation system (Speece 1969). The total tube length is 20 m. The inner and outer tube diameters amount to 175 and 350 mm, respectively. The difference in height between inlet and outlet is 1 m. Each tube satisfactorily aerates 80 m³/h. The heating unit with a total capacity of 1.7 × 10⁶ kcal/h is placed after the aeration in order to optimize the partial oxygen pressure in the water. A part of the total water intake is heated and cold and hot water lines run into the hatchery, in which the water temperature is thermostatically controlled, offering possibilities for 13 independent different rearing temperatures.

The water discharged from the fish basins is re-collected within the same unit and can be reused after passing again through the water processing plant mentioned. By means of an adjustable valve between the return pipe and the suction pipe of the pump, the intensity of water reuse can be varied at desire between 0 and 100% minus evaporation and water spoilage. A scheme of the water processing is given in Figure 2.

Cage Facilities

In the near vicinity of the pond farm a conventional electrical power station of the Electrical Power Company of the Province Gelderland (PGEM) is located on an artificial island in the Lake IJssel. Total installed capacity of the station is 846 MW and the discharged water flow at full load amounts to 92,000 m³/h (average velocity in the cooling water canal: 16 cm/s). Because of Dutch legislation concerning cooling water discharges, the temperature of the cooling water can be raised 8°C above the ambient water temperature with an absolute maximum of 30°C, while during winter the ΔT may be 12°C.

By consequence the temperature of the discharged cooling water increases from 18°C in April to 26-30°C during midsummer and decreases again to 18°C at the end of September. During winter the water temperature drops to a minimum of about 12°C during January. The situation of the power plant and the site of the cages are indicated in Figure 3.

The above mentioned temperature profile of the discharged cooling water enables a thermal complementary fish culture, offering opportunites for warmwater cyprinid fish culture from April till September-October and for coldwater trout culture from October till March.

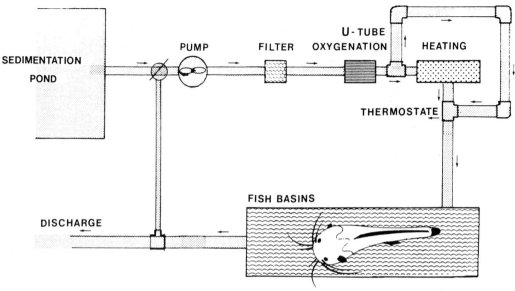

FIGURE 2. — Scheme of water processing in the hatchery.

This type of fish culture is practiced by the Organization for Improvement of Inland Fisheries in floating rafts, each unit containing 4 cages of 6.5-m³ volume each (Fig. 4). The floating units are constructed and linked together in a way enabling access for a small tractor to facilitate the distribution of fish and food as well as the harvesting procedures. Moreover the cages are constructed in such a way that they can be lifted up and turned over easily for harvesting/removal of the fish. A plasticized metal wire with a mesh size of 1.3 by 2.6 cm used in mink farming is ap-

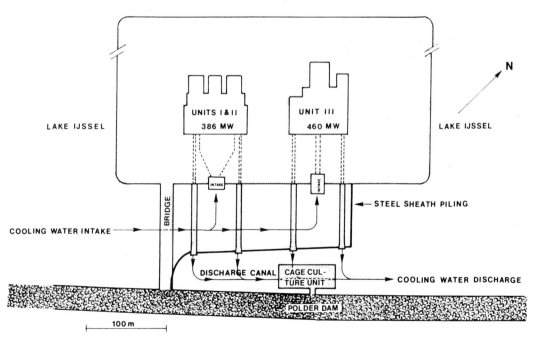

FIGURE 3. — Ground plan of the power station and the cage culture unit in the cooling water discharge canal.

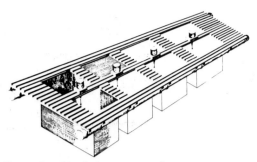

FIGURE 4. — Floating cage construction.

plied as cage material, and for small fish a nylon inner net of suitable mesh size is used in addition. In order to reduce when necessary the water current in the cages, adjustable "Venetian blinds" are mounted to the upstream side of the floating rafts.

Feeds are administered to the fish by means of demand feeders or — and more especially for small fishes — automatic feeders operating like conveyor belts.

This type of fish culture was started in 1972 and has been steadily expanding since. At the moment 80 cages are in use at the mentioned location. An annual labor input of 3 man-years is sufficient to manage the cage culture unit the whole year round.

Results of Fish Culture Management for Carp and Grass Carp

Hatchery and Nursery Operations

Artificial reproduction of carp and grass carp is carried out in the hatchery during winter to ensure the availability of 1.5-2.0-g fingerlings at the start of April to be stocked in the cage culture unit.

For the purpose of reproduction in December-January adult fishes are taken from the outdoor ponds and transferred into the hatchery basins. Subsequently the water temperature in the hatchery is raised by 1 °C/day from about 4 °C (depending on outdoor weather conditions) to 23 °C. The latter temperature is maintained for 2 weeks in the case of carp and for 4 weeks in the case of grass carp. After these respective periods hypophysation can be carried out successfully. When the temperature

regime mentioned is given at a later time (for instance, spring), the 2-weeks (or 4-weeks) period can be shortened. Detailed data on the calculation of the temperature regime prior to hypophysation have been given for carp and grass carp by Huisman in 1974 and 1979, respectively.

Hypophysation and stripping of spawner fishes, as well as fertilization and incubation of eggs, are conducted as published earlier (Huisman 1976).

Fertilization and hatching rates under production scale conditions vary between 63-87% and 25-47% respectively, without big differences between the two species.

Hatched larvae are transferred by the overflow of water out from the incubation jars into the fine mesh stainless-steel wire cages. These cages float in the hatchery basins supplied with running water and are used only during the sac-fry period and the very first hours of exogenous feeding. A suspension of hard-boiled egg yolk is used to determine the moment at which the fry are able to take up food. This is shown by the yellowish color of the belly.

At that moment the fry are stocked in 120-l aquaria in densities of about 40,000 fry an aquarium. These aquaria are used during the first 5-7 days of exogenous feeding, when freshly hatched nauplia of *Artemia salina* are fed exclusively. Afterwards the *A. salina* are steadily replaced by a pelleted trout starter or by small zooplankton, if the latter can be caught in the outdoor ponds — sometimes impossible because of an ice cover. Afterwards the fry are stocked in the hatchery basins in densities up to 100,000 fry a basin.

During this whole process the water temperature is maintained at 23-25 °C. The water supply is adjusted at least twice a week according to the increasing biomass to provide a minimum dissolved oxygen content of 4 mg/l at the discharge of the basins.

In this way millions of fry and fingerlings have been produced and the efficiency of this type of propagation of cyprinid fishes can be illustrated by the following data.

In Table 1 the growth pattern of grass

TABLE 1. — Growth of grass carp fry under controlled hatchery conditions.

After	Body weight in mg[a]
1 week	30-55
2 weeks	65-137
3 weeks	127-270
4 weeks	234-484
5 weeks	425-764
6 weeks	722-1168
7 weeks	1180-1652

[a]Weight range depends on differences in fish densities.

carp fry under controlled hatchery conditions is given. Growth of carp fry is in the same range.

Both for carp and grass carp on the average every hundred of stripped eggs will result in 20-30 fingerlings of 1.5-g body weight within a 50-day feeding period.

The whole operation from the intake of spawners into the hatchery until a fingerling of 1.5-g size is reached takes approximately 3 (carp) to 3.5 months (grass carp) during the winter season and can be shortened in spring and summer according to the previous outdoor temperature regime.

During the phase of nursing feed conversions using pelleted trout starter vary between 1.1 and 1.4.

Nursery results compare favorably with those reported by other authors (Appelbaum 1978; von Lukowicz 1979; Schlumberger, Anwand, and Mende 1976). In this respect, two major items may be of importance.

First, feeding young fry with artificial feeds supposedly starts with a learning process in which the fry has to recognize a dead particle as a palatable prey. This process will be accelerated by high densities of both fry and prey, thus maximizing the chance for a fry to encounter a prey (pellet). Fry densities in the aquaria therefore have to amount to several hundreds of fry per liter volume. Moreover, such a learning process takes time, during which energy expenditure (maintenance metabolism, activity, etc.) inevitably takes place. The bigger the fry will be, the more learning time will be available before the point of no-return because exhaustion will be reached. Therefore, the changeover from life food to pelleted feeds has to take into account the size rather than the age of the fry.

Secondly, it is clearly illustrated in Table 1 that the specific growth rates during the very first days of exogenous feeding are extremely high, up to 100% of body weight per day. As a consequence of this it should be strongly stressed that during this very first period the daily food ration has to be several times the total fry biomass.

Reviewing the literature of artificial feeding in fry rearing experiments it is striking how often these goals are not accomplished.

Cage Culture Operations

Hatchery-reared fingerlings of carp and grass carp are transferred to the cages at the beginning of April preferably. Initial stocking densities are up to 2,500 fingerlings per cubic meter cage volume. Initial stocking densities are adjusted to the increasing body weight of the fish. Detailed data on cage culture of grass carp are given by Huisman (1979).

Both carp and grass carp are fed exclusively with pellets using demand feeders. The pellet used has been compounded on the basis of a commercial trout feed mixed with an equal portion of wheat flour before pelleting and the proximate analysis is given in Table 2.

Under the Dutch circumstances grass carp is stocked at a body weight of ± 350 g in the early spring for recreational and/or water managerial purposes. Carp is used in recreational fisheries management at a weight of ± 350 g and ± 1250 g to be stocked in the autumn.

Since during winter no growth is realized, the goal of cage culture is to produce a marketable product before the end of the propagation period, which is climatologically set at September-October. To reach this goal the fish stocking density is used to control the individual growth rate within the limits of prevailing water temperature, water quality, etc.

Disease problems so far encountered in

TABLE 2. — Proximate analysis of the feed used for carp and grass carp production.

Dry weight (%)	Composition of dry matter (%)					cal/g dry matter
	protein	fat	fibre	carbohydrate	ash	
87.18	33.90	6.80	2.10	47.53	9.72	4377

the cage culture are mostly mild parasitic infestations, which were easily cured, and infestations with *Myxobacteria,* as for instance *Flexibacter columnaris.* The latter infection proved to be more serious for carp than for grass carp. Routine therapeutic measures comprise the use of Terramycine® in the feed and/or short term baths with Furanace® (Bootsma 1976).

As an overall characterization of carp and grass carp culture in cages several relevant fish cultural parameters are given in Table 3. These figures are based on production data collected for the periods April to October during the last five years, in which several hundreds of tons of fishes were produced at different stocking densities. Apart from these average production data, the production capacity of this very intensive type of fish culture is illustrated by some data on maximum production figures achieved under experimental management in the same facilities. These results indicate that the level of intensification of grass carp culture in running water facilities as used by Shiremen, Colle and Rottmann (1977, 1978) can be increased successfully for large scale production purposes.

Pond Utilization for Carp and Grass Carp Culture

The use of the outdoor ponds in the integrated approach of fish culture as described here is limited but still of major importance.

For grass carp culture — even during summer — the outdoor ponds do not meet the temperature requirements for a satisfactory growth performance. Still they serve three main purposes.

First, the ponds are well suited as holding facilities for spawner animals.

Secondly, the ponds can serve as a source of zooplankton for the fry rearing phase in the hatchery.

And third, during winter the ponds are used as holding facilities for the grass carp raised in the cages, as the growing period ends at October but stocking in Dutch waters takes place not earlier than March for logistic reasons. Even sometimes in summer the grass carp are to be transferred from the cages into the ponds when the desired body weight is already reached in the cages before the end of the growing period. In this way the ponds are used to slow down or stop the growth rate.

TABLE 3. — Average and maximum production figures in kg/m³ cage volume, feed conversion values and losses of carp and grass carp in cage culture (1974-1978).

	Carp		Grass carp	
	Average	Max.	Average	Max.
Fish density at harvest for 350-g fish	169	317	143	166
Fish density at harvest for 1250-g fish	191	349		
Weight gain Apr.-Oct. for 350-g fish	156	284	136	150
Weight gain Apr.-Oct. for 1250-g fish	170	265		
Feed conversion (kg feed/kg gain) for 350 g-fish	1.55		1.80	
Feed conversion (kg feed/kg gain) for 1250 g-fish	2.19			
Percentage of losses	14.3		13.8	

The above mentioned functions of the outdoor ponds are also valid for carp. However, under the Dutch climatological conditions carp still have an acceptable — although marginal — growth rate in outdoor ponds, while grass carp have not. This implies that reproduction, fry rearing, and ongrowing of carp in outdoor ponds is still possible. But this process of carp production can be accelerated considerably in time by integrating the hatchery and cage culture procedures in addition.

Based on the results of a three years monitoring of labor input, food utilization, amortization of investment and other cost determining parameters the best strategy for the culture of carp and grass carp at the "Lelystad" fish farm is likely to be reproduction and nursing in the hatchery from December till March/April, followed by a cage culture phase during summer up to an individual body weight of about 350 g, with a consecutive wintering in outdoor ponds.

Conclusions and Remarks

Based on the results mentioned the following conclusions and remarks can be made:

(1) Reproduction under environmentally controlled conditions can be carried out during winter as well as summer in a successful way both for carp and grass carp.

(2) Intensive cage culture in warm water effluents using pelleted feeds and demand feeders is an adequate fish husbandry technique for both carp and grass carp. For the latter species this may be the only feasible type of production under cool climatological conditions.

(3) The results mentioned raise doubt on the strict herbivoral character of grass carp and indirectly support the ideas of Fischer (1973), that optimal diets for grass carp must include relatively high percentages of animal matter to ensure adequate growth performance.

(4) The integration of hatchery, cage and pond culture procedures for carp and grass carp production offers great possibilities for a flexible intensive culture of both species, especially in climatologically moderate regions of the world.

References

APPELBAUM, S., AND U. DOR. 1978. Ten day experimental nursing of carp *(Cyprinus carpio* L.*)* larvae with dry feed. Bamidgeh 30 (3):85-88.

BOOTSMA, R. 1976. Studies on two infectious diseases of cultured freshwater fish. Thesis, State University of Utrecht, The Netherlands. 90 pp.

FISCHER, Z. 1973. The elements of energy balance in grass carp *(Ctenopharyngodon idella* Val.*)*. Part IV. Consumption rate of grass carp fed on different type of food. Pol. Arch. Hydrobiol. 20(2):309-318.

HUISMAN, E.A. 1974. (A study on optimal rearing conditions for carp *(Cyprinus carpio* L.*))*. Thesis State Agriculture University of Wageningen (The Netherlands). 95 pp. (In Dutch with English summary).

———. 1976. Hatchery and nursery operations in fish culture management. Pages 29-50 in E.A. Huisman, ed. Aspects of fish culture and fish breeding. Misc. Papers, No. 13. State Agriculture University, Wageningen, The Netherlands.

———. 1979. The culture of grass carp *(Ctenopharyngodon idella* Val.*)* under artificial conditions. Proc. EIFAC-Symposium on finfish nutrition and feed technology. Hamburg 1978. Heenemann Verlaggesellschaft mbH. Berlin. (in press).

SCHLUMPBERGER, W., K. ANWAND, AND R. MENDE. 1976. Erfahrungen bei der Brutaufzucht von Amurkarpfen *(Ctenopharyngodon idella* Val.*)* und Silberkarpfen *(Hypophthalmichthys molitrix* Val.*)* mit verschiedenen Trockenfuttermitteln. Zeitschr. Binnenfisch. DDR. 23(6):164-174.

SHIREMAN, J.V., D.E. COLLE, AND R.W. ROTTMANN. 1977. Intensive culture of grass carp, *Ctenopharyngodon idella,* in circular tanks. J. Fish Biol. 11(3):267-272.

———. 1978. Growth of grass carp fed natural and prepared diets under intensive culture. J. Fish Biol. 12(5):457-463.

SPEECE, R.E. 1969. U-tube oxygenation for economical saturation of fish hatchery water. Trans. Am. Fish. Soc. 98:789-795.

VON LUKOWICZ, M. 1979. Experiments with first feeding of carp fry with Alevon and freeze-dried fish. Proc. EIFAC Workshop on mass rearing of fry and fingerlings of freshwater fishes. The Hague.

Bio-Engineering Symposium for Fish Culture (FCS Publ. 1):274-281

A Perspective on the Salmonid Enhancement Program in British Columbia

ALAN F. LILL, P. ENG.

Salmonid Enhancement Program
Department of Fisheries and Oceans
1090 West Pender Street
Vancouver, B.C. V5E 2P1

ABSTRACT

The Salmonid Enhancement Program in British Columbia has received a seven-year commitment of $157.5 million by both Federal and Provincial governments. It is designed to reestablish anadromous salmonid populations at historic levels, thereby resulting in significant economic and social benefits.

This paper discusses the planning and implementation processes in the Program, which are governed by the three major criteria areas of stock manageability, enhanceability, and socioeconomics. It outlines some of the enhanceability methods currently in use and the results to date.

A program of this size has, in addition to its successes, its problems, some of which are important concerns of the bio-engineer. From experience with this Program, it is apparent to the author that the bio-engineer must look both beyond and within the fish culture facility to assess his effectiveness and explain his mission to those that pay the bills. If he fails to do either, the future of artificial enhancement as a means of maintaining and expanding anadromous salmonid stocks is in doubt.

Introduction

The Salmonid Enhancement Program (S.E.P.) in British Columbia was approved by the Canadian Cabinet in May 1977 in two phases. Phase One was to last five years (now seven years) and cost $150 million. If the economic, social, and environmental benefits projected in the Program justification document (DFO 1978) appear to be coming on-line, the Cabinet agreed in principle to the continuation of the Program into Phase Two.

In March 1978 a Federal-Provincial Agreement was signed and Provincial participation confirmed at a rate of five percent of the Federal contribution (or $7.5 million).

The Program is now in its third year and has reached an expenditure rate of about $26 million per year, which corresponds roughly to the annual average commitment to the end of Phase One.

The Plan

From study of historic levels of the British Columbia commercial salmon catch (Fig. 1), it is believed that the marine environment can support approximately twice the present stock size, and this level was chosen as a long-term overall objective. In a two-year planning period prior to the start-up of the S.E.P., integrated multidisciplinary geographic working groups were formed. They established preliminary long-term targets and a development strategy for every species (broken down by principal geographic areas and fisheries). The potential enhancement capability of many of the 1,800 salmonid-producing rivers in British Columbia was reviewed in the process, and computer project proposals prepared using "standard" costs and production rates. The projects were priorized, and the most desirable early projects were included in the Program justification document.

The geographic working groups had to keep both "enhanceability" and "manageability" in mind, while developing their plans. From Figure 1, it is apparent that the present salmonid stocks in British Columbia are of a highly natural origin, and it was fundamental that the "natural" stocks of lower productivity be protected from high exploitation in mixed stock fisheries designed to harvest "enhanced" salmonids (Wood 1979).

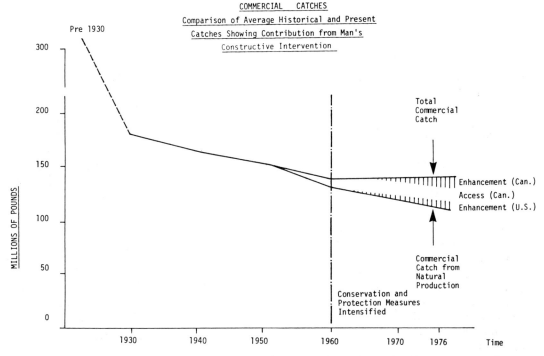

FIGURE 1. — Ten-year average annual commercial catch of salmon in British Columbia with recent sources of production indicated.

Project scale, order, and location became exceedingly important in developing the Program, with projects generally falling into two categories: "prerequisites" or "opportunities." A "prerequisite" project involves upgrading weak, natural stocks before enhancement of target species in a given fishery can take place. "Opportunity" projects are ones that can deal with a total fishery management unit and can be fished without impact on unenhanced natural stocks. Bio-engineering reconnaissance and previous experience yielded "enhanceability" information for each alternative, which was then assembled for socioeconomic evaluation.

A complex socioeconomic evaluation is undertaken for all project proposals, and these evaluations are recycled as often as necessary until a feasible project proposal has been developed.

It is important for all participants in the planning process to have a basic understanding of all three criteria areas, so that viable proposals are developed with a minimum of recycling and confrontation.

The socioeconomic criteria fall into two major categories. The Federal Treasury Board has stipulated guidelines and set discount rates by which projects must be evaluated in dollar terms (Planning Branch 1976). All measurable benefits and costs are considered in this category called "National Income." Unfortunately, these standards tend to mitigate against long-term primary resource production (such as salmonid enhancement) because multiplier effects are not allowed and the primary evaluative discount rate is ten percent. (However, the Program is currently projecting a benefit/cost ratio of approximately 1.8:1, which is very healthy by these standards).

The second major area is in the social and environmental considerations, which do not lend themselves easily to evaluation in dollar terms. Projects are assessed subjectively and relatively, with respect to employment, regional economic development, Native people, and the environment (Barclay 1978; DFO 1978).

Obviously, past experience with various technologies and the promise of new methods coloured the original projections. However, it was recognized at the outset that the Program needed to be flexible and innovative. Therefore, a fixed, long-term plan was not appropriate.

Methods and Results

Enhanceability Analysis

In choosing technique(s) for a given situation, it is usually helpful if the bio-engineer follows a checklist or "prescription" approach to the watershed. He should attempt first to assess the limiting factors on production in the current situation. After a given solution is tried, he should reexamine the impact of the carrying capacity of the environment at all other stages in the life cycle.

During this exercise, the bio-engineer must project survival at each stage of the life cycle for each alternative. As is common practice elsewhere, in S E.P. we have developed production standards for use in consistent evaluation of all project proposals. These standards are regularly revised and updated when new knowledge of survival rates becomes available.

To facilitate costing for each alternative, the Program engineers developed cost "standards." While these were used early in the S.E.P., current practice for major facilities is to draw up a conceptual design and make a cost estimate on latest experience by elements for a given location.

Our older cost standards are still used as a check, but the trend to mixed technologies within a single facility has made our original standards largely obsolete.

Fishways, Obstruction Removals, and Access

Prior to the S.E.P., approximately twenty major fishways were constructed in British Columbia to overcome migration problems at man-made and natural obstructions. All but two were considered successful, and considerable potential re-

mains for more, although most opportunities are on smaller streams or on rivers requiring transplants because of barriers near the mouth.

Jim Wild of the S.E.P Engineering Division has developed a new technique for building fishways in rock in isolated areas. A trench is provided by preshearing the wall line and then pattern-blasting to remove the rock. Precast, vertical-slot baffles are set in the trench and anchored by poured-in-place concrete pilasters (DFO 1978b). The technique has significantly reduced site time and costs over those of a conventional fishway.

Habitat Upgrading

The S.E.P. is investing heavily in a variety of pilot-scale projects involving side channels, boulder placement, cover provision, flow enhancement, etc. Many of these projects are being done on a volunteer basis (DFO 1978b, 1979).

Spawning Channels

The spawning channel remains an important tool for chum, pink, and sockeye

FIGURE 2. — Little Qualicum River spawning channel: length, 4400 m: width, 8 m; capacity, 50,000 chum spawners.

FIGURE 3. — Capilano Hatchery.

enhancement, where adequate land and water supplies are available. It has been most recently applied at the Little Qualicum River (Fig. 2) in a three-million-dollar project for chum salmon.

Fish Cultural Techniques

The greatest variety of techniques with the largest potential impact on salmonid survival is in this area. Classification of facilities by technique is difficult at best, because many facilities encompass many techniques at one location, and many life stages are covered. In general, it is best to consider alternatives at each life stage and assemble the best combination for all species under the circumstances.

Egg Incubation

The three major techniques are in common use — upwelling incubation boxes (usually used in isolation and for small projects), modified Atkins boxes (used for large batches of eggs as a part of the "Japanese system" [Hokkaido 1976]), and the well known "Heath" tray (used for "low-risk" sites, smaller individual egg takes, and areas with seasonal siltation

where washing prior to hatching is desirable).

Alevin Incubation

It has been found advantageous, in many circumstances, to handle the eggs after eyeing, and move them to a secondary incubator. Eyed eggs held in Atkins boxes are moved to alevin raceways, as in the "Japanese system" (Moberly and Lium 1977).

Late eggs are also being moved to Heath trays at Capilano Hatchery, where the timing of earlier eggs permits, thereby doubling the use of the trays. A unique application to S.E.P. is at our Pallant Hatchery, where we opted to start chum eggs in Heath trays and then moved them to shallow-matrix incubation boxes upon eyeing. It was felt that the gravel box would provide a better quality fry, but that is prone to clogging from siltation. At Pallant, it was possible to wash the eggs when eyed and then plant them in the gravel box after the time when siltation from fall storms had passed.

Freshwater Rearing

Conventional "Burrows" type ponds are in use at our pre-S.E.P. facilities and are producing well for us at the Capilano Hatchery (Fig. 3). However, an interesting experiment (Sandercock and Stone 1979) indicates that half the normal loading by surface area will produce the same returns. We have long felt that environmental conditions during rearing have an impact on ultimate survival, and have experienced excellent returns from seminatural, gravel rearing channels, with relatively low densities (Fig. 4). Our thinking, at present, is to use these channels (wherever adequate water supplies and topography permit) for rearing coho, chinook, and steelhead. We often will first pond our fry into troughs, and then subsequently transfer them to the channels or to more conventional concrete raceways. Lately, we have been designing our raceways for both adult holding in late summer and autumn, and for rearing chinook fry in the spring. This system usually works quite well, and we have in-

FIGURE 4. — Puntledge Hatchery, upper site. Note the gravel rearing channel with automatic feeder stations.

corporated it into our Puntledge Hatchery (Fig. 5).

For chum salmon, we have generally been following standards proven in Japan until such time as Canadian experience is available. The Program has constructed several facilities on groundwater supplies, so that egg and alevin development is accelerated, thereby allowing time for freshwater rearing to increase the size at release, without changing the normal timing.

Centralization and Satellites

In keeping with the manageability restraints placed on our developments, we have opted for a central hatchery with satellite release/adult capture sites such as at Tlupana Inlet (Fig. 6). The program is adopting a conservative attitude with respect to maintaining independent stock strengths in a common fishery zone, and strives to return fry from the donor streams and avoid transplanting, wherever stock levels warrant.

The collection of eggs from many streams for the central hatchery is an expensive and often difficult proposition. The Program has resorted to all of the conventional techniques and has experimented with various types of fences, such as permanent overflow/flow-through combination units with or without downstream fry-trapping gear (Fig. 7 [Lill, A.F. and P. Sookachoff 1974]). Usually, formal fences and holding facilities are prohibitively expensive in our flashy coastal rivers, but often there is no alternative if an egg supply is to be secured.

Lake Enrichment

Perhaps the most exciting development in recent years is the phenomenal success

FIGURE 5. — Puntledge Hatchery, lower site.

of our lake fertilization program on improving sockeye fry survival in certain lakes, primarily on Vancouver Island (Stockner 1979). This tactic is now being employed on virtually every feasible lake that meets Program economic and stock manageability criteria.

Transition to the Marine Environment

It is generally accepted that substantial losses do occur after emergent chum and pink fry leave freshwater and enter the marine environment. A pilot project in Japan (Iioka et al. 1977) has overcome this problem by a strategy of taking chum salmon directly from freshwater to saltwater pens. In 1979, this strategy was first tried on a production scale in Canada at our Pallant Hatchery on the Queen Charlotte Islands, without any serious fish cultural problems.

Marine Environment

The real success of the Program is most often tested in the marine environment where the production to fisheries predominantly takes place. It is crucial to be able to evaluate our strategies in the freshwater environment against this benchmark. The complexity of environmental factors and jurisdictions gives little potential to date for artificial manipulation during this phase of the life history, except in the areas of predation control and in fishing strategies.

Discussion

The Salmonid Enhancement Program is unique for a government program, in that it is funded in a controlled allotment from the federal government, and is subjected to far-ranging evaluation procedures, in order to determine its continued long-term

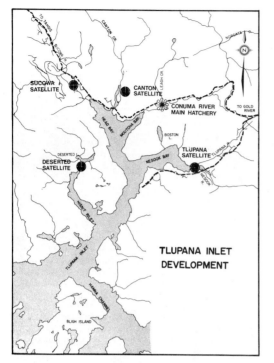

FIGURE 6. — An example of a central hatchery with satellites serving several stocks fished in a common area.

viability. It follows that there are the formal economic evaluations undertaken by government economists, but perhaps more importantly, there are informal evaluations ongoing with our client fishermen and the general public.

This atmosphere is very much one of "results-motivation" and puts considerable pressure on the technical staff responsible for design, construction, and operations of artificial enhancement facilities. From the outset, all staff members have been encouraged to measure output in terms of catch and escapement, not in terms of numbers or weight of smolts leaving the facility.

In order to understand our effectiveness in economic terms, it is very important for all bio-engineers to understand basic economics, particularly the principles of "present worth" of a stream of investments or benefits, over time. Considerable confusion exists in this area, and it often causes unnecessary conflict between disciplines in the assessment of alternative strategies of investment.

FIGURE 7. — Little Qualicum River diversion fence. The sketch shows the fence with gratings removed. During storms the flood waters and debris can pass over the fence. Salmon are diverted into the fishway on the left.

Another significant factor in terms of acceptance of the Program by our clients is the government commitment to "cost recovery." Commercial and recreational fishermen are expected to pay for the S.E.P. (in terms of increased special taxes), so they must receive an overall benefit from the investment.

Among the general public of British Columbia (especially steelhead fishermen), there is a vocal body of opinion against hatcheries and for "emulation of nature." The S.E.P. has conducted two province-wide public inquiries, and this opinion is often voiced at these gatherings (DFO 1980). The staff bio-engineering groups responsible for hatcheries have begun an educational process with a province-wide public advisory body to the S.E.P. It will be a long struggle to convince these people of the strides that are being made in improving the reliability of hatcheries, and any failures will be seen as a confirmation of the worst fears of the concerned public.

The mixture of social and economic concerns that shape policy are represented in the Board of Directors of the Program, which reports to the federal and provincial ministers. This Board has both client and government representation, and fulfills a very essential role in keeping the S.E.P. on track. They have grown to expect high performance from the S.E.P. professionals, and for the most part, this performance has been achieved to date.

From experience with the S.E.P., I have concluded that the challenges that face the bio-engineer are immense, but potential rewards are great. To succeed, he must watch his costs, maximize his benefits (from *added* production), and be in tune with the public and client groups.

References

BARCLAY, J.C. 1978. Benefits and costs of salmon enhancement — The British Columbia case. Presented to the Joint Annual Conference of the Western Association of Fish and Wildlife Agencies and the American Fisheries Society, San Diego, California.

DEPARTMENT OF FISHERIES AND OCEANS. (DFO). 1978. Salmon Enhancement Program. Vancouver, British Columbia.

——. 1978b. Salmonid Enhancement Program. Annual Report 1977. Vancouver, B.C. 84pp.

——. 1979. Salmonid Enhancement Program. Annual Report 1978. Vancouver, B.C. 92 pp.

——. 1980. Salmonid Enhancement Program. Update '78. Public inquiry submissions. Vols. 1 and 2. Vancouver, B.C.

HOKKAIDO SALMON HATCHERY. 1976. An outline of the management of salmon hatching and rearing. Fisheries and Marine Service, Canada. Trans. Ser. No. 4320. 9:1-43.

IIOKA, C., ET AL. 1977. An examination of the release of seawater-released fingerlings. Iwate Ken Suisan Shikenjo Hokoku-Sho. Dept. Fish. Oceans, Canada. Trans. Ser. No. 4384. 59 pp.

LILL, A.F., AND R. LIUM. 1977. Japan Salmon Hatchery Review. Alaska Fish and Game, Juneau.

LILL, A.F., AND P. SOOKACHOFF. 1974. The Carnation Creek counting fence. Fisheries and Marine Service Tech. Rep. No. PAC/T-74-2. 23 pp.

MOBERLY, S.A., AND R. LIUM. 1977. Japan salmon hatchery review. Alaska Fish and Game, F.R.E.D., Juneau. 124 pp.

SANDERCOCK, F.K., AND E.T. STONE. 1979. The effect of rearing density on subsequent survival of Capilano coho. Proceedings of the Northwest Fish Culture Conference. (abstract.)

PLANNING BRANCH, TREASURY BOARD SECRETARIAT. 1976. Benefit cost analysis guide. Information Canada, Ottawa. 73 pp.

STOCKNER, J.G. 1979. Lake fertilization as a means of enhancing sockeye salmon populations: The state of the art in the Pacific Northwest. Revised edition. Fisheries and Marine Service Tech. Rep. No. 714. 14 pp.

WOOD, F.E.A. 1979. Policy implication of the salmonid enhancement program. Fisheries course presentation, University of British Columbia. (mimeo.) 33 pp.

Bio-Engineering Symposium for Fish Culture (FCS Publ. 1):282-295

Design Features of a Carp Hatchery in Bangladesh[1]

JOHN R. SNELL

Snell Environmental Group, Inc.[2]
Lansing, Michigan 48906

A.Q. CHOWDHURY

Directorate of Fisheries
Dacca, Bangladesh

WILLIAM M. BAILEY[3]

Snell Environmental Group, Inc.
Dacca, Bangladesh

KERMIT E. SNEED

Snell Environmental Group, Inc.
Lansing, Michigan 48906

ABSTRACT

An irrigation and flood control embankment, which encircles a land area of 560 km², has been constructed in Bangladesh. The embankment, along with the river and floodwater regulators, pumping stations, and locks and dams, effectively limits the free-flow and major flooding of the South Dakatia River. These structures also inhibit the migration and dispersal of important economic species of fishes and prawns that formerly entered the area during the rainy season and were dispersed and left as natural stock in ponds, canals, and rivers after floodwaters receded.

The loss of natural recruitment of fry and juvenile fish from outside the irrigation and flood control area and the reduction in flooded areas that formerly produced foodfish indicated a need for a fish hatchery to produce fry and fingerlings of the most important fish species. These hatchery-reared fish are to be stocked in controllable waters to encourage managed fish culture in the 28 thousand small ponds in the project area to compensate for the deficit in former natural production.

The need for a flexible, multispecies hatchery was the primary design criterion for a faclty to produce 50 million fry. Additional criteria established requirements for a land area of 20 or more hectares for buildings and adjunct support facilities, including ponds to rear 25 million fingerlings of the major carp species, *Catla catla, Labeo rohita,* and *Cirrhinus mrigala.* Most of the ponds were to be drainable and relatively seepage-free in spite of variable soil textures. Local construction materials and technology were to be used to avoid expensive imports. A hatchery complex was designed that comprises a hatchery/laboratory building; nursery, rearing, and broodstock ponds; and support structures and facilities, such as deep and shallow wells, water processing plant, and water storage reservoirs. The heart of the hatchery is a system of circular tanks for spawning fish and a "jar" egg-hatching and fry-recovery facility.

Other design features include aeration and iron removal via multimedia filters and methods used to assure a continuous supply of water during power outages. Economy features concern filling and draining of ponds, earth moving and compaction methods, and use of local construction materials. Since 15-cm PVC pipe is the largest available in Bangladesh, optimum looping of the water supply network by computer analysis permitted passage of the required 4000 l/min to the ponds with low head loss.

[1]The Irrigation Fisheries Development Project is funded by the Government of Bangladesh and the International Development Association through the world Bank.

[2]Other Snell Environmental Group contributors to design and implementation are Thomas Powers and Sam Nalluswami.

[3]Formerly with the Arkansas Game and Fish Commission.

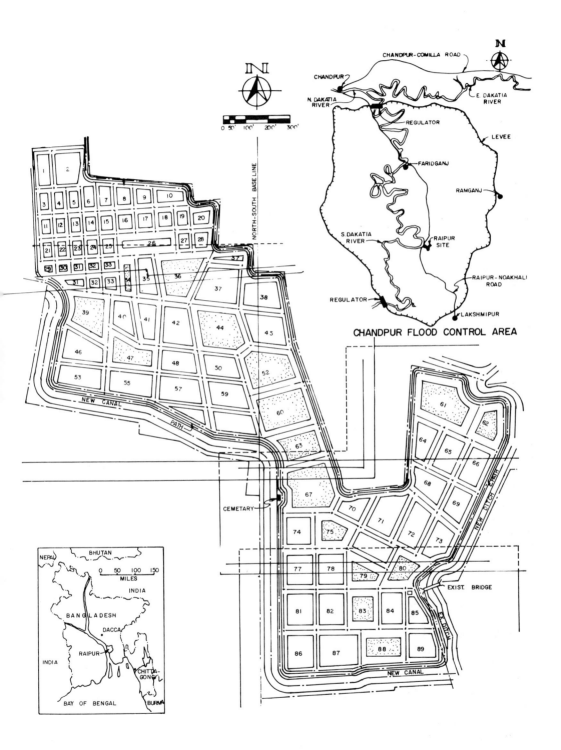

FIGURE 1. — Location maps of the Chandpur irrigation and flood control area and pond layout of the Raipur Fish Hatchery and Training Center. Those stippled are dugout, nondrainable ponds. All others are elevated and drainable.

Design Features of a Carp Hatchery in Bangladesh

The occurrence of frequent destructive floods in Bangladesh, particularly since 1954, motivated the Bangladesh Water Development Board (BWDB) to develop a master plan in 1964 to control future floods, with the aim also to provide irrigation facilities to increase agricultural production (Fig. 1). By June 1977 the BWDB had completed 34 flood control and irrigation projects that help control the waters on more than 800 thousand hectares. At present there are 109 ongoing irrigation and flood control construction projects, intended to bring an additional 360 thousand hectares under control.

Initially, for the first five or six years, little thought was given to one of the most important resources of the country — fish, prawns, and other aquatic food organisms. The Department of Fisheries (DOF) first expressed its serious concern in the year 1970 after the BWDB had initiated the construction program in the Chandpur Irrigation and Flood Control Project. The possible detrimental effects of the engineering design for irrigation and flood control to the existing important prawn and carp fisheries of the Chandpur area were identified. Subsequently, a study plan was prepared and proposed by the DOF to determine the status of fisheries in the area to be enclosed by a 96-km-circumferential embankment. This area includes approximately 360 km² of fertile agricultural land, as well as numerous irrigation canals and ponds, many of which can be managed to increase fish production.

The engineering and construction plan involved the construction of two dams (regulators) on the Dakatia River — one at Char Baghadi (includes a pumping station) just south of the forking of the river into the cast, south, and north branches, and the other located farther south, at Hajimara (Fig. 1, upper right). These two dams and regulators divert the monsoon rainwater to the other branches and on to the main stem of other river systems. During the monsoon season, excess rainwater falling inside the embanked area can be pumped out, or the excess water can be released downstream through the regulators if outside river water levels permit.

The embankment and regulators effectively cut off the free flow of the south branch of the Dakatia River inside the project area. At the same time these structures blocked the migratory and dispersal pathways of important economic species of fishes and prawns. These species spawn in free-flowing water or floodwater habitats, in brackish water farther downstream, or in the Bay of Bengal. Then the juveniles migrate to upstream river habitats where they complete their life cycle. Examples of such species are the giant freshwater prawn, *Macrobrachium rosenbergii;* hilsa, *Hilsa ilisha;* and the major carps: catla, *Catla catla;* rohu, *Labeo rohita;* and mrigala, *Cirrhinus mrigala.*

In addition to the design and construction of the fish hatchery complex, an extensive biological and fish and prawn recruitment and economic assessment program was conducted to determine the present status and future needs for the economic and important foodfishes of the various irrigation and flood control areas. Particular attention was devoted to the Chandpur project, which was completed and put into operation in February 1978. A survey was conducted to determine the number and size of ponds and other manageable water bodies that might be used for fish farming by private owners or by the DOF. This program also included a survey of the needs for extension services and the anticipated requirements for training of professional personnel and pond owners in the methods of aquaculture.

Design Criteria

Design criteria were formulated based on data from early surveys and feasibility reports, the assessment of the economics of the fishery, the surveys to determine recruitment of fishes from outside the irrigation area (before and after closure of the dams), and the number and area of water bodies that could be used in the future for managed fish culture. These baseline data are in *Working Documents,* printed by the DOF, Bangladesh, and listed in "References." Some of the most pertinent criteria established for design, construction, and operation of the hatchery were the following:

1. The pond survey revealed that there were as many as 28,000 small ponds in the project area, most of which can eventually be renovated and made suitable for improved fish culture. These ponds, together with irrigation canals and borrow ditches, offer about 5000 hectares of water area for fish management, and more ponds are being built.
2. Since these ponds, canals, and the now cut-off portion of the river would no longer be adequately stocked from natural sources, a flexible, versatile fish hatchery was needed with capability and capacity to handle as many as 100 million eggs to produce 50 million hatchlings or advanced fry or 25 million fingerlings of three major Indian carps. The hatchery required the flexibility to spawn and rear fry or fingerlings of other species; for example, tilapia, *Sarotherodon* spp; common carp, *Cyprinus carpio;* the river catfishes such as *Pangasius pangasius;* and particularly the grass carp, *Ctenopharyngodon idella,* and silver carp, *Hypopthalmichthys molitrix.*
3. The operational flexibility required to satisfy the second criterion indicated that a moderate level of sophistication be built into the facility, especially the hatchery building itself, even though the operational personnel will need training in the use of complicated mechanical and electrical equipment.
4. The very high initial cost plus custom duty and taxes on imported supplies, materials, and equipment dictated the use of locally available building materials, labor capabilities, and construction methods.
5. It was necessary to select a site that was already in the public domain or one that could be purchased by the DOF. The

subsoil structure should have physical properties for support of two-story, concrete-masonry buildings, and permit economic construction of reasonably watertight ponds. Both surface and groundwater of good quality for fish culture and in sufficient quantity (4000 l/min minimum) had to be available. It had to be located in the project area, be accessible via all-weather road, and have reasonable availability to infrastructure such as vehicle parts and repair shops, fuel, electricity, food markets, and telephone. The lack of some or all of this infrastructure in most rural areas of Bangladesh mandated the design, construction, and equipping of a facility to be as self-sufficient as possible in the way of parts and repair, standby and backup units of equipment, as well as for quarters for all supervisory and operational personnel.

Site Selection

In order to fulfill the biological, operational, and production requirements of the hatchery, as established by fishery biologists and fish culturists in the DOF and Snell Environmental Group (SEG), it was necessary to select a suitable site with a usable area of 20 or more hectares. An area of that size was calculated to be sufficient for the hatchery building and adjunct support structures and facilities, levees, roadways and drainage canal, leaving an area large enough on which to build ponds having a water area of about 13 ha. This pond area was not considered sufficient to produce the required 25 million fingerlings, even though the hatchery (i.e., jar and circular tank system) could produce the necessary 50 million fry. However, DOF plans envisioned the recruitment and training of owners of the largest and best ponds to rear fry to desirable stocking size for sale to other pond owners, a procedure that is very profitable in Bangladesh and is presently practiced on a small scale.

After investigation of numerous sites, a tract of land (approximately 21 ha) was selected near the town of Raipur within the

FIGURE 2. — Artist's concept of main hatchery building.

project area (Fig. 2). This land was judged to meet most of the stated criteria, even though certain deficiencies in soil texture, the existence of an irrigation canal traversing the center area, and the presence of squatters and two graveyards (socio-religious problem to relocate) presented several engineering, social, and public relations complications.

Design Features

Fish cultural facilities must be designed and adapted to the requirements of the species to be produced. The physical plant must be efficient and of adequate size to meet the projected production goals, with realistic views of the expected level of management of operating personnel. The facility should also be adaptable to modification to permit production of other species as aquacultural practices and management needs change. The design of the proposed facility in the Chandpur Irrigation Project area is primarily for production of flowing water spawners with semibuoyant eggs, but the rearing facilities are, for the most part, all-purpose ponds that may be used for pond spawning of species that do not require hormone inducement or flowing-water, such as common carp or tilapia. The facilities were designed with production foremost in mind, with training of fishery workers and pond owners a related but equally important activity.

Site Plan

A casual layout of ponds and structures on the site was impossible. The size, configuration, and type of pond had to be designed and sited to conform to the existing characteristics of the land, i.e., soil types, the scarcity of soil and clay to build the elevated, drainable ponds, public footpaths, and the encumbrance of an irrigation canal that meandered through the center of the property.

It was necessary to design and route a new canal around the west and south borders in compliance with BWDB specifications (Fig. 2). However, to some degree, this impediment was a disguised aid to pond construction. A large amount of earth was needed to build up the levees of the "elevated," drainable ponds in order to retain pond bottoms at nearly original ground elevation, which was only 30 cm to 50 cm above highest expected operating water level in the canal. Additional excavated material was used to build an elevated perimeter pathway for pedestrian and small vehicle travel. Also, the course of the old canal was re-excavated (and some sections filled) to form "dugout," non-drainable ponds (shown stippled, Fig. 2) to store broodstock or grow large-sized fish.

All of the buildings were situated on the highest ground near the main access road. Pathways and bridges for pedestrians and small vehicles had to be either retained or rerouted around and outside this perimeter wall, a very necessary security measure wherever fish are cultured.

Main Hatchery Building

The main hatchery building and all other buildings were designed with aid from an associated local engineering firm to comply with the requirements of simplicity and the restraints imposed by the necessity to use locally available materials and workmanship. Practically all first-class housing and public buildings in Bangladesh are constructed with walls of brick covered over with cement plaster. Usually footings and foundations are also brick masonry, but support columns and beams are of steel-reinforced concrete. All of the materials are available in Bangladesh, even though steel and cement are sometimes in short supply.

The simplicity of external lines and elevations and the architectural restraint in ornamentation is shown in Figure 2. The design of other structures — residential quarters, training center, and office building — was in keeping with this basic structural concept.

The hatchery's ground floor plan (Fig. 3) includes circular tanks for spawning or for hatching eggs, a 5-table jar egg incubation system comprising 100 jar units, and 20 rectangular fish holding troughs. A small laboratory was also included for routine analytical work normally associated with fish cultural activities, such as disease diagnosis, water chemistry, plankton counts, and larval development observations. Research was not to be a primary function of the hatchery; however, it will support practical, applicable experiments and testing to further its own effectiveness in production, training, extension activities, and management procedures.

Jar Egg-Hatching Units

Most of the fish species to be produced in the hatchery spawn naturally only in flowing water at limited, specific, well-defined sites. The site is usually oxbow bends of major rivers during the monsoon season when extensive flooding occurs. These species do not spawn in captivity, even though they mature sexually in ponds. In order to secure viable eggs, the females of these species must be injected with fish pituitaries or anterior-pituitary-like materials extracted from human pregnancy urine (APL, HCG, etc.). After injection, eggs and sperm are obtained by hand-stripping and mixed to achieve fertilization. On the other hand, spawning and fertilization can be achieved (after injection) by placing males and females together in a spawning tank or hapa. In either case, an egg-hatching facility is necessary.

There are many different types of hatching facilities in use around the world today. Those used for semibuoyant eggs have one thing in common: a water flow-through system to buoy the eggs and keep them in motion and in contact with abundant dissolved oxygen.

In Japan and Taiwan, the hatching facility usually consists of a water container in which a screen-cloth funnel is suspended to confine the eggs. Water flows

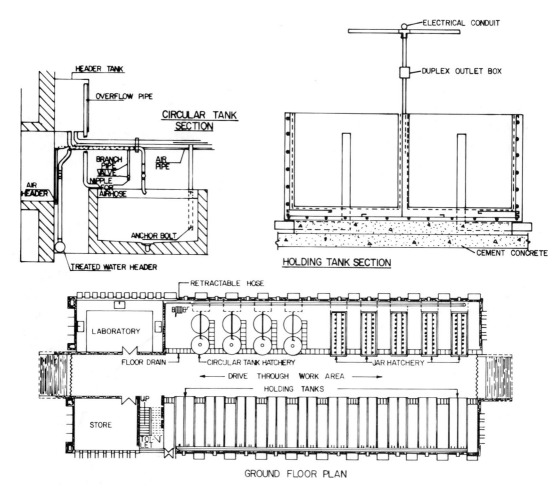

FIGURE 3. — Ground floor plan of hatchery and cross-sections of circular spawning tank, *upper left,* and holding tanks, *upper right.*

up into the bottom of the funnel to keep the eggs in motion. This apparatus works reasonably well as indicated by the mass production of carp fry in these countries. However, this device does not permit close observation of egg movement or easy separation of live and dead eggs or shells and larvae.

In India, hapas have been used successfully. This apparatus consists of a screen-cloth egg retainer placed inside a close-woven cloth larvae container suspended in a pond. This is a simple and useful method of hatching eggs where other more efficient hatching facilities are not available, but mass egg mortality is common. India has been using a glass jar

hatchery during recent years with more efficiency.

Any apparatus that serves to confine the eggs and maintain motion and oxygen levels might serve to hatch the eggs of the species that this facility is expected to produce. However, in many countries where the Indian or Chinese carps are cultured, the jar hatching system is most commonly used.

The glass (or plastic) jar hatchery is an efficient system for hatching semibuoyant eggs. The jars are clear, allowing close observation of egg circulation and development. Disease control and the separation of live and dead eggs are convenient. Fry are not held in confinement with the slower developing eggs or dead eggs, as they can be quickly flushed out of the jar into less crowded containers. The water inside a container during the hatching process is at its lowest quality because of the breakdown of egg chorions, and it is advantageous to free the larvae as quickly as possible. It is also wise to reduce handling (with dipnet or siphon) as much as possible at this stage. Hence, a simple fry recovery system (Fig. 5) is a necessary addition to the jar hatchery. It was important to develop a hatching facility that was not overly technical or complicated. The jar hatchery is as easy to operate as the funnel hatchery, and the jar system allows more control over hatching processes to assure greater efficiency and productivity.

The jar hatchery consists of five separate units, each connected to an overhead water supply tank (Fig. 4). Each table unit is equipped with 20 jars, a receiving trough, fry chutes, and a concrete holding vat with cloth hapas. The header tank is used to maintain constant water pressure to facilitate the regulation of water flow. The tank is equipped with a manual valve and overflow pipe that can be adjusted to maintain a constant water level. This tank stores enough water to operate all jar and circular tank hatching units for 1/2 hour in case of power failure. Water flows by gravity into a main supply pipe 5 cm in diameter that is fitted with 22 valves. Two valves regulate water to shower heads that spray water over the fry vat, and 20 valves supply water through surgical tubing to the 20 hatching jars. The jars are of approximately 8-liter capacity and each will hold approximately 50,000 eggs during each incubation period. Waterflow through the jars should be just enough to keep all the eggs in very slow constant motion without causing them to overflow, approximately 2 l/min. The overflow from the jars falls into a receiving trough that is divided lengthwise by two removable small-mesh nylon screens. Two screens are provided so that one can be removed and cleaned if clogging occurs, while the other remains in place and continues to function. As eggs hatch, the egg shells and larvae are discharged into the receiving trough, where the egg shells and dead eggs are retained by the screen, while larvae pass through and are swept down the chute by the water flowing into the concrete vat, which is partitioned by cloth hapas into four sections, one for each section of the receiving trough. Fry are held in these vats until they begin feeding; then they are moved to nursery ponds. If the vats are needed for another crop before the first reaches the postlarvae stage, the newly hatched larvae may be moved to the other concrete holding tanks on the other side of the building for care until feeding begins.

Circular Tank System

The circular tank spawning and hatching system is not commonly used anywhere except in China. For Bangladesh and for the species of fish to be spawned, the tank system in combination with the jar hatchery should offer considerable versatility to the spawning methodology of the several major carps.

A 2-m-diameter circular tank can be used with air screen and proper waterflow pattern to incubate about 2 million eggs. These tanks also offer the opportunity for "natural" spawning if hand-stripping methods are not necessary. Eggs may be moved to the jar hatchery for incubation after natural spawning occurs to make room for more brooder fish, or they may be incubated and hatched in the tanks without further handling. This method of incubation does not offer good control over eggs and fry, but often hatching success is greater than in jars, because of the lower stress of natural release of eggs and less crowded conditions.

A circular tank system was designed to accommodate both "free spawning" and egg incubation (Fig. 5). These units are also connected to header tanks. Influent water is supplied at an angle to create a circular current in the tank, which helps stimulate the broodfish to spawn "naturally" and which will keep the semibuoyant eggs in motion. An air screen (air supplied by blower) around the standpipe screen also helps to keep the eggs buoyant. When used for natural breeding, one female and two males are placed in the tank after

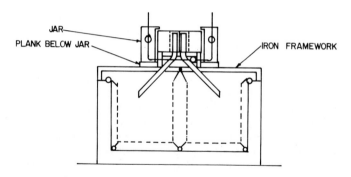

SECTION OF FRY RECOVERY SYSTEM

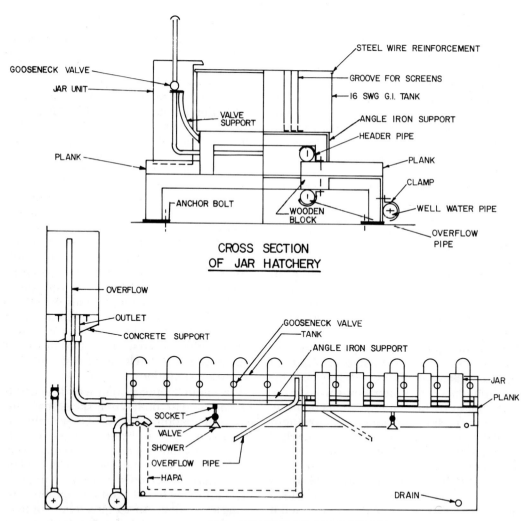

FIGURE 4. — Sectional diagrams of the jar-hatching and fry-recovery system.

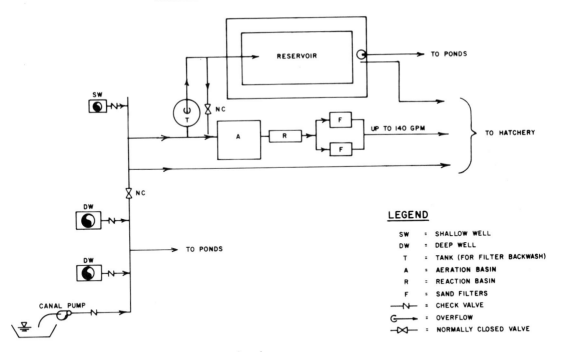

FIGURE 5. — Schematic diagram of water processing plant.

pituitary injection and left undisturbed until spawning occurs. The adults are then removed. The eggs continue to incubate in the tank until hatching. The larvae may be held there until the post-larval stage when they are moved to nursery ponds, or they may be moved to other holding tanks.

These tanks are equipped with a screw-in, inside standpipe 5 cm in diameter and a turn-down, outside standpipe 5 cm in diameter, which allows easy removal of fry. During incubation the water level is regulated by the length of the inside standpipe. The standpipe screen is approximately 20 cm in diameter and has a rubber gasket at the bottom, which fits tightly over the standpipe. When the fry are to be removed the inside standpipe is replaced with one that has a 7.5-cm-long solid lower end, and the water is drained off by lowering the outside standpipe. After the water level is low and the fry are concentrated, the screen and the inside standpipe are removed and the fry and remaining water are slowly allowed to drain from the tank into a bucket or bag for transfer to ponds or other holding tanks.

Water Process Plant

Since it was known that most of the groundwaters in the Raipur area contained dissolved iron and that some surface water from the canal would be used in the hatchery and need filtering, a water processing plant was necessary. A schematic flow diagram is shown in Figure 5. The design consists of a basin where four or more agitators aerate the well or surface water. The aeration basin is followed by a reaction section to permit time for oxidation of dissolved iron.

Two gravity-flow, multimedia sand filters are used alternately to remove iron and other particulate matter. The filtered water then flows by gravity to the indoor hatchery facilities.

The process plant water can also be supplied from an elevated earthen reservoir filled by rain or well water. The elevation level of water in this reservoir will be about 6 m above ground level, permitting gravity flow to the process plant or into the hatchery building. Additionally, a concrete reservoir on top of the earthen reservoir levee will provide the necessary 2500 l/min of water for backwashing the filters.

This system provides four different kinds of water to the hatchery — untreated or processed well water and unfiltered or filtered reservoir water. This flexibility permits each supply to be used separately or variously mixed by valve adjustment outside and inside the hatchery building. Also, each water supply can act as a backup in case of failure of another. This is comforting assurance to the fish culturist, and is a substantial aid to diversified fish cultural activities.

Water Supply

Plans for the hatchery water supply include two deep tube wells to furnish 4000 or more l/min for the ponds and one shallow well to supply 1000 l/min for the hatching and holding facilities. These plans were based on data from groundwater maps of the water resources of the alluvial plains area and oral information concerning a few nearby deep and shallow wells.

A minimum of 4000 l/min of water was considered necessary to fill or flush the ponds, many of which are located long distances from the planned well sites and the canal water intake. Calculations of head and friction loss revealed that 15-cm-diameter PVC pipe, the largest size manufactured in Bangladesh, would not carry the needed volume if the customary single-line header pipe were run along main levees between each two rows of ponds. Therefore, in order to obtain a workable pipeline layout, the calculated and established parameters were put to computer analysis using a program based on Cross's analysis of flow-in networks (Cross 1936). The analysis resulted in a design for optimum looping of the pipe network to permit passage of the required volume with low head loss.

Ponds

The most applicable criteria for pond design and construction required that most of the ponds, particularly the nursery and rearing ponds, be small, completely drainable, and reasonably watertight. The water table in the Raipur area is very high, being about 2.5 m below ground level during the dry season and 1 m or less in the wet season. Even though soil tests indicated mixed clays with sand and silt with moderate water-holding capacity, it was thought that any substantial excavation of earth inside the pond area would put the pond bottoms below the water table, permitting significant loss of water during the dry season, and preventing complete draining for harvest of fish. Hence most of the ponds were designed to be elevated with bottoms above groundwater level, with clay-cored levees built from materials mostly excavated from both the old and new canals. The excavations form "dugout" nondrainable ponds suitable for broodstock and other large fish that can be adequately captured by seine, or the ponds can be pumped dry for complete harvest.

The pond and perimeter levees not only were clay-cored but were also thoroughly compacted to improve their waterholding ability. Equipment for earth moving or compaction (bulldozers, rollers, draglines, or loaders) is scarce in Bangladesh. Ordinarily, earth is moved with grubhoes and headbaskets by hundreds or even thousands of people digging and carrying earth to the proper site. Because of high unemployment, low wages, and excessive population, it is actually more economic to use hand labor than to import heavy equipment that is very expensive to operate and maintain in Bangladesh. Fortunately, the construction contractor was able to rent a dozer and sheepsfoot roller to be used mostly for earth compaction, but it was also used to grub out deep-rooted tree stumps and to move large piles of earth that workmen carried out of the canal where the dozer could not work effectively. To move the earth by dozer for more than a few feet was more expensive than the hand-and-head method. In addition, this method gave much-needed jobs to a large number of people, an important consideration for any emerging nation (Fig. 6).

In other ways, pond design and construction were typical and standard practices as ordinarily used in the United

FIGURE 6. — The hand-head basket method of pond construction in Bangladesh.

States. A typical cross-section of the levees of both the drainable and dugout ponds is shown in Fig. 7. The catch structures are similar to many in use in the United States at state, federal and private fish hatcheries. The catchments consist of a concrete basin and drain structure ("monk"). Pond water levels are controlled by "dam boards." Three dam board channels are provided in the monk to accommodate a screen, baffle, and weir boards. The front channel secures the screen, the middle channel holds the baffle boards to permit draining bottom water (by leaving out a bottom board), and the back channel boards regulate the pond water level and are withdrawn to drain or lower the pond water.

Discussion

All of the many irrigation and flood control projects will significantly affect the freshwater fisheries of Bangladesh. Some aquatic species, those favored by stable, static water habitat, may increase in numbers and growth; but in many instances, adverse effects will produce severe, irreversible alterations in the species composition and reduction in productive potential of the former aquatic communities. These changes will be particularly noticeable in those species of fish and prawns that are migratory or anadromous. These adverse impacts on foodfish production can be mitigated partially by stocking hatchery-reared fish in manageable ponds and canals where modern aquaculture methods can be practiced to boost pond fish harvests high enough to compensate for the loss in former natural production. However, mitigatory hatcheries alone will not ameliorate the adverse impacts. The hatchery program of fish stocking must be accompanied by training and education for both professional aquaculturists and prospective fish farmers to assure even modest success. As a result of this reasoning, the

293

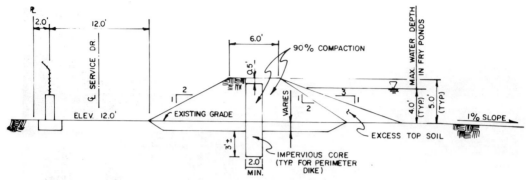

TYPICAL FRY AND REARING POND SECTION

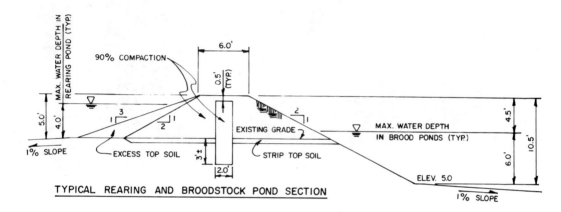

TYPICAL REARING AND BROODSTOCK POND SECTION

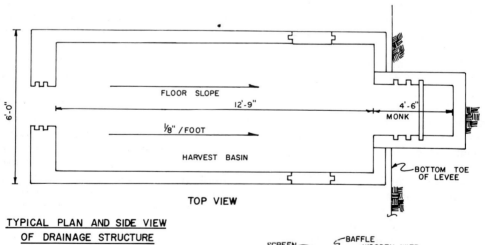

TOP VIEW

TYPICAL PLAN AND SIDE VIEW
OF DRAINAGE STRUCTURE

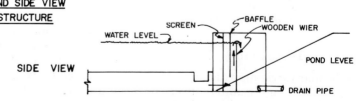

SIDE VIEW

FIGURE 7. — Typical sections of levees, catch basin, and water-level control structure.

Raipur Fish Hatchery and Training Center was designed and construction was begun in late February 1979.

Many impediments face the design engineer who works in a developing nation. Faced with the necessity of utilizing crude equipment, local materials, and sometimes primitive construction methods, he must usually avoid innovative designs not previously proven functional in that country. Even social and religious mores and practices must be considered in the design and, sometimes, the orientation of certain buildings.

In order to meet the requirements of simplicity with operational versatility and adaptability, the designers incorporated or combined several functional features which have been used and proven successful in other countries, particularly in Arkansas, U.S.A. At the same time, it was intended that the facility be modern and fully supplied with up-to-date instruments, equipment, and materials so that the hatchery can fulfill the urgent need of fish farmers for quality fry and fingerlings.

References

BAILEY, W.M., JR. 1978a. Design of the fish production center, Chandpur irrigation project. W.D. No. 8. Mimeo rep. 71 pp.

———. 1978b. Management of the fishery resources and operational scheme for the fish production center in the Chandpur irrigation project area. W.D. No. 10. Mimeo rep. 59 pp.

CROSS, HARDY. 1936. Analysis of flow in networks of conduits or conductors. Univ. Ill. Eng. Exp. Sta. Bull. No. 286.

DIRECTORATE OF FISHERIES AND SNELL ENVIRONMENTAL GROUP. Working Documents. Dacca, Bangladesh.

PANTULU, V.R. 1977. Establishment of fish hatchery and rearing facilities in Chandpur irrigation project area. W.D. 3. Mimeo rep. 58 pp.

RAINBOTH, WALTER J. 1978. The fish and prawn stocks of the Chandpur region and the impact of the Chandpur irrigation project on these stocks and their fisheries. W.D. 16. Mimeo rep. 69 pp.

TSAI, CHU-FA. 1978. Impact of the South Dakatia River regulator and pumping station on the fishery stocks inside the Chandpur irrigation project area. W.D. 11. Mimeo rep. 42 pp.

Bio-Engineering Symposium for Fish Culture (FCS Publ. 1):296-307

Evaluation of a Closed Recirculating System for the Culture of Tilapia and Aquatic Macrophytes[1]

JAMES E. RAKOCY[2] AND RAY ALLISON

Department of Fisheries and Allied Aquacultures
Agriculture Experiment Station
Auburn University
Auburn, Alabama 36830

ABSTRACT

Aeration, sedimentation, biofiltration, and aquatic macrophyte production were the wastewater treatment processes employed in three identical units of a closed recirculating system in outdoor concrete tanks. Each unit consisted of one fish production tank and three wastewater treatment tanks and had a water volume of 61,600 liters. The mean flow rate was 68 l/min and the mean retention time in the fish rearing tank was 3.8 hours. Average daily makeup water amounted to 1.1% or less of the total water volume. *Tilapia aurea* fingerlings were stocked at rates of 1,000, 2,000, and 3,000 per unit and fed for 109 days. The standing crop of fish at harvest was 234 kg, 437 kg, and 596 kg in the low, intermediate, and high density units, respectively. Dissolved oxygen appeared to be the only limiting water quality variable. The efficiency of the wastewater treatment system was evaluated. Through competition for light and nutrients, the macrophytes prevented the establishment of phytoplankton and the water remained clear. Two floating species (*Eichhornia crassipes* and *Spirodela oligorhiza*) and two submerged species (*Egeria densa* and *Vallisneria* sp.) were evaluated for growth and nitrogen uptake. The combined plant populations removed 15.8%, 13.4%, and 12.0% of the waste nitrogen in the low, intermediate, and high density units, respectively.

Introduction

Recirculating systems have generally been designed to discharge 10% of the total water volume per day (Liao and Mayo 1974). This represents a substantial loss of nutrients that could contribute to the production of by-products. Attempts have been made to recover these nutrients through the use of hydroponics. Tomatoes have been successfully grown in a recirculating system for catfish culture (Lewis et al. 1978). Makeup water in this system amounted to only 6.6% per day. but nutrient levels were not high enough for optimum tomato growth and nutrient supplements had to be added to the system. In the present study, a slow rate recirculating system was designed where makeup water amounted to 1.1% or less of the total system volume per day. Aquatic macrophytes were used to recapture waste nutrients.

[1]Research supported by Hatch Project Alabama Number 399.

[2]New address: College of the Virgin Islands Agriculture Experiment Station, P.O. Box 920, Kingshill, St. Croix, U.S.V.I. 00850.

Methods

Three identical units of the recirculating system were established in outdoor concrete tanks measuring 7.32 m long × 2.74 m wide × 0.76 m deep. The water volume of each tank was 15,400 liters. Each unit consisted of one fish production tank and three wastewater treatment tanks (Fig. 1). The total volume of each unit was 61,600 liters.

A diffusion aeration system was used in the production tank and consisted of three lines of perforated polyvinyl chloride pipe (PVC), 2.5 cm in diameter, that were anchored to the bottom of the tank. A 5-horsepower blower supplied air to all the units. Near the end of the growing season, an additional aeration system was used that consisted of a submersible pump spraying water into the air (Figs. 2, 3).

The wastewater treatment tanks were divided into a number of compartments for sedimentation, macrophyte production, and biofiltration. The position and relative size of these compartments were partially dictated by the structure of the fixed con-

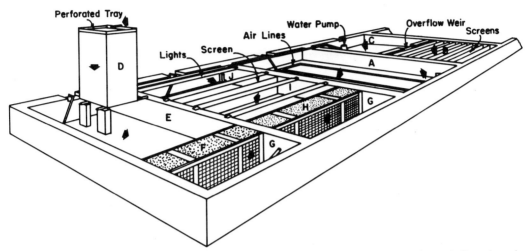

A Tilapia Production
B Clarifier
C Hyacinth Production
D Trickling Filter
E Duckweed and Clam Production

F Coarse Limestone Filter
G Duckweed Production
H Fine Limestone Filter
I *Egeria* Production
J *Vallisneria* Production

FIGURE 1. — A closed recirculating system for the culture of tilapia and aquatic macrophytes. Arrows indicate the path of water flow.

FIGURE 2. — Experimental production facilities, View 1.

FIGURE 3. — Experimental production facilities, View 2.

crete tanks, especially the openings between tanks, and did not necessarily reflect optimum configuration.

The first treatment tank was equally divided by a block wall into a clarifier and a water hyacinth *(Eichhornia crassipes)* production compartment. The clarifier consisted of twelve plastic window screens, 30 cm apart, suspended across the width and depth of the tank. The screens were anchored to the bottom of the tank by bricks, but they were not attached to the sides so that water could flow around them if they became clogged. Water passed into one end of the clarifier from the bottom of the fish rearing tank and was discharged at the opposite end into the water hyacinth compartment through an overflow weir in the block wall. A submersible pump at the opposite end of the water hyacinth compartment continuously pumped water to the top of a trickling filter tower above the second treatment tank.

The influent of the second treatment tank was uniformly distributed through the tower by a perforated plastic tray at the top of the tower that measured 1.22 m by 1.22 m. The tower was 2.44 m in height and 3.63 m³ in volume, and consisted of plastic modules of B.F. Goodrich Koro-Z media. At the opposite end of the second tank was a submerged biofilter of crushed limestone averaging 2.5 cm in diameter. The limestone was retained by a crib of wood (0.61 m wide) and vinyl-coated hardware cloth (2.5-cm × 1.3-cm mesh) that extended across the width and depth of the tank. The volume was 1.27 m³. The filter was aerated by a stream of bubbles from two lengths of perforated PVC pipe at the bottom of the filter. All available surface area in this tank was used for duckweed *(Spirodela oligorhiza)* production. The compartment between the biofilters (5.49 m long) had a sand bottom for the production of clams *(Corbicula sp.)*. After passing through both biofilters, water flowed from mid-depth into the third treatment tank (Fig. 1).

The third treatment tank contained a biofilter of crushed limestone with aggregates ranging from the size of sand to 1.3 cm in diameter. In every other respect, it was identical to the first submerged

FIGURE 4. — Vegetation growth.

biofilter. Duckweed was grown in the small compartment (1.22 m in length) between the limestone biofilter and the influent end of the tank. The remaining section of the tank was used for the production of submerged macrophytes. An aluminum window screen, secured to the tank walls and bottom, divided this section into two equal compartments (2.75 m in length) for the growth of *Egeria densa* and *Vallisneria* sp. The plants were rooted in a substrate, 8 cm deep, consisting of 50% clay soil and 50% builders sand (Fig. 4). Two sets of fluorescent grow-lights were suspended across each compartment. Each set of lights contained two, 73-watt bulbs that were 2.44 m in length. The lights were turned on at night to promote continuous photosynthesis. These compartments were covered at night with a sheet of black polyethylene to avoid attracting insects. Surface water from the *Vallisneria* compartment flowed into the fish rearing tank through a stand-pipe to complete the purification cycle.

The rate of water flow within the units decreased over time resulting from the buildup of biological growth in the water pipes. Flow rates were calibrated frequently to 76 liters/min. Mean flow rates for the growing season were 70, 67, and 66 liters/min in Units I, II and III, respectively. Mean retention times ranged from 3.6 to 3.9 hours in the fish production tanks and from 10.9 to 11.7 hours in the wastewater treatment tanks. Water flowed by gravity through the system from the trickling filter tank to the pump compartment. Makeup water was added frequently, but not daily, to replace water lost by evaporation, seepage, accidental overflow, and sludge removal. Mean daily makeup water was 1.0% of the total volume of Unit I, 1.1% of Unit II, and 0.8% of Unit III. During the entire experimental period, water loss by overflow and sludge removal was 5.0% of the total volume of Unit I, 9.5%, of Unit II and 16.9% of Unit III.

On 31 May and 1 June 1978, a mixed population (males and females) of *Tilapia aurea* fingerlings, averaging 22 g, were stocked in the production tanks at a density of 1,000 fish in Unit I, 2,000 fish in Unit II, and 3,000 fish in Unit III. Growth rates were determined at monthly intervals by weighing and counting a sample of fish from each unit. On 18 September through 20 September, the fish were harvested, sorted into groups by length, weighed, and counted.

The fish were started on sinking Purina Trout Chow (#4), fed twice daily at 4% of their body weight per day. After 10 days, they were given floating Purina Trout Chow (#5) for the remainder of the experiment. At the end of the first month, the feeding rate was reduced to 3% of their body weight per day in three feedings per day, spaced 6 hours apart. At the end of the second month, the fish were fed to satiation in three feedings per day until the end of the experiment. The fish were fed for a total of 109 days.

The stocking and final harvest dates of the aquatic macrophytes are shown in Table 1. As the plants became crowded, a partial harvest was made to keep the remaining plants in a state of rapid growth.

TABLE 1. — Production of aquatic macrophytes used in the wastewater treatment process of a closed recirculating system for tilapia culture.

Macro-phyte	Unit	Stocking dates	Final harvest dates	Dry weight (%)	Nitrogen[a] (%)	Crude protein[a] (%)	Yield (kg/m²) Wet weight	Dry weight	Productivity (g/m²·d⁻¹) Wet weight	Dry[b] weight	Nitrogen[b] weight
Water	I	28 May	22 Sept.	5.7	4.0	25.1	44.0	2.37	374.2	20.2	0.81
hyacinth	II	28 May	22 Sept.	4.9	4.6	29.0	54.2	2.92	461.0	24.9	1.14
	III	28 May	26 Sept.	5.3	4.7	29.1	64.8	3.50	533.7	28.8	1.35
Duckweed	I	30 May 5 June	5 Sept.	10.1	5.5	34.3	2.9	0.27	38.6	3.6	0.20
	II	30 May 5 June	22 Sept.	9.2	5.8	35.9	5.0	0.46	52.0	4.8	0.28
	III	30 May 5 June	22 Sept.	8.8	5.6	35.1	5.8	0.54	62.0	5.8	0.32
Egeria	I	27 May 5 June	1 Oct.	7.6	4.1	25.9	13.0	0.96	102.1	7.6	0.31
	II	27 May 5 June	30 Sept.	7.0	4.4	27.4	15.1	1.12	119.6	8.8	0.39
	III	27 May 5 June	30 Sept.	7.7	4.2	26.6	17.3	1.28	136.4	10.1	0.42
Vallisneria	I	20 May	1 Oct.	4.6	2.8	17.7	6.9	0.32	51.5	2.4	0.07
	II	20 May	30 Sept.	4.7	2.8	17.8	8.9	0.41	66.5	3.1	0.09
	III	20 May	30 Sept.	4.7	3.2	19.8	9.5	0.45	71.3	3.4	0.11

[a]Percent of dry weight.
[b]Based on the mean dry weight of that species over all three units.

When *Egeria* and *Vallisneria* reached the surface, the upper 0.3 m of the plants were cut off and removed without disturbing their roots. Excess water was drained from the harvested plants and they were weighed. The following areas in each unit were devoted to macrophyte production: 10.0 m² for water hyacinth; 21.7 m² for duckweed; 7.5 m² for *Egeria;* 7.5 m² for *Vallisneria.* A total of 58% of the surface area of each unit was devoted to macrophyte production.

On 5 July, 2,000 clams *(Corbicula* sp.*)* were weighed and stocked into the compartments under the trickling filters. They were harvested, weighed, and counted on 14 November in Units I and II and on 16 November in Unit III.

Sludge was removed from the clarifier periodically as dissolved oxygen concentrations decreased to low levels in the fish production tank. Sludge was either removed from the surface with fine-mesh net or siphoned from the bottom of the clarifer with a tube. The sludge was placed into a fine-mesh basket, allowed to consolidate for several days, and then weighed. A sample was collected for solids determination. Water lost from the system during sludge removal was recorded.

A number of water quality parameters were measured using standard methods (APHA 1975). Dissolved oxygen (DO) and temperature were measured daily with a YSI Model 54 oxygen meter. Weekly measurements were made of pH (glass electrode) and total alkalinity (by titration with dilute acid to the methyl red-bromcresol green end point). Total ammonia-nitrogen and nitrite-nitrogen were measured weekly using the phenate method and diazotization method, respectively. Five-day biochemical oxygen demand (BOD₅) was determined weekly using Hach Kit manometer bottles. Infrequent determinations were made of

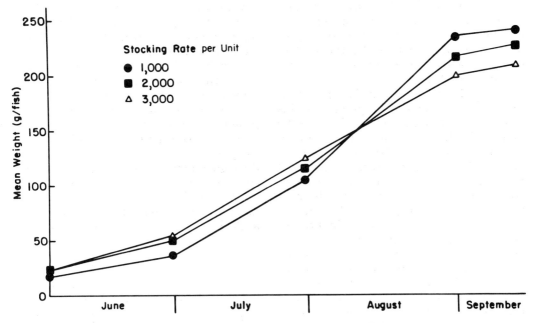

FIGURE 5. — Growth rates of *Tilapia aurea* stocked at three rates in a closed recirculating system.

nitrate-nitrogen (phenoldisulfonic acid method), total hardness (EDTA titration), and suspended solids. Water samples were collected shortly after sunrise before the first feeding of the day with the exception of BOD_5 samples, which were collected at midday.

At the end of the experiment, samples of fish, feed, plants, and sludge were analyzed for percent dry weight and for nitrogen content using a macro-Kjeldahl technique (AOAC 1970).

Results

The total fish weight at stocking was 18 kg in Unit I, 44 kg in Unit II, and 66 kg in Unit III. At harvest the standing crop was 234 kg in Unit I, 437 kg in Unit II, and 596 kg in Unit III. The standing crop of the production tank at harvest was 15 kg/m³ in Unit I, 29 kg/m³ in Unit II, and 39 kg/m³ in Unit III. In terms of the mean flow rate, the final standing crop was 3.3 kg/l·m⁻¹ in Unit I, 6.5 kg/l·m⁻¹ in Unit II, and 9.1 kg/l·m⁻¹ in Unit III. Survival was 99% in Unit I, 98.6% in Unit II, and 97.5% in Unit III. The feed conversion ratio (feed weight divided by fish weight gain) was

1.22, 1.18, and 1.16 in Units I, II, and III, respectively.

The most rapid growth occurred during July and August (Fig. 5). Growth rates decreased as early morning water temperatures declined from a mean of 24.7 °C in August to a mean of 22.9 °C in September. After satiation feeding started in August, growth rates were lower in the high density units. The maximum daily feeding rate of 10 kg occurred in Unit III.

After harvest, the fish were marketed (Fig. 6). The minimum marketable size was found to be 20 cm (182 g). The percen-

FIGURE 6. — Harvest size, *Tilapia aurea.*

tages of marketable fish in Units I, II, and III were 89, 80, and 73, respectively. The number of young fish resulting from reproduction ranged from 6 to 50 in the production tanks and 140 to 303 in the wastewater treatment tanks.

Macrophyte production data are summarized in Table 1. The greatest yields were obtained in Unit III, which had the highest feeding rate and nutrient levels. Water hyacinth was the fastest growing plant and removed the greatest amount of nitrogen per square meter of growing area. Although duckweed had the highest nitrogen content, it grew slowly and was infested by insects. A combination of aphids and moth larvae completely destroyed the duckweed in several of the smaller compartments. *Egeria densa* grew at one third of the rate of water hyacinth but grew faster and removed more nitrogen than duckweed. *Wallisneria* sp. grew slowest and removed the lowest amount of nitrogen. *Vallisneria* increased from 12 plants at stocking to 771 plants (greater than 25 cm in height) in Unit I, 815 plants in Unit II, and 939 plants in Unit III at harvest.

Clams had a survival rate of 97% in each unit but grew very slowly. During 132 days or more of growth the mean weight gain was only 7%.

The dry weight of sludge removed from the clarifier was 0 kg in Unit I, 13 kg in Unit II, and 31 kg Unit III.

The water quality variable that limited production in the system appeared to be DO. The mean DO concentration during early morning for the months of July through September was 6.7 mg/l in Unit I compared to 4.7 mg/l in Unit III. During the day, DO levels decreased in the production tanks. Mean DO concentrations at dusk for the same time period were 5.8 mg/l in Unit I compared to 2.8 mg/l in Unit III. If the DO was less than 2.0 mg/l, the fish were not fed until DO levels increased. The fish continued to feed at a DO concentration of 2.0 mg/l but increased respiration caused a rapid decrease in the DO concentration to about 1.0 mg/l. At that point, feeding activity ceased and the fish came to the surface for oxygen. Mean DO concentrations in Unit II ranged from 5.3 mg/l shortly after dawn to 3.7 mg/l at dusk during July through September.

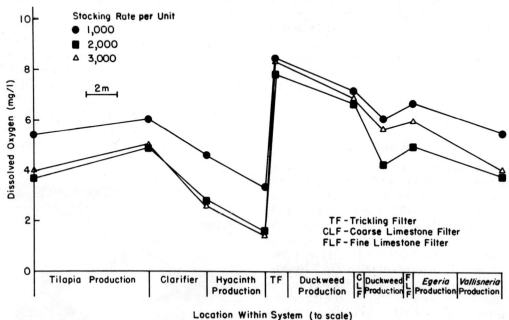

FIGURE 7. — Means of early morning dissolved oxygen concentrations at treatment process stages within a closed recirculating system stocked with *Tilapia aurea* at three rates.

The trickling filter contributed significantly to the oxygen budget of the system (Fig. 7). The mean DO levels increased by at least 6 mg/l upon passage through the filter in Units II and III. The mean hydraulic loading rate and total surface area of the trickling filter were 66 $m^3/m^2 \cdot d^{-1}$ and 648 m^2, respectively. The submerged macrophytes were another significant source of oxygen to the system. On sunny days the DO concentrations in the effluent from the submerged macrophyte compartments were frequently greater than the saturation value. There was no evidence that the lights increased DO levels at night in the submerged macrophyte compartments.

As the feeding rate and standing crop of fish increased, total ammonia-nitrogen increased in the production tanks. The highest level recorded was 2.96 mg/l in Unit III on 13 September. In Unit III, pH values remained at 6.6 or 6.7 near the end of the growing season and prevented un-ionized ammonia-nitrogen concentrations from going higher than 0.007 mg/l. The highest un-ionized ammonia-nitrogen

level recorded in a production tank was 0.014 mg/l. This value was well within safe limits. Redner and Stickney (1979) reported that 2.40 mg/l of un-ionized ammonia-nitrogen was the 48-hour median lethal concentration for *Tilapia aurea* not previously exposed to ammonia. *T. aurea* tolerated concentrations as high as 3.4 mg/l without mortality after exposure to sublethal concentrations for 35 days.

The capacity of the biofilters for ammonia removal was not reached. After nitrifying bacteria became established, in approximately 1 month, ammonia-nitrogen concentrations in the production tank influent were not higher than 0.32 mg/l and were typically below 0.20 mg/l. The average reduction of ammonia-nitrogen in the wastewater treatment tanks was 67.3% in Unit I, 78.8% in Unit II, and 85.8% in Unit III (Table 2). Based on the means of three samples collected over a 9-week period, the greatest ammonia reduction occurred in the coarse limestone filter of Unit III and trickling filter of Unit II (Fig. 8). The mean hydraulic loading rate of the limestone filters was 46.7

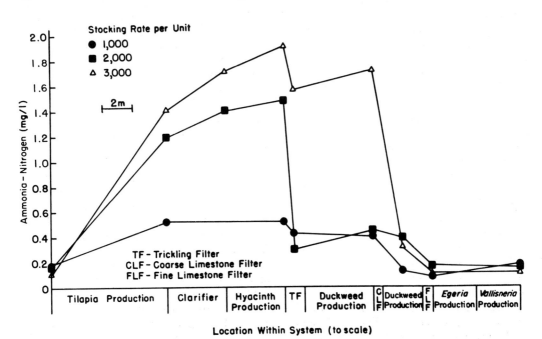

FIGURE 8. — Mean ammonia-nitrogen concentrations at treatment process stages within a closed recirculating system stocked with *Tilapia aurea* at three rates.

TABLE 2. — Efficiency of the wastewater treatment process in a closed recirculating system for the culture of tilapia and aquatic macrophytes.

Unit	Location in production tank	Mean NH$_3$-N[a] (mg/l)	Percent reduction	Mean NO$_2$-N (mg/l)	Percent reduction	Mean BOD$_5$[b] (mg/l)	Percent reduction	Mean suspended solids[c] (mg/l)	Percent reduction
I	Effluent	0.52		0.20		8.5		7.8	
	Influent	0.17	67.3	0.14	32.7	3.1	63.5	0.8	89.7
II	Effluent	1.04		0.44		12.6		10.6	
	Influent	0.22	78.8	0.34	23.0	4.8	61.9	0.8	92.5
III	Effluent	1.27		0.36		16.2		28.8	
	Influent	0.18	85.8	0.30	17.3	4.1	74.7	0.8	97.2

[a]Total ammonia-nitrogen.
[b]Five-day biochemical oxygen demand.
[c]Means based on two morning samples collected near the end of the experiment.

$m^3/m^2 \cdot d^{-1}$. One cubic meter of combined biofiltration media handled the nitrogenous waste of 96.6 kg of fish at the highest standing crop.

Nitrite-nitrogen levels increased sharply during August in the production tanks of Units II and III and reached a peak of 1.35 mg/l in Unit II on 30 August. Concentrations of nitrite-nitrogen decreased in September as the water temperature and feeding rate decreased. During August and September, the mean concentration of nitrite-nitrogen was highest in Unit II. The wastewater treatment process was less efficient at removing nitrite-nitrogen (Table 2). Nitrite-nitrogen levels increased through the trickling filter, remained about the same through the coarse limestone filter, and decreased sharply in the fine limestone filter (Fig. 9).

Mean nitrate-nitrogen concentrations at the end of the growing season were 41.8 mg/l in Unit I, 51.2 mg/l in Unit II, and 68.6 mg/l in Unit III.

The wastewater treatment process was very effective at removing suspended solids and BOD. Suspended solids reduction was 90% or greater (Table 2). The mean retention time and overflow rate of the clarifiers were 113 min and 9.7 $m^3/m^2 \cdot d^{-1}$, respectively. The BOD levels in the production tank influent did not increase significantly over time or at the higher stocking rate (Table 2). BOD levels decreased gradually throughout the entire treatment process.

The pH in the production tanks ranged from 6.6 to 7.5. Total alkalinity was initially around 80 mg/l as $CaCO_3$ in all the units. As the growing season progressed, total alkalinity gradually decreased to about 40 mg/l and stabilized at that value. Total hardness increased from less than 20 mg/l as $CaCO_3$ in the water supply to 214 mg/l in Unit I, 325 mg/l in Unit II, and 411 mg/l in Unit III. This indicates that a considerable amount of alkalinity was produced from the solution of limestone and was used to neutralize acids in the system or was used by nitrifying bacteria as a carbon source.

The principal nitrogen input to the systems was feed, of which the fish assimilated between 36.5 and 38.3% (Table 3). The combined plant populations in Units I, II, and III removed 15.8%, 13.4%, and 12.0%, respectively, of the waste nitrogen. Sludge removal accounted for 2.5% of the waste nitrogen in Unit II and 3.8% in Unit III. Nitrogen remaining in the water at the end of the experiment in the forms of ammonia, nitrite, and nitrate equaled 25.1%, 18.8%, and 17.9% of the waste nitrogen in Units I, II, and III, respectively. The percentage of unaccountable waste nitrogen was 58.9 in Unit I, 65.4 in Unit II, and 66.3 in Unit III.

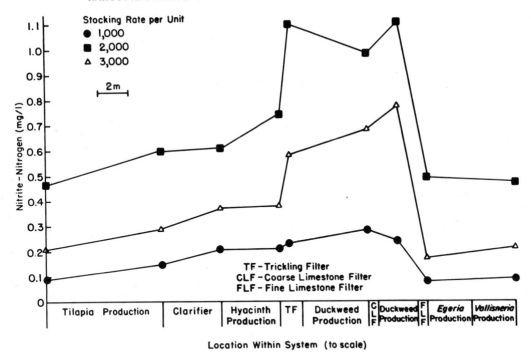

FIGURE 9. — Mean nitrite-nitrogen concentrations at treatment process stages within a closed recirculating system stocked with *Tilapia aurea* at three rates.

TABLE 3. — Nitrogen mass balance in a closed recirculating system for the culture of tilapia and aquatic macrophytes.

	Unit I Nitrogen		Unit II Nitrogen		Unit III Nitrogen	
	(kg)	(% of input)	(kg)	(% of input)	(kg)	(% of input)
INPUT	16.21		28.53		37.92	
OUTPUT						
Tilapia	5.91	36.5	10.75	37.7	14.51	38.3
Hyacinth	0.95	5.9	1.34	4.7	1.64	4.3
Duckweed	0.32	2.0	0.58	2.0	0.65	1.7
Egeria	0.29	1.8	0.37	1.3	0.40	1.1
Vallisneria	0.07	0.4	0.09	0.3	0.11	0.3
Sludge			0.44	1.5	0.88	2.3
Total	7.54	46.6	13.57	47.6	18.19	48.0
REMAINDER						
NH_3-N^a	0.04	0.2	0.08	0.3	0.10	0.3
NO_2-N	0.03	0.2	0.04	0.3	0.10	0.3
NO_3-N	2.53	15.6	3.22	11.3	4.08	10.8
Total	2.59	16.0	3.34	11.7	4.20	11.1
UNACCOUNTABLE	6.07	37.5	11.62	40.7	15.52	40.9

[a]Total ammonia nitrogen.

Discussion

The growth of tilapia was reduced when low DO concentrations developed in the high density production tanks at the end of the season. Tilapia, which can tolerate very low DO concentrations, grew fastest in Unit I, where the average DO was greater than 6.0 mg/l. Low dissolved oxygen levels in the production tank were not a function of the buildup of oxygen demand throughout the system, since influent BOD values remained relatively constant, but rather they were a function of the standing crop, feeding rate, and water temperature. One of the most important of these variables was the feeding rate. The maximum rate was 10 kg/d, but the data indicate that at higher oxygen levels a higher feeding rate would be possible.

Intensive culture of tilapia proved to be an effective method of preventing overproduction of bisex populations. Massive recruitment has been a major obstacle of tilapia culture in ponds.

Although aquatic macrophytes removed a surprisingly small percentage of the waste nitrogen, they are essential to the successful operation of a closed recirculating system located outdoors in warm climates. Through competition for light and nutrients, though nutrients did not appear to be limiting, macrophytes prevented the establishment of phytoplankton. There was virtually no phytoplankton in the system and water in the system remained clear. Dense algal blooms in a recirculating system interfere with the treatment process by clogging biofilters, impose the risk of oxygen depletion, and may reach lethal concentrations. Parker and Simco (1974) reported catfish mortalities, resulting from mechanical obstruction of the gills, when a bloom of *Scenedesmus dimorphus* reached a concentration of 0.05 g/l in an outdoor recirculating system. In the present study, the phytoplankton concentrations equivalent to the entire macrophyte yields would have been 0.65 g/l in Unit I, 0.85 g/l in Unit II, and 1.00 g/l in Unit III. Macrophytes are also more easily managed (harvested) than phytoplankton.

Snails, particularly *Helisoma* spp., were important in controlling the growth of periphyton on the submerged macrophytes. In the absence of predators, large populations of snails (*Helisoma* spp., *Physa* spp., and others) developed in the submerged macrophyte compartment and the clarifier, where they kept the screens relatively free of periphytic growth.

Water hyacinth and *Egeria densa* were the most efficient plants of the ones tested for nitrogen removal. The accumulation of nitrates in the system indicates that larger areas could have been used for plant production. It would be difficult to extrapolate the nitrogen removal rates to find the plant production area needed to reduce dissolved inorganic nitrogen to a desired level because nitrogen removal rates decrease as nutrient levels decrease. For warmwater fish culture, the percentage of the surface area used for floating plants should not be greater than is necessary to control phytoplankton because they reduce the water temperature by shading it. The area devoted to submerged plants like *Egeria* should be maximized because they remove nutrients from the entire water column and produce oxygen during the day.

With the exception of dissolved oxygen, all other water quality parameters were within acceptable ranges from warmwater fish culture. The wastewater treatment process was very efficient at removing suspended solids and ammonia. The clarifier was well within the design limits of overflow rate and detention time set by Davis (1977). Moreover, all the wastewater treatment tanks acted as clarifiers. The clarifier performed satisfactorily on an experimental basis, but on a larger scale a conventional clarifier would be required for easier and more efficient sludge removal. The three-stage biofiltration process of various media was promising, but its overall efficiency needs to be evaluated in terms of ammonia and nitrite removal rates of single-stage biofilters of one medium. The fine limestone filter clogged in Unit III shortly before the end of the experiment and water had to flow over the top of the filter. Larger grades of limestone and some mechanism for periodic cleaning of the submerged filters will be required to prevent clogging and channelization.

We cannot fully explain what happened to all the unaccountable waste nitrogen. Some of it was in the form of dissolved organic nitrogen, which was not measured, or was lost with the sludge water or overflow water. The assimilation of nitrogen by snails or insect larvae, though not measured, was probably minor. The pH did not favor the volatilization of gaseous ammonia on passage through the trickling filter tower. Denitrification may have accounted for the removal of a large portion of the waste nitrogen.

For a number of reasons recirculating systems are very desirable, but certain technologically based economic problems have hindered their development (Weaver 1979). Aquatic macrophyte production is a potential tool in the bio-engineering and economics of recirculating systems for warmwater fish culture.

Acknowledgment

We thank Leonard Pampel for his help in designing the system.

References

AOAC (ASSOCIATION OF OFFICIAL ANALYTICAL CHEMISTS). 1970. Methods of analysis of the Association of Official Analytical Chemists. Washington, D.C. 1015 pp.

APHA (AMERICAN PUBLIC HEALTH ASSOCIATION), AMERICAN WATER WORKS ASSOCIATION, AND WATER POLLUTION CONTROL FEDERATION. 1975. Standard methods for the examination of water and wastewater. 14th ed Am. Public Health Assoc., New York. 1193 pp.

DAVIS, J.T. 1977. Design of water reuse facilities for warmwater fish culture. Ph.D. dissertation. Texas A&M University, College Station, Texas. 100 pp.

LEWIS, W.M., J.H. YOPP, H.L. SCHRAMM, AND A.M. BRANDENBURG. 1978. Use of hydroponics to maintain quality of recirculated water in a fish-culture system. Trans. Am. Fish. Soc. 107:92-99.

LIAO, P.B., AND R.D. MAYO. 1974. Intensified fish culture combining water reconditioning with pollution abatement. Aquaculture 3:61-85.

PARKER, N.C., AND B.A. SIMCO. 1974. Evaluation of recirculating systems for the culture of channel catfish. Proc. Southeast. Assoc. Game Fish Comm. 27:474-487.

REDNER, B.D., AND R.R. STICKNEY. 1979. Acclimation to ammonia by Tilapia aurea. Trans. Am. Fish. Soc. 108:383-388.

WEAVER, E.W. Technical and economic aspects of closed cycle aquaculture. Presented at the World Mariculture Society Annual Meeting, Honolulu, Hawaii, Jan. 22-26, 1979. 23 pp. (mimeo.)

NOTES

NOTES

NOTES